MARKETS FOR WATER
Potential and Performance

NATURAL RESOURCE MANAGEMENT AND POLICY

Editors:

EDITORIAL STATEMENT

There is a growing awareness to the role that natural resources such as water, land, forests and environmental amenities play in our lives. There are many competing uses for natural resources, and society is challenged to manage them for improving social well being . Furthermore, there may be dire consequences to natural resources mismanagement. Renewable resources such as water, land and the environment are linked, and decisions made with regard to one may affect the others. Policy and management of natural resources now require interdisciplinary approach including natural and social sciences to correctly address our society preferences.

This series provides a collection of works containing most recent findings on economics, management and policy of renewable biological resources such as water, land, crop protection, sustainable agriculture, technology, and environmental health. It incorporates modern thinking and techniques of economics and management. Books in this series will incorporate knowledge and models of natural phenomena with economics and managerial decision frameworks to assess alternative options for managing natural resources and environment.

There is a growing body of evidence showing that water markets can help improve water allocation and produce substantial gains from trade, at the same time there are also concerns that water markets are not efficient and create inequity problems. This book documents, through a series of studies, how water markets have emerged and are actively working in various parts of the world. It illustrates the key factors that have fostered the development of water markets and supported their operation. The book highlights the potentials and problems facing the expansion of water markets, world wide, and in water scarce regions.

The Series Editors

Recently Published Books in the Series

Antle, John, Capalbo, Susan and Crissman, Charles:
Economic, Environmental, and Health Tradeoffs in Agriculture: Pesticides and the Sustainability of Andean Potato Production
Spulber, Nicolas and Sabbaghi, Asghar:
Economics of Water Resources: From Regulation to Privatization
Second Edition
Bauer, Carl J.:
Against the Current: Privatization, Water Markets, and the State in Chile

MARKETS FOR WATER
Potential and Performance

edited by

K. William Easter
University of Minnesota

Mark W. Rosegrant
International Food Policy Research Institute

Ariel Dinar
The World Bank

KLUWER ACADEMIC PUBLISHERS
Boston / Dordrecht / London

1998

Distributors for North, Central and South America:
Kluwer Academic Publishers
101 Philip Drive
Assinippi Park
Norwell, Massachusetts 02061 USA
Telephone (781) 871-6600
Fax (781) 871-6528
E-Mail <kluwer@wkap.com>

Distributors for all other countries:
Kluwer Academic Publishers Group
Distribution Centre
Post Office Box 322
3300 AH Dordrecht, THE NETHERLANDS
Telephone 31 78 6392 392
Fax 31 78 6546 474
E-Mail <orderdept@wkap.nl>

Electronic Services <http://www.wkap.nl>

Library of Congress Cataloging-in-Publication Data

A C.I.P. Catalogue record for this book is available
from the Library of Congress.

Printed on acid-free paper.

Printed in the United States of America

Table of Contents

List of Figures

List of Tables

Preface

There are many myths concerning water markets, including that they do not exist, or if they do exist, that they are somehow unfair. In this volume, we dispel many of these myths and give the reader a comprehensive picture of how informal and formal markets have developed and are operating in different parts of the world. There are, of course, potentials for market failures and excessive transaction costs in water markets. It also may be politically difficult to establish institutional arrangements such as water rights that are important for the effective operation of water markets. On the other hand, there is a growing body of literature indicating water markets are an efficient way to introduce market signals for water users and to reallocate water.

Too often water trading is banned because the water resources have been developed with public funds and the water agencies do not want to lose control over water. There is also a concern that poor farmers or households will be disadvantaged by water trading. These concerns about public resources and the poor are not very different from those that have been voiced in the past about land sales. The problem is that, in many cases, the poor already have limited access to resources, but this limit is not due to water trading. In fact, as shown in the chapter on India, in many cases, water trading is likely to expand access to water for small scale farmers.

Not only does the book have examples of existing water markets, it also shows the importance of institutional arrangements in the development and performance of water markets and provides guidelines for those countries interested in promoting formal water markets. Several chapters illustrate how water markets have been established while other chapters provide estimates of the economic gains accruing from water markets. Finally, three chapters use economic analysis and experimental economics to determine how new water markets might perform. Hopefully, the readers will find the results illuminating and instrumental in changing some of their myths concerning the potential for water markets.

K. William Easter, Mark W. Rosegrant, and Ariel Dinar

Acknowledgments

The editors want to thank all the authors who have contributed chapters for this volume. It is the thoughtful contributions of these leading scholars that is the real strength of the book. We want to especially thank Linda Littrell, who has done the word processing for the manuscript and helped guide it through to completion. Without her conscientious help the manuscript would still be just a good idea in the minds of the editors. We also want to thank a number of people who have read the manuscript and suggested editing changes, including Cynthia Fang, Claudia Ringler, and Nancy Leach. In addition, we would like to thank Cynthia Fang and Anthony Mieloch for helping with the index and for creating the figures for several chapters.

About the Authors

Sandra O. Archibald, Professor, Humphrey Institute of Public Affairs, University of Minnesota, Minneapolis, Minnesota, 55455.

Bonnie G. Colby, Professor, Department of Agricultural and Resource Economics, The University of Arizona, Tucson, Arizona, 85721.

Ariel Dinar, Senior Economist, Rural Development Department, the World Bank, 1818 H. ST. N.W., Washington, D.C., 20433.

K. William Easter, Professor, Department of Applied Economics, University of Minnesota, St. Paul, Minnesota, 55108.

Alberto Garrido, Associate Professor, Departamento de Economía y Ciencias Sociales Agrarias, Universidad Politécnica de Madrid, Ciudad Universitaria, 28040, Madrid, Spain.

Ronald C. Griffin, Professor, Department of Agricultural Economics, Texas A&M University, College Station, Texas, 77843.

Robert R. Hearne, Research Associate, International Institute for Environment and Development, 3 Endsleigh St., London, WC1H 0DD, U.K.

Theodore M. Horbulyk, Associate Professor, Department of Economics, The University of Calgary, Calgary, Alberta, Canada, T2N 1N4.

Charles W. Howe, Professor, Environment and Behavior Program, Institute of Behavioral Science, University of Colorado at Boulder, Boulder, Colorado, 80309.

Richard E. Howitt, Professor, Department of Agricultural Economics, University of California, Davis, California, 95616.

Marie Leigh Livingston, Professor, Department of Economics, University of Northern Colorado, Greeley, Colorado, 80639.

Lynda J. Lo, Economist, The Royal Bank of Canada, Calgary, Alberta, Canada.

Ruth S. Meinzen-Dick, Research Fellow, International Food Policy Research Institute, 1200 17th St. N.W., Washington, D.C., 20036.

S.J. Rassenti, Associate Director of Economics Science Laboratory, University of Arizona, Tucson, Arizona.

Mary E. Renwick, Research Associate, Humphrey Institute of Public Affairs, University of Minnesota, Minneapolis, Minnesota, 55455.

Mark W. Rosegrant, Research Fellow, International Food Policy Research Institute, 1200 17th St. N.W., Washington, D.C., 20036, and Fellow, International Irrigation Management Institute, P.O. Box 2075, Colombo Sri Lanka.

R. Maria Saleth, Reader, Institute of Economic Growth, University Enclave, Delhi, 110 007 India.

Mateen Thobani, Principal Economist, The World Bank, Washington, 1818 H. St. N.W. D.C. 20433.

Vernon L. Smith, Professor and Director, Economics Science Laboratory, University of Arizona, Tucson, Arizona.

1 Water Markets: Transaction Costs and Institutional Options

K. William Easter, Ariel Dinar, and Mark W. Rosegrant

1.1 Introduction

In the past, water has been considered an abundant resource that should be available to everyone at little or no charge. However, the world's growth in population and increase in per capita water use has dramatically changed this situation over the past half century. Today the outlook for future water use and development is very different from what it was when USAID and the World Bank first began investing in water development during the 1950s.

Projections of water withdrawals by sector show the dramatic increase in pressure on water resources over the next three decades. Rosegrant, Ringler, and Gerpacio (1997) project domestic water demand as a function of population growth, per capita income growth, and the income elasticity of domestic water demand. Industrial use of water is based on water use intensity per unit of gross domestic product (GDP) and the growth in GDP. Agricultural use is a function of irrigated area and water use intensity. Globally, water withdrawals are projected to increase by 35 percent by 2020, to 5,060 billion cubic meters (Bm^3), with growth in developing countries much faster than in developed countries. Developed countries as a group will increase water demand by 22 percent to 1,710 Bm^3, more than 80 percent of which will be for industrial uses. The most serious pressure on water resources will be in the developing world, where water withdrawals are projected to increase by 43 percent, from 2,347 Bm^3 in 1995 to 3,350 Bm^3 in 2020. In sharp contrast to past growth patterns in developing countries, the absolute increase in domestic and industrial water demand will be greater than the increase in agricultural water demand, projected at 589 Bm^3 and 415 Bm^3, respectively, from 1995 to 2020. The combined share of domestic and industrial use in total water demand in developing countries will more than double from 13 percent to 27 percent, representing a significant structural change in water demand in developing countries (Rosegrant et al., 1997).

1.2 Meeting Future Water Demands

Can the rapid growth in water demand, particularly in the domestic and industrial sectors, be met without massive transfers of water out of agriculture that could derail the growth in crop production needed to effectively meet the food demand in the coming decades? Development of irrigation and water supplies has become increasingly expensive. In India and Indonesia, for example, the real costs of new irrigation have more than doubled since the late 1960s and early 1970s; costs have increased by more than 50 percent in the Philippines; they have tripled in Sri Lanka; and increased by 40 percent in Thailand (Rosegrant and Svendsen, 1993). The cost of supplying water for household and industrial uses is also increasing rapidly. In Amman, Jordan, the average incremental cost of water from groundwater has been \$0.41 per m^3 (all values are in U.S. dollars unless noted otherwise). However, with shortages of groundwater, the city has begun to rely on surface water, pumped with a lift of 1,200 meters (m) from a site 40 kilometers (km) from the city, at an average incremental cost of \$1.33 per m^3. In Shenyang, China, the cost of new water supplies will nearly triple from \$0.04 to \$0.11 per m^3 between 1988 and 2000 because pollution of the current groundwater source will require a shift to water conveyed by gravity from a surface source 51 km from the city. In Mexico City, water is currently being pumped over an elevation of 1,000 m into the Mexico Valley from the Cutzamala River through a pipeline about 180 km long, at an average incremental cost of \$0.82 per m^3, almost 55 percent more than the previous source, the Mexico Valley aquifer (World Bank, 1993). Non-traditional sources of water are unlikely to be a major component of new water supplies. Desalination offers an infinite supply of freshwater, but at a high price, and will not be a significant source in most regions. The reuse of wastewater will similarly make an important contribution only in arid regions such as the Middle East where the cost of new water supplies is very high. Water harvesting (the capture and diversion of rainfall or floodwater to fields to irrigate crops) will be important in some local and regional ecosystems but again, will not have a significant impact on global food production and water scarcity (Rosegrant, 1997; Rosegrant and Meinzen-Dick, 1996).

With high costs of development limiting the expansion of water supply, the rapidly growing household and industrial demand for water will need to be met increasingly from water savings from irrigated agriculture. A particularly difficult challenge will be to improve the efficiency of agricultural water use to maintain crop yields and output growth while at the same time allowing reallocation of water from agriculture to rapidly growing urban and industrial uses. How this will be managed could determine the world's ability to feed itself.

A clear place to start in seeking water savings, improving water use efficiency, and boosting crop output per unit of water is through the reform of existing water policies that have contributed to the current predicament. Both urban and rural water users are provided with large water subsidies. In many areas, irrigation water is essentially unpriced and in urban areas, the price of water does not cover the cost of delivery. These water-wasting policies can be attacked through comprehensive reforms to improve the incentives at each level of the water allocation process. Institutional and legal reforms must empower water users to make their own decisions regarding resource use, while at the same time providing a structure that reveals the real scarcity value of water.

1.3 Gains from Water Markets and Transaction Costs

In areas with irrigation and scarce water supply, research suggests that there is a large potential gain in economic efficiency from conservation and the reallocation of water among users (Hearne and Easter, 1995; Vaux and Howitt, 1984). Traditionally, economists have argued that what is needed is a water pricing system that reflects the scarcity value of water. Their approach has been to harangue irrigation departments and water utilities to charge much higher prices for water (Rogers, 1992; Repetto, 1986). The problem is that users have been able to use their political power to prevent major increases in water prices, especially for irrigation water. Given this dilemma, a number of economists such as Rosegrant and Binswanger (1994) have suggested that there may be a better way to achieve an efficient allocation of water by changing the incentives so that users support the efforts to reallocate water. They argue that if water markets are established where water is scarce, then users can buy and sell water and the whole incentive system is changed. When users can benefit from the reallocation, they are willing to sell water or pay a higher price for new supplies. The numerous informal water markets that have evolved suggest that water users will buy and sell water even if it is illegal or discouraged by governments (Shah, 1993; Renfro and Sparling, 1986).

Problems arise when governments are asked to encourage or help develop formal markets or allow informal markets to procreate, particularly within government constructed irrigation projects. Many government officials feel that since these are subsidized projects, the users should not be able to sell the water at a profit and that "poor farmers" will be disadvantaged by such water markets.[1] Thus, even though water markets can change the incentives for water users and improve water allocation, there are a number of constraints that prevent the introduction of water markets.

For formal water markets to work, one key component is for users to have some type of water right or use right they can buy or sell. This may create a serious problem, not only because water rights may be difficult to establish, but also because public water agencies feel they will lose a great deal of power if they allocate water rights to users. If users have rights, this means that system operators (the government officials) have the responsibility to deliver water to the users, based on user demands. In contrast, if a government agency holds the water rights, they can dictate to farmers the conditions under which they will receive water, including in some cases, the necessary side payments from farmers. If the water rights are made tradable, this creates an even greater dilemma for government water agencies. To prevent losing control over tradable water rights, the National Water Commission (CNA) in Mexico and some of the water districts in the western U.S. limit trading among water districts. In Mexico, a water user must obtain special government approval to sell water outside the district or jurisdiction and any profits from the sale must accrue to the district and not to the seller. This, of course, discourages interdistrict trading but it reduces the chance that trades will have third-party effects. In addition, during the drought in the mid-1990s, it allowed CNA to reallocate water from agriculture to urban users without giving markets a chance to work (See Hearne, Chapter 11).

Even if the water management unit that allocates water is set up and hired by the users, the unit may have an incentive to discourage water trades with other jurisdictions. The fear is that if too much water is transferred out of the district, it may reduce economic activity and make it difficult to effectively operate the irrigation system or, at least, parts of the system. Once most of the farmers along a canal have sold their water, it may be difficult and expensive to serve the few remaining farmers who own water rights on the canal. This means that although water markets may change water user incentives and encourage water use efficiency, the management of the system may prevent trades or, at least, raise the transaction costs of interdistrict or interjurisdictional trades.

Similarly, Cooter (1996) argues that:

> an organization that seeks to maximize the wealth of its members will behave both monopolistically and efficiently. By fixing prices, establish-

> ing territories, and withholding information from the public, the organization will seek to create monopoly power for its members in dealing with nonmembers. The organization will also seek to minimize agency costs that members incur in dealing with each other. To minimize agency costs, the organization must establish efficient property rights and contracts among members. Thus a general principle of motivation for an intermediate organization can be stated as: monopoly for outsiders, efficiency for insiders (p. 211).

These motivations may affect water user organizations that can use their infrastructure as a monopolistic tool to block trades with outsiders. They can either say that their canals are used to capacity or charge such high transmission fees that the trades become unprofitable.

Consequently, the organizational problem appears to involve two important aspects. First is the resistance to water trading between or among districts or jurisdictions (Cooter's outsiders). Second is the problem of establishing water rights and giving the users more control over water. Other problems which raise the transaction costs of water trading include legal challenges by third parties who feel they might be damaged by a transfer, the lack of sufficient infrastructure to transfer water among potential buyers, and the lack of an effective means for verifying and enforcing water rights. The question is whether or not it is in the best interest of a country to take action to reduce these transaction costs. If the answer to this question is, at least, a tentative "yes," then the second question is "how can these transaction costs be lowered?"

The problem is illustrated in Figure 1.1, where transaction costs raise the market price of water and reduce the amount of water transferred. The question is whether these transaction costs move us closer to the social optimum. Let us start with a quantity, Q_1, as the initial allocation of water to the urban sector, while the rest of the water is allocated to agriculture. When water becomes tradable, farmers can increase returns by selling water to the urban sector. The farmers' supply curve, S_1, measures the benefits farmers forego when they sell water. The urban water demand with no transaction costs is D_1. In the market allocation without transaction costs, the urban demand, D_1, equals S_1 at Q_2. If farmers sell water until their opportunity cost of water equals the market price, the urban sector can obtain an additional quantity of $Q_2 - Q_1$ units of water.

When the transaction costs of administering the trades are included, the urban demand drops to D_2 which is the urban demand minus the transaction costs (this assumes the buyer pays the administratively-induced transaction costs, which is the case in countries such as Chile). The administratively-induced transaction costs are the shaded area, AIC, between the two demand curves. The quantity going to the urban sector, when transaction costs are included, becomes Q_3, which is the quantity where S_1 equals D_2. The purchase price of water, that includes AIC, will increase. Yet this solution does not include policy-induced transaction costs, which, if included, would further increase water prices and reduce transactions. When these policy-induced transaction costs (PIC) are paid by the sellers, as they are in California, then the supply curve would be shifted up and to the left. Assuming that the shift is to S_2, then the new equilibrium is Q_4 where $S_2 = D_2$. These policy-induced transaction costs result from the political action by third parties who are concerned about instream water use, environmental quality, and the economic impacts on regional sales caused by reduced agricultural production. If, through political action, third parties are able to impose high policy-induced transaction costs on sellers, then trading may be minimal and the equilibrium allocation will be close to Q_1 and not the social optimum.

Figure 1.1. Transaction Costs in Water Market Allocation Among Sectors

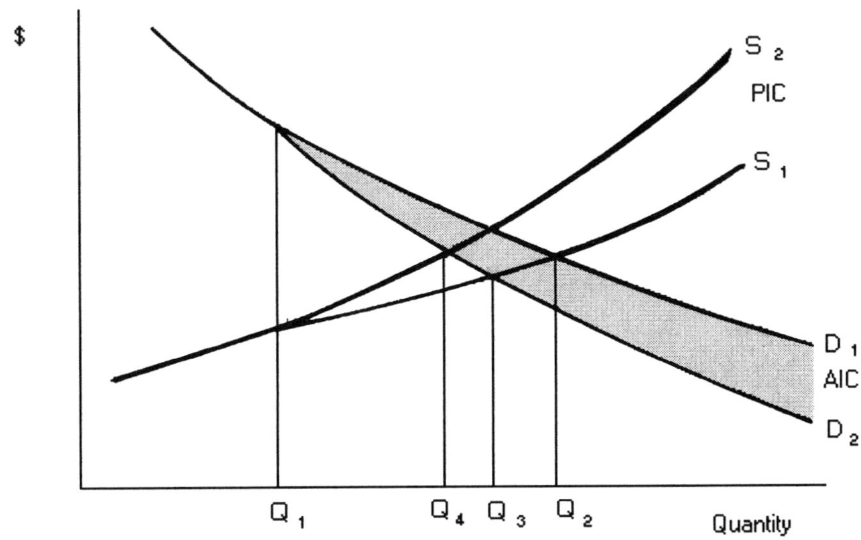

Adapted from Colby, 1990

1.4 Institutions and Establishing Water Markets

Before we determine if transaction costs should be reduced and how they can be reduced, it is important to understand the larger social structure of a country in which such economic exchanges might take place. If a country has little experience with markets for allocating scarce goods and services, then it is unlikely that water will be one of the first goods to be exposed to market forces. In contrast, in a country that is exploring new ways to use the market to improve the allocation of publicly managed resources, scarce water resources might be a good candidate for market trading. Formal water markets developed during the 1990s in California, not only because of the general trend toward the wider use of markets to allocate public resources, but also because of the high economic and environmental costs of developing new sources and the resulting potential high benefits from trades of existing water supplies. In contrast, one would not expect formal water markets to be developed very soon in the Central Asian Republics of the former Soviet Union because of their lack of experience and institutions that would facilitate market development.

There are two distinctly different views concerning the institutional setting required for efficient market exchanges. One is the neoclassical view that a legal system is required to foster efficient market exchanges. The other is the more pragmatic view that emphasizes the importance of informal contract enforcement to foster market exchanges.

> This neoclassical view that places the legal system at the center of contract enforcement in market economies has recently been criticized on the basis of evidence indicating that many contemporary exchange relations in the West and elsewhere are informal. The associated contract enforceability is not provided by the legal system but is based on reputation, general morality, and personal trust within social networks. Empirical evidence indicates the importance of two distinct systems of informal contract enforcement: the individualistic system of informal contracts enforcement prevalent in the West, under which the reputation and morality of the individuals are key, and the collectivist system of contract enforcement prevalent in most other societies, under which personal trust within the social network is critical (Greif, 1997, p. 239-240).

Greif (1997) argues that, "the neoclassical theory does not imply that a legal system is necessary for anonymous exchange, while the history of the West does not support the view that the legal system is key for the operation of market economies. . . . Furthermore, 'asymmetric information' and incomplete contracts limit the extent to which legal systems can facilitate exchanges" (p. 240-241). He goes on to argue that the informal contract enforcement systems are also not costless and can be more expensive than enforcement through a legal framework. Furthermore, in the collectivist model, market exchanges can take place among members of a community, but not between members of different communities because the exchanges are based on social ties and personal trust within each small community.

Cooter (1997) comes to a similar conclusion in reviewing the problem of contracting and establishing a rule-of-law that is consistent with a country's social norms.

> Although enforcement of contracts typically requires coercion by a third party such as a state, the problem can also be solved without recourse to third-party enforcement. Investment in a business network usually occurs among people who maintain enduring relationships. . . . Contracts solve the problem of cooperation through law, and enduring relationships solve the problem of cooperation through repetition. In tentative relationships, the problem of cooperation is solved by social norms (p. 194).

At least in the case of markets for irrigation water, it appears that both the formal neoclassic-legalistic system and the informal system are at work. The transfer of permanent water rights appears to require the certainty that is provided by a legally-based approach where water rights are recorded and can be defended in court. In addition, if water is being transferred between districts and the sale will change return flows, a more formal market may be required with safeguards that prevent losses to those using return flows. In contrast, if the sales are temporary, i.e., for one season or less, and do not change return flows, then informal water markets based on "informal" water rights can suffice (See Howitt, Chapter 8). These sales will likely be among farmers in the same water district and in many cases, among farmers served by the same canal. In addition, these sales are not likely to be anonymous and enforcement of the contracts will not be provided by the legal systems, but will be based on reputation and personal trust within the water user group. This is another reason why trades among districts or jurisdictions may be limited. It also suggests that to obtain more interdistrict or interjurisdictional water trades, a country will have to develop legally-based water rights that can be easily verified and defended in court at a reasonable cost. However, if interdistrict trades yield only small efficiency gains, while significant gains are possible with intradistrict trades as illustrated in Chapter 15, then informal water markets based on collective enforcement may be all that is needed.

Informal water markets have worked fairly well for groundwater as long as recharge is adequate and there is a sufficient number of sellers in the market (Shah, 1993; Palanisami and Easter, 1991; Saleth, Chapter 12). The "tit for tat" game theory enforcement strategy appears to work. If farmers do not pay, they will not receive water the next time, or if a seller does not or will not deliver, then the buyer can go to another supplier. In one area of Gujarat, India, farmers have pipelines from three to four different suppliers coming to their fields (Shah, 1993). They can buy from the supplier who offers the best price and service. Shah (1993) also found that, "while the main beneficiaries of the private investments in pipelines have been the buyers of water, early operators in the water business were motivated mainly by the desire to establish monopoly positions and to overcome topographical constraints in supplying water to a large command" (pp. 61-62).

In many cases, a buyer's choices may be quite limited either because there is not an abundance of suppliers or because the conveyance system is limited by capacity or location (Saleth, Chapter 12). In some groundwater-served areas, there are few canals between farmers. Even in an area with canals connecting most farmers, these canals may have limited capacity or may be limited by the irrigation authorities to delivering only water from within the irrigation system. The control of the canal system has been used as a way to reduce competition and trading among districts. An inadequate canal system can also limit trades within a district. However, where water is being traded, users have a strong incentive to see that a canal is well maintained since it is transporting "their" purchased water (Palanisami and Easter, 1991).

If a country decides to establish a more formal or legalistic approach to water trading, care must be taken in establishing and allocating the water rights to users. The law must be written so that the resulting allocation of rights is considered "fair." For water markets to work, especially in transferring water among sectors or districts, the community of users need to support the concept. When the economic rents from water trading are concentrated in the hands of a few individuals, or negative third-party effects are large and not mitigated, then the community is not likely to obey the law. Cooter (1997) argues that this will hinder future economic development. "A modern economy needs effective laws to promote cooperation among people. Yet states enact many laws that few people obey. People tend to disobey, or obey out of fear, laws that are not consistent with social norms and to obey laws that reflect social norms" (p. 191).

A good example of a water law that farmers tend to disobey or obey out of fear is the law in Pakistan against trading canal water. In contrast, the 1981 Chilean water law that establishes private water-use rights is widely obeyed because not only does Chile have a long record of private water development, but it also allocated the water rights based on past use (Hearne, 1995). Yet problems have arisen because the law permitted a hydroelectric power company to obtain non-consumptive water-use rights free of charge and then to change the timing of return flows for downstream users.

1.5 Experience with Water Markets

As discussed above, there are cases of both formal and informal water markets. The formal sales can be for a specified volume or share of water for a set period of time, or they can be for the transfer of permanent water rights. Informal markets usually involve the sale of unmeasured flows of surface water in a canal for a set period of time or from a well based on the number of hours water is pumped. Although the unit sold in the informal market may not be metered, both the buyer and the seller usually have good information about the volume transferred. The key difference between the formal and informal markets is the enforcement of the trade. If the enforcement is by the users with no recourse to the legal or administrative system, then the market is informal. Enforcement in the informal markets is based on the reputation of the individuals or personal trust within the irrigation community.

It is important that we use our experience with both formal and informal markets to help decide when and how water markets might be effectively introduced and how the transaction costs of trading can and should be reduced. Country experiences with water markets are wide-ranging with most of the formal markets found in North and South America while the informal markets are concentrated in the irrigated areas of South Asia. For example, in several areas of Chile, the legalizing of water trading and recording of water-use rights were sufficient to foster an active formal water market with relatively low transaction costs. The combination of effective water user associations and well developed irrigation infrastructure allowed these farmers to easily trade water throughout the system, including trades with the urban sector (Hearne, Chapter 9).

Informal Water Markets

In the case of informal markets in Asia, Shah (1993) finds that in areas with dependable groundwater recharge, the selling of tubewell water has benefitted a wide range of farmers by allowing them to increase crop production and income. Meinzen-Dick, in Chapter 13, finds that the economic gain from groundwater markets can come from:

> improving utilization of tubewell capacity, increasing access to irrigation water supplies (especially among farmers with small or fragmented holdings), and lowering water tables in areas of waterlogging. . . . Access to groundwater makes it possible for those without wells to use groundwater for irrigation. This improves the equity of resource distribution because it is the smaller scale farmers who do not own tubewells. The opportunity to sell groundwater can make it profitable for farmers to invest in wells even if their own holdings are too small to use the full pumping capacity. . . . Even some farmers with wells purchase groundwater from others. This may be to irrigate fragments located away from their own well, or to take advantage of the more reliable

> water supply from wells located closer to the canals or in pockets of fresher groundwater.

The groundwater markets are important for agricultural production and the distribution of water in much of the irrigated area of South Asia. Saleth, in Chapter 12, estimates that 20 percent of the 14.2 million pumpsets are likely to be involved in water trading. This means that water markets are providing water for about 6 million ha, or 15 percent of the total area irrigated by groundwater. In Pakistan, a survey of well owners found a similar percentage of well owners selling water, amounting to 21 percent (NESPAK, 1991).

In one of the few studies that has tried to estimate the economic returns from having access to informal water markets, Meinzen-Dick, in Chapter 13, found that water markets increased the availability of water and the reliability of supplies. With greater water supplies and reliability, yields and income increased for those who purchased water, particularly for those who also had access to canal water. However, the highest yields and income were still found among farmers who owned tubewells and had access to canal water.

Even though groundwater markets appear to be growing rapidly and providing a more assured source of water to many farmers in South Asia, a number of concerns have been raised about the performance of informal groundwater markets. Given that markets for the sale of groundwater draw on an open-access resource, it is not surprising that problems arise in areas with high water demands and limited groundwater supplies. Farmers have an incentive to ignore the scarcity value and buffer stock value of the groundwater and pump until their cost of pumping equals the market price of water (Ramasamy, 1996).[2] Over time, the cost and price rise as the groundwater level declines. For example, the overdraft in Coimbatore district of India is almost 5 thousand m^3. Ramasamy (1996) estimates that if the over-pumping continues, it will mean a drop in the total net returns to farmers of $42 to $69 million, due to higher energy costs and excessive well investments. Here is a case where informal markets may exacerbate the overdrafting or depletion problem. Yet formal markets may not work any better unless water rights can be established and enforced in strict quantity terms. The problem is not the water markets but the lack of exclusive property rights for groundwater. Consequently, the market places a zero value on the buffer stock and scarcity values of water. To establish exclusive water rights, the number of wells and the volume of pumping would have to be agreed upon and restricted. Such restrictions are probably unrealistic unless there is strong support in the irrigation community to implement pumping limitations and strict enforcement. However, if exclusive water rights can be established, then the water market should reflect the scarcity value of water and help restrain over-pumping.

Blomquist (1995) reports on one of the cases where the demand for water is increasing and the community of water users have been able to stop the depletion. In the Los Angeles-Orange County-Riverside metropolis, pumping is metered and taxed. He finds that:

> Water users have shifted at least some of their reliance on local groundwater to more expensive but more plentiful imported water. Surface and imported water are impounded for basin replenishment. In the coastal basins, saltwater

> intrusion from the ocean has been halted. In some basins, pumping rights have been defined, limited to the basin safe yield, and made transferable within the basin (p. 43).

A more typical case is reported by Shah (1993) in coastal Gujarat, India. Here, the overdraft of coastal aquifers has caused declining groundwater supplies in some areas and saltwater intrusion in others. Shah contends that it is likely to be difficult to effectively reduce overdrafting without good local leadership and the involvement of water user groups. He argues that "legal, quasi-legal, and organization instruments of public policy will not, on their own, succeed in securing the compliance of farmers unless they are accompanied by measures aimed at affecting private returns to irrigation...or unless the structure of property rights on the water resource itself is drastically reformed" (p. 147). Similarly, in Pakistan, Meinzen-Dick, in Chapter 13, doubts "whether government would have the institutional capacity to regulate sales among hundreds of thousands of private tubewells, and if it had such capacity, it is unclear what such direct intervention could achieve."

Yet, in both India and Pakistan, the impact of water markets on the depletion of groundwater is much smaller than the impact of the very low or zero per unit charges for electricity. The zero or near zero marginal cost of pumping in most states of India and Pakistan means that farmers have an incentive to use water up to the point where the marginal value produce is close to zero. This, of course, encourages farmers who can sell water to use their wells at close to full capacity. The low power rates not only create overdrafting problems, but they also lead to an inefficient use of electricity in countries without adequate electricity supplies.

As noted above, water markets can actually help solve the depletion problem by increasing the incentives for efficient water use and making it possible to purchase water from areas not faced with an overdraft. The ability to find another source of water, but at a higher marginal cost, can help promote community action for self-regulation and demand management. Shah (1993) cites a case in coastal Gujarat where self-regulation became possible when additional new supplies were piped into the area.

In large areas of northern India and Pakistan, over-pumping is not a problem. The groundwater depletion problem tends to be concentrated in coastal regions and the hard rock regions of southern India. For many of the northern areas, pumping actually improves crop growing conditions by lowering the water table to below the root zone (Shah, 1993; Meinzen-Dick, Chapter 13). In areas with high water tables or rapid recharge rates, water markets are not likely to cause negative externalities, except possibly for temporary well interference due to the closeness of neighboring wells.[3] Where these negative externalities are small, personal trust and repetition may be enough to foster efficient informal water markets. This is particularly true where farmers own a number of separate plots that cannot be served by the same well. In such cases, most water sellers are also buyers, since most farmers who own a well are able to irrigate only their larger plots and must purchase water to irrigate the rest (Shah, 1993; Saleth, Chapter 12; Meinzen-Dick, Chapter 13). Furthermore, their wells are likely to be underutilized unless they can sell water. Yet because of the conveyance costs and the need to get the cooperation of neighboring farmers when water is to be transported any distance, transaction costs can be high. High transaction costs not only can limit trades but they can also prevent water markets from being competitive.

This brings us to the other concern that has been raised about water markets, the potential for monopoly pricing and discrimination. Groundwater markets are somewhat confined by the physical limits of the location and supply of groundwater. Still, pipelines can extend markets as they have in parts of Gujarat and the costs of investing in wells should put a limit on monopoly power. An abusive monopolist who raises prices too high will find others investing in wells and undercutting the monopoly price. Shah (1993) notes a lack of balance between the numbers of buyers and sellers in areas with high capacity wells, where there may be as many as 70 or 80 buyers served by one seller. Yet, he does not say how many sellers an individual buyer can access. If the buyers can purchase water from four to five sellers, this may be enough to prevent monopoly pricing as long as the sellers do not collude.

The evidence on monopoly pricing is mixed. Meinzen-Dick, in Chapter 13, found in a 1991-92 survey, that the price of water was not much higher than the cost of pumping.

> The two most common ways of charging for groundwater are a flat charge per hour of pumping (ranging from \$0.57 to \$3.27 per hour, depending on the pump type, capacity, and location) and . . . arrangements whereby the buyer supplies the diesel and motor oil for the pump, and pays an additional fee of \$0.16 to \$0.24 per hour to the well owner to cover the wear and tear on the engine.[4] Water sellers with diesel pumps are apparently only recovering their own costs under either type of contract.

Saleth, in Chapter 12, suggests that in India, monopoly rents may be high in some areas. He cites as evidence the variation in water charges compared to pumping costs in different areas. For example, water charges are 1.3 to 2.0 times higher than operating costs in the Indo-Gangetic region but 2.5 to 3.5 times higher in the water-scarce hard rock regions of south India. Another explanation for the difference in rates is the difference in water scarcity. Since groundwater scarcity is greater in most hard rock areas than it is in the Indo-Gangetic plain, the rates may be directly related to the scarcity value of groundwater.

The degree of monopoly power may also be related to the terms of the transaction or contract for water. Not surprisingly, some of the contracts for water are quite similar to contracts for land. Contracts for water include crop sharing, crop and input sharing, and labor arrangements. If the payment is cash-based, then the buyers have more freedom to take their business to another well owner anytime during the season. When the transaction is a contract in kind, especially if it is based on a crop sharing or a crop and input sharing contract, then the buyer is tied to the seller for at least one season, if not longer. Similarly, when compensation is in terms of labor, it may be difficult to change suppliers, depending on the length of time a buyer provides labor to the seller. Yet within the village setting of South Asia, it appears that informal markets do not face extreme cases of monopoly rent taking. In fact, if monopoly power restrains trading in areas with serious problems of declining groundwater, this may help reduce over-extraction. When suppliers are taking advantage of their monopoly position and there are adequate groundwater supplies, the best strategy is to encourage (legalize) trading and increase competition through support of community and/or private well development (Palanisami and Easter, 1991).

Thus, informal water markets are ways to improve water use and incomes in irrigated areas where the water rights are not well-defined or recorded. They also may be a good option if third-party challenges are likely to be common and result in excessively high transaction costs for formal water markets. Finally, informal markets would work well in traditional or community based irrigation systems where the farmers manage the irrigation system and would be able to maintain a relatively modest level of transaction costs.

Formal Markets

In situations where informal markets can work well, it may not be necessary to incur the extra expense of establishing formal water markets. However, formal markets may be required to provide the certainty necessary for permanent water transfers or for transactions between different sectors and jurisdictions. For example, formal markets in the Middle East may be one way to solve some of their critical interjurisdictional water allocation problems (Becker, 1995; Becker et al., 1996).

Since the need for permanent trades and interjurisdictional water exchanges are likely to become more important with the rapid growth in non-agricultural water demands, we are likely to see a growth in formal markets. The growing demand in water scarce regions has been a driving force behind the new interest in water markets. A number of studies have already illustrated the benefits that are possible from interjurisdictional trading (Vaux and Howitt, 1984). These markets can be for the water rights (permanent) or they can be spot markets like the California market discussed by Howitt in Chapter 8.

Water markets in the Rio Grande Valley of Texas have been used primarily to transfer water out of agriculture. Griffin, in Chapter 4, indicates that over 99 percent of the trades in the valley have been from the agriculture sector to non-agriculture. Of the municipal water rights in the valley that existed in 1990, 45 percent had been purchased during the two preceding decades. While water markets are not active in other areas in Texas, Griffin feels that the surface water law has evolved to the stage where trading will be more ubiquitous in the future. In contrast, the groundwater law is just beginning to evolve.

Garrido, in Chapter 14, shows that when trading is restricted to an individual water district or community, the economic gains may be relatively modest in the Guadalquivir Basin of southern Spain. In contrast, if trades are allowed among communities that are subject to different supply constraints and drought conditions, the gains from trade can be substantial. In his analysis of intra-community trading in four irrigation communities, he estimates that the total welfare gains would exceed those under the current water allocation by no more than 10 percent. However, with inter-community trading, the economic gain for one of the older irrigation communities could reach almost 50 percent. He also shows that both inter-community and intra-community trading are very sensitive to the level of transaction costs. If transaction costs exceed 8-12 percent of the market price, trading and the gains from trading would be too small to justify the expense of establishing the formal markets. Yet Garrido may underestimate the potential gains since he only considers the crops traditionally grown in the region, i.e., cotton, wheat, corn, oilseed and sugar beet, and excludes any potential transfers to non-irrigation uses. In Chile, there were significant changes in cropping patterns that accompanied water trading (Hearne and Easter, Chapter 10).

In contrast, Horbulyk and Lo, in Chapter 15, found that most potential gains from introducing water markets in Alberta Province of Canada were likely to come from intrasub-basin trades. They considered four sub-basins and compared the current situation with the water allocation under four separate markets (one in each sub-basin), as well as with a market encompassing the total basin. The four separate market scenarios created 90 percent of the welfare gains that were obtained when unrestricted trading was allowed among the four sub-basins. The urban sectors purchased most of the water, except on the South Saskatchewan River where the agricultural sector purchased additional water, when market trading was allowed among the sub-basins.

In their analysis of selected water markets of Chile, Hearne and Easter, in Chapter 10, found trading both within and between sectors. For permanent transactions either within or between sectors, well established water-use rights that were recorded and recognized by the government were critical. La Serena, a growing vacation destination located on the coast in the dry region some 400 km north of Santiago, has found it difficult to meet water demands, particularly during the summer tourist season. The opening of water markets allowed the city to purchase water-use rights and lower costs by delaying the need to develop new sources of water. Starting in 1992, the city's water company, ESSCO, purchased 292 shares which added 28 percent to its water supply. In addition, upstream from La Serena, 63 shares were purchased by households for domestic uses and 351 shares were acquired by farmers for irrigation.

Another study area in Chile with significant trading was the Limarí Valley where trading was very active but restricted to only the agricultural sector (locally, the urban sector has adequate water). To estimate the benefits from this water trading, a farmer survey was conducted in 1993-94. It included 37 farmers who had sold water-use rights and 19 farmers who had purchased water-use rights, for the combined total of 9.2 million cubic meters (m^3). The average gains-from-trade were \$2.47 per m^3 (\$3,045/acre-ft),[5] with a transaction cost of \$0.07 per m^3 (\$86/acre-ft). This was not a random sample or an attempt to obtain a complete sample. The numbers surveyed were large enough to indicate that the water market was very active and has created significant gains-from-trade. The largest gains were for water purchased by three large grape producers who purchased 5.8 million m^3 of water (63 percent of the total amount traded in the sample).

In the active water markets in Chile, the transaction costs were low and did not seem to constrain trading. However, in some other areas in Chile, such as the upper section of the Maipo River that supplies the southwestern Santiago area and irrigates about 100,000 ha, the transaction costs are high and trading is quite limited. The Maipo River is divided into three sections for management and water trading. Uncertain water rights and the lack of adjustable control structures both limit trading (Hearne, Chapter 9).

Similar concerns about transaction costs are raised by Archibald and Renwick in Chapter 7. Employing a model analogous to the one used by Hearne and Easter (1997) they found that a large number of potential trades in California had not occurred, even though substantial gains-from-trades were possible. These trades did not occur, largely because of the transaction costs related to possible third-party effects and legal action.

Archibald and Renwick's analysis of the California water market suggests that two types of transaction costs have limited potentially profitable trades. Administratively-induced transaction costs (AIC) are explicit and included in the sales price of water sold through the California Water Bank. Policy-induced transaction costs (PIC) are attributed to existing legal requirements such as no injury to vested water rights, no unreasonable effects on fish and wildlife, and the mitigation of third-party effects. The AICs included the costs of search for buyers and sellers and negotiating quantities, timing, and other terms of transfer and were $0.04/m^3 ($50/acre-ft) in 1991 and $0.01/m^3 ($17.50/acre-ft) in 1992 and 1994. In effect, what has happened is that when PICs were added, the supply curve (S_1) for water shifted up to S_2 (Figure 1.1) which made many potential transfers of appropriative water rights in the Central Valley of California unprofitable. Colby (1990) found PICs in the West ranging from $0.15/m^3 ($187/acre-ft) in Colorado to $0.04/m^3 ($54/acre-ft) in New Mexico, all states with less stringent state and federal transfer requirements than California. Policy-induced transaction costs in this range would be as much as or more than the potential gains for trading in the California Water Bank.

Because of the high transaction costs in Colorado, Howe, in Chapter 5, recommends a shift away from water courts as a means for administering water transfers, to administration by the State Engineer's Office. He also recommends that water be reserved or acquired for "public good" type uses such as recreation. [6] He argues that these and other changes are necessary to allow water to be allocated more freely in Colorado's water markets, as it is in neighboring states.

Colby, in Chapter 6, suggests that water claims by Native Americans on many western rivers have the effect of imposing high transaction costs on water trading. The problem may be that markets will not be able to work under such high transaction costs. However, she argues that when the transaction costs of water markets are compared with the cost of litigated solutions to the Native American water rights claims, water markets look like a much better alternative.

Howitt, in Chapter 8, reports that spot markets and options markets performed well during California's droughts in the 1990s. Even though these markets are a fairly recent phenomenon, he thinks that they can go a long way in stabilizing stochastic water supplies such as those in California. However, for permanent shifts in demand, a much more active water market for permanent water rights may be needed.

A more recent development of water markets has occurred south of the California border in Mexico. The 1992 Mexican National Water Law provides a strong institutional framework for water markets based on transferable water-use concessions. The law also contains a strong role for the Federal government through CNA. Hearne, in Chapter 11, raises concerns about how CNA has used this power during periods of drought to impose command and control regulations and effectively eliminate the security offered by the water-use concessions. The law also fails to provide water users with any incentive to sell outside their irrigation districts since the benefits from such sales accrue to the district.

1.6 To be or Not to be Formal

Thus, as we consider the use of more formal water markets, the question of transaction costs must be addressed directly. What are the legitimate uses that need to be protected in water transactions? Clearly, return flows have to be considered, as do third-party economic effects and instream uses, but should these interests be able to block all transfers because they might be affected adversely? On the other hand, should third parties have to take into account the economic losses caused by not allowing the transfers? The 1981 legislation establishing water markets in Chile did not include safeguards for possible third-party or return flow impacts. In more recent legislation, they have tried to correct for some of these potential problems by including provisions for minimum stream flows and other restrictions on water transfers. Is the 1981 Chile model a better one for developing countries to use as they approach the question of water reallocation? If they use this model, does it pose potential future problems, especially when they are dealing with return flows in international waters?[7]

In regions with numerous small-scale water users, it may be better to encourage and remove barriers to the creation of informal markets. If this can be done, many of the costs of establishing and enforcing water rights can be avoided. In addition, the third-party effects can be dealt with within the irrigation community. Yet as markets expand into areas with limited groundwater supplies, regulations may be necessary to prevent serious market inefficiencies due to the lack of well-defined water rights. Furthermore, informal markets will not be adequate to take advantage of potential gains from trading among different jurisdictions. Also, instream uses and environmental damages are likely to be ignored with informal markets.

1.7 Organization of the Book

The idea of water markets has many detractors. Some critics argue that markets cannot work because of market failures, while others say that there are no existing examples of water markets. Thus, they conclude that the government must step in and take water from some users and give it to others. The problem is that this provides no incentives for users to cooperate or conserve water. If they conserve, the government will just take away the water they save.

This book documents, through a series of studies, how water markets have emerged and are actively working in various parts of the world. It illustrates the key factors that have fostered the development of water markets and supported their operation.

Chapters 2 and 3 list and discuss the important institutional and organizational arrangements that are required to support active water markets. These are followed by five chapters which explain how water markets have helped improve the allocation of water in the western U.S. In these chapters, it is illustrated how water markets can play an even more important role in the future. The strengths and weaknesses of these markets are also discussed, highlighting the importance of government policies that reduce transaction costs. Chapter 8 also shows how spot water markets and options markets can significantly improve water allocation in arid areas like California.

Chapters 9 through 13 review the successes and failures of formal and informal water markets in Chile, Mexico, India, and Pakistan. In the two former countries, the governments are formally establishing and recording tradable water-use rights and supporting the development of water markets. In the two latter countries, informal water markets are very active, especially for groundwater, even though the governments prohibit trading in public irrigation projects.

In Chapters 14 and 15, the authors estimate the economic gains that can be achieved from developing water markets in selected regions of Spain and Canada. Both examples indicate that significant economic gains can be obtained from water trading. The Canadian example includes transfer of water from agriculture to the urban sector while the Spanish model does not. Transaction costs are particularly important in determining whether or not net economic gains will be achieved with the proposed water markets in Spain.

Chapter 16 completes the critique with a discussion of 'smart markets' and the application of experimental economics to test their use for water allocation in California. Experimental economics allows one to test various institutional arrangements for markets and to estimate their impact on transaction costs and market performance. The authors conclude that water networks, using a computer-coordinated smart allocation mechanism, will yield essentially competitive outcomes and should be further tested. This is followed by the concluding Chapter that highlights the potentials and problems facing the expansion of water markets, world-wide, in water scarce regions.

Endnotes

[1] It is not clear that the "poor farmers" will be disadvantaged by water markets. Both Meinzen-Dick and Saleth in Chapters 12 and 13 find that small-scale farmers benefit from water markets.

[2] Scarcity value is the intertemporal opportunity costs of water. It is the present value of the marginal sacrifices imposed on the future by using the resource now. Buffer stock value is the value of groundwater in stabilizing water supplies when the supply of surface water is uncertain. (Tsur, 1990, p. 816).

[3] Problems also arise in areas where deep tubewells, used for irrigation, interfere with shallower domestic wells. Such interference has been reported, for example, in some areas of Bangladesh.

[4] The 1995 exchange rate of 24.5:1 was used to convert Pakistan rupees to U.S. dollars.

[5] An acre foot equals approximately 1,233 m^3.

[6] Public goods have two characteristics. First, there is no conventional consumption of the resource during its use and the resource will continue to provide the same level of service after its use to other consumers (the marginal resource cost of supplying another consumer is zero or nonrival). Second, consumers cannot be excluded from using the goods (the cost of exclusion is too high) (World Bank, 1993).

[7] The problem in international waters is that upstream countries have a strong incentive to ignore the impacts their water use has on countries downstream. Yet as Becker (1995) suggests, water markets may offer a means for giving these upstream countries an incentive to conserve and trade water downstream.

References

Becker N., N. Zeitouni, and M. Shechter, 1996. "Reallocating Water Resources in the Middle East through Market Mechanisms," International Journal of Water Resources Development, 12(1):17-32.

Becker, Nir, 1995. "Value of Moving from Central Planning to a Market System: Lessons from the Israeli Water Sector," *Agricultural Economics,* 12:11-21.

Blomquist, William, 1995. "Institutions for Managing Groundwater Basins in Southern California," in *Water Quantity/Quality Management and Conflict Resolution: Institutions, Processes, and Economic Analyses*, edited by Ariel Dinar and Edna T. Loehman, Westport Connecticut, Praeger Publications, pp. 43-59.

Colby, Bonnie G., 1990. "Transactions Costs and Efficiency in Western Water Allocation," *American Journal of Agricultural Economics*, 72(5):1184-1192.

Cooter, Robert D., 1997. "The Rule of State Law and the Rule-of-Law State: The Economic Analysis of the Legal Foundations of Development," in *Annual World Bank Conference on Development Economics, 1996*, edited by Michael Bruno and Boris Pleskovic, Washington, D.C., pp. 191-217.

Greif, Avner, 1997. "Contracting, Enforcement, and Efficiency: Economics beyond the Law,"in *Annual World Bank Conference on Development Economics, 1996*, edited by Michael Bruno and Boris Pleskovic, Washington, D.C., pp. 239-265.

Hearne, Robert R., and K. William Easter, 1997. "The Economic and Financial Gains from Water Markets in Chile," *Agricultural Economics,* 15: 187-199.

Hearne, Robert R., and K. William Easter, 1995. *Water Allocation and Water Markets: An Analysis of Gains-from-Trade in Chile,* World Bank Technical Paper Number 315, Washington, D.C., 75 p.

Hearne, Robert R., 1995. "The Market Allocation of Natural Resources: Transactions of Water-use Rights in Chile," Ph.D. Dissertation, Department of Applied Economics, University of Minnesota.

NESPAK (National Engineering Services of Pakistan), 1991. "*Contribution of Private Tubewells in the Development of Water Potential,* (Field Report) Pakistan: Government of Pakistan, Ministry of Planning and Development, Planning and Development Division.

Palanisami, K., and K. W. Easter, 1991. "Hydro-economic Interaction between Tank Storage and Groundwater Recharge," *Indian Journal of Agricultural Economics,* 46(2):174-179.

Ramasamy, Chellappan, 1996. "Intertemporal Allocation of Groundwater Resources with Stochastic Surface Supplies: The Case of Southern India," Ph.D. Thesis, Department of Applied Economics, University of Minnesota, 101 p.

Renfro, Raymond Z.H., and Edward W. Sparling, 1986. "Private Tubewell and Canal Water Trade on Pakistan Punjab Watercourses," in *Irrigation Investment, Technology and Management Strategies for Development,* edited by K. William Easter, Boulder, Westview Press, pp. 193-210.

Repetto, Robert C., 1986. "Skimming the Water: Rent-Seeking and the Performance of Public Irrigation Systems," Research Report 4, World Resources Institute, Washington, D.C.

Rogers, Peter, 1992. "Comprehensive Water Resources Management: A Concept Paper," Working Paper Series 879, World Bank, Infrastructure and Urban Development Department, Washington, D.C.

Rosegrant, Mark W., 1997. "Water Resources in the Twenty-First Century: Challenges and Implications for Action," Food, Agriculture, and the Environment Discussion Paper 20, International Food Policy Research Institute, Washington, D.C.

Rosegrant, Mark W., Claudia Ringler, and Roberta V. Gerpacio, 1997. "Water and Land Resources and Global Food Supply," Paper prepared for the 23rd International Conference of Agricultural Economics on "Food Security, Diversification, and Resource Management: Refocusing the Role of Agriculture," Sacramento, California, August 10-16.

Rosegrant, Mark W., and Ruth S. Meinzen-Dick, 1996. "Water Resources in the Asia-Pacific Region: Managing Scarcity," *Asia-Pacific Economic Literature*, 10:32-53.

Rosegrant, Mark W., and Mark Svendsen, 1993. "Asian Food Production in the 1990s: Irrigation Investment and Management Policy," *Food Policy*, 18(2): 13-32.

Rosegrant, Mark W., and Hans P. Binswanger, 1994. "Markets in Tradable Water Rights: Potential for Efficiency Gains in Developing Country Water Resource Allocation," *World Development,* 22:1613-1625.

Shah, Tushaar, 1993. *Groundwater Markets and Irrigation Development: Political Economy and Practical Policy,* Bombay, Oxford University Press, 228 p.

Tsur, Yacov, 1990. "The Stabilization Role of Groundwater When Surface Water Supplies are Uncertain: The Implications for Groundwater Development," *Water Resources Research,* 26(5): 811-818.

Vaux, Henry J., and Richard E. Howitt, 1984. "Managing Water Scarcity: An Evaluation of Inter-regional Transfers," *Water Resources Research,* (20):785-792.

World Bank, 1993. "Water Resources Management, A World Bank Policy Paper," World Bank, Washington, D.C.

2 Institutional Requisites for Efficient Water Markets

Marie Leigh Livingston

2.1 Introduction

The Role of Institutions in Water Markets

Efficient water use is critical to the economic well-being of many regions and countries around the world. In many cases, technical approaches to expanding water resources are not economically or politically viable. Increasingly, policy makers and water managers are looking to institutional innovation, in general, and water markets, in particular, as a means to achieve economic and environmental goals.

Institutional arrangements set the ground rules for resource use. Institutions are "ordered relationships among people which define their rights, exposure to rights of others, privileges and responsibilities" (Schmid, 1972). They determine who has the right to claim the benefits from resource use and who must bear the costs (Bromley, 1982). Therefore, institutions establish the incentives, information, and compulsions that guide behavior and influence economic outcomes.

Institutional arrangements establish the foundation for efficient water markets. At best, institutions create order and relative certainty for water users which facilitate the achievement of economic and social goals. At worst, they establish impediments to efficient resource use and significant resources must be expended by individuals to compensate for their obsolete or poor design (North, 1987).

The peculiar characteristics of water resources pose special challenges for institutional design. Water resources are prone to market failures that must be addressed by institutions in order to yield efficient allocation and use. Water markets may allow allocations to change in order to adapt efficiently to changing physical and economic circumstances (Howe et al., 1986; Vaux, 1986). In order to facilitate sound economic development, the institutions that underlie markets must create sufficient security and flexibility (Ciriacy-Wantrup, 1956).

Preventing Market Failures

Economic theory suggests that under specific conditions, markets will yield accurate incentives and foster efficient resource use. These conditions are restrictive but are met to a greater or lesser degree in specific resource and goods markets. When these particular conditions are not met, markets do not yield the appropriate incentives and "fail" to achieve efficient resource use (Randall, 1983).

Three conditions necessary for efficient resource allocation to occur, that are particularly relevant to water are: 1) the resource user must be certain of the quantity, quality, location, and timing of resource availability, 2) resource use must not affect, or be affected by, utilization of the resource by another party, i.e., externalities must not exist, and 3) it must be possible to reallocate the resource over time in response to changing conditions.

The unique physical nature of water makes the resource prone to market failure and inefficient use (Livingston, 1995). Water supplies are typically uncertain. Extreme within-year variability in precipitation and stream flow tends to be the rule, rather than the exception in water supply, especially in arid regions where efficient water allocation is more critical. The fugitive (mobile) nature of water also creates physical uncertainty. Without credible institutions, deferred use carries no guarantee of future availability.

Finally, water use can be rife with externalities. Because water supplies are concentrated, withdrawal, consumption, and return flows by one individual, in all likelihood, affects the quality, quantity, and timing of supply for other water users along the stream or within the same aquifer. Return flow issues are minimized on short rivers (like those found in Chile, Chapters 9 and 10) but can be very significant under other circumstances.

Without well-designed institutions, uncertainty about the physical quantity of water available at particular times and locations impedes efficient resource use by lessening the expected value of engaging in water-related activities. Moreover, when externalities exist, private costs and benefits diverge from social costs and benefits, which may create serious distortions in allocation and prevent the formation of markets altogether.

Fortunately, institutions governing water use, in general, and water markets in particular, can be structured to accommodate these peculiarities. To a considerable extent, market failures can be prevented in order to generate the security and flexibility that provide the foundation for efficient use.

Water markets can be structured to address these fundamental characteristics of water in order to create a fairly secure and flexible system of water use. Of course, it is not realistic to expect institutions to correct for every potential market failure; indeed, perfect markets rarely exist. For example, markets cannot entirely eliminate uncertainty nor is complete certainty necessary. The certainty and flexibility created by market institutions must simply be sufficient, rather than ideal, in order to be effective.

Water markets themselves tend to evolve over time and there is a variety of water markets operating around the world, ranging from rudimentary to sophisticated systems. Even very simple and/or informal water markets, if structured well, can be a stimulus for more efficient water use and can spawn continual refinements in the trading system.

The discussion in this chapter begins with a focus on individual water users, then progressively broadens in scope. Section two describes how the nature of the water supply (surface, ground or conjunctive) influences inherent externalities and, in turn, the structure of a successful water market. Part three treats the role of private and public water organizations in facilitating or hampering efficient markets. In section four, new and changing water interests, especially environmental interests (i.e., instream rights and water quality), are presented along with a discussion of the way in which they can be integrated into markets.

2.2 Water Sources, Externalities, and the Implications for Markets

Water markets must generate sufficient certainty and flexibility in order to yield efficient resource use. At a minimum, water must be tradable and separate from land. Beyond this minimum, structuring markets so that they reflect and accommodate externalities is arguably the most critical factor in institutional design. If markets enable opportunism and cost shifting, the system cannot function well in terms of economic efficiency and institutional breakdown is the likely result. On the other hand, if markets recognize, reflect, and internalize external effects on other water users when trades occur, water users will find the system beneficial and the market will flourish.

If the goal is to maximize flexibility, while recognizing the inherent interdependence between water users, then the nature of the water source becomes critical in designing a market. The scope and kind of interdependence and, therefore, the potential externalities, are largely defined by the physical connection among water users via the water itself. (The discussion here concerns only externalities on other water users.) Surface water in rivers and streams and groundwater in aquifers entail specific kinds of connections between water users that markets must accommodate in order to be efficient. These differences are explored below.

Appurtenancy

Appurtenancy between water and land constitutes an example of a poorly designed institution, in terms of flexibility based on legitimate connection. There is no economic reason that water must be connected to a particular parcel of land. Indeed, one would expect the location of water demand to change over time. Nonetheless, some countries have adopted institutional arrangements that make water appurtenant to land, which effectively establishes a ban on permanent water transfers without accompanying land transfers.

This principle pervades many areas of the developed and developing world (Maass and Anderson, 1978). The water allocation system in Valencia, Spain, is typical of many around the world. Water rights are appurtenant to specific tracts of land. In the U.S., Elwood Mead was very influential in advancing the notion that water should be appurtenant to land. This system is embedded in Wyoming water law as well as elsewhere in the U.S.. However, with changing times, there has been pressure to change this rather rigid water system.

Surface Water Variability in Streams

Much of the world that needs vigilant water management faces extreme variability in precipitation and stream flow throughout the year. In an analysis of 35 developing countries, Cestti (1989) reports that nearly half of these countries experience 40 percent variability in rainfall. Typically, this physical uncertainty has been dealt with through a combination of technological and institutional accommodations, in order to produce relatively secure water rights. The technical aspect of creating security involves developing storage facilities to capture water during periods of high stream flow (usually due to spring snow melt or high rainfall periods) and preserving it for times of lower stream flow and higher demand. This technical approach can be augmented by institutional arrangements that enhance the security of water availability and enable trades.

Priority by time of diversion. One approach to the security issue that can facilitate markets is represented by the concept of priority in time, as used in the Western U. S. The doctrine of prior appropriation (in contrast to riparian water rights) addresses the market failure of uncertainty in supply. The prior appropriation doctrine stipulates that senior (prior) water rights are fulfilled before junior rights, which have been acquired later in time. Thus, based on temporal priority, a water user can be relatively certain about the prospects for receiving water in a particular year. The more junior the right, the less likely that the right is to be fulfilled. In a market for water rights, the most senior rights would have the highest price.

In this system, a water right must be continually used in order to be valid (the loss doctrine). Non-use results in "abandonment" or "forfeiture." This principle has a two-pronged effect on security. Some economists argue that in order to protect the security of a water right, right holders use their maximum entitlement, regardless of need. This incentive results in overuse of water resources by those with senior rights. On the other hand, the loss doctrine negates "sleeper rights" (unused) which increases certainty in the overall system.

In the Western U.S., the number of senior rights expected to be fulfilled in a given year is determined by snowpack measurements and satellite monitoring of stream flows. Water commissioners are "river cops" who also monitor water withdrawals to ensure actual use and compliance with priority. Violations of priority and absence of damage rules are institutionally enforced either administratively, or through a system of water courts.

Drought poses a particular challenge in the design of institutional arrangements. In the interest of economic efficiency, the most profitable and drought-sensitive enterprises need to be assured continued water supply, while less sensitive and less economically important endeavors need to be sacrificed.

Adaptation to drought in priority based systems is achieved by combining priority with tradability. There is no guarantee that within the initial allocation of water, the most senior right carries the greatest net economic benefit. So, under the priority system, one of two approaches may be taken in order to ensure water supply during drought years. First, a sufficiently senior water right may be purchased from an existing water user. More senior water rights are more valuable in terms of greater certainty of supply. Only those enterprises with large profit margins or with high ability to pay (e.g. municipalities) will be in a position to buy senior water rights. Second, it may be possible to buy a "drought year option" (Michelsen and Young, 1990). In this case a contract is drawn up wherein a right holder with a low-valued use agrees, for a price, to forego water use and transfer it to a high-valued use under conditions of drought. This alternative does not require permanent water transfers and is, in all likelihood, less disruptive to local economies than the first approach.

While the priority system creates substantial security that undergirds water markets, the security of water rights, and therefore markets, in the Western U.S. is somewhat compromised by sleeper "reserved" water rights. Reserved water rights are those accruing to Indian or other Federal reservations of land. These lands carry an implicit right to the water necessary to achieve the designated purpose of the land and carry a priority date equal to the reservation date. Unlike other water rights, reserved rights are not lost from non-use. To date, reserved rights are largely unexercised and unquantified, which creates great uncertainty. Under most scenarios, the full and eventual use of these reserved rights will likely supplant a significant fraction of existing water uses.

Proportionality. When institutions governing water allocation are based on the principle of proportionality, the insecurity inherent in variable water supplies is shared among all water users of a given supply. Shares may be defined in many ways. They may be based on the number of rights held, the amount of land owned, or on some other grounds. When water markets are in operation, shares, rather than a specific amount of water, are traded.

In many water systems that utilize the principle of proportionality, water rights are defined as a percentage of the water available in any given year. For example, some localities in Australia use the concept of "capacity sharing" to reduce uncertainty (Dudley, 1991). Share holders in a dam project receive a certain percentage of dam capacity and stream flow. While the percentage is constant, the actual quantity of water received will vary from year to year. In this way, the structure of water rights spreads the risk of variability equally among share holders.

Drought year water allocation in Valencia, Spain, is representative of many privately organized water distribution systems. Private institutional arrangements specify that in times of shortage, water is to be allocated in proportion to land holdings among irrigators (Maass and Anderson, 1978). The result is greater time intervals between water deliveries. It appears that the primary goal is equity rather than economic efficiency. In Pakistan, proportionate allocation is found to be ineffective or unworkable due to conveyance losses. By default, because of this physical problem, water allocation is prioritized based on the proximity of the user to the head of the canal (Chaudhry and Young, 1990). Those at the end of the canal may receive very limited water.

Effective water markets are certainly not precluded in the setting of proportional water rights. Market prices will simply adjust to reflect the relative uncertainty of the rights. Firms with inelastic water demands will buy more rights than they need for a normal year or purchase options to use in dry periods.

Absence of damage doctrine. In parts of the Western U.S., notably Colorado and Texas, market transfers of surface water are possible within the overall institutional framework (see Chapters 4 and 5). In Colorado, transfers cannot adversely affect the stream flow available to senior or junior right holders. Any party who claims to be injured may bring suit to modify or stop the proposed transfer. This provision prevents potential market failures by ensuring that external impacts on other water users are incorporated into transfer decisions.

The problem of interdependence and externalities is addressed by the far-reaching doctrine of "absence of damage." The "absence of damage" rule has been translated into the notion that only "consumptive" water use can be permanently transferred, rather than the full amount withdrawn from the stream. Return flows must remain intact in order to prevent injury to downstream water users.

Establishing "absence of damage" often requires substantial time and money. Transaction costs have been cited as a reason why more water transfers do not take place, even when large differences in the value of water in alternative uses exist (Young et al., 1986). (See Chapter 7 for examples from California). Ingram and Oggins (1990) note that a transfer of 15,000 acre feet of water in Nevada required 15 months of transaction time. In Utah, 7 of the 85 million dollars spent on water rights for the Intermountain Power Project went to lawyers and engineers (Saliba, 1987).

As discussed in the context of market failure, efficient resource use requires that the externalities imposed on other water users are accounted for in decisions to transfer water. However, this stipulation does not take into consideration that there may be other parties, who are not water users, who are adversely impacted. These "secondary impacts" tend to be on people who supply water users and the community at large (Howe and Easter, 1971). For example, the Owen's Valley purchase of land and water by southern California has had major secondary impacts on the region. Some argue that it has prevented growth in the valley.

Theoretically, secondary impacts on suppliers and the community tax base are negligible if full employment exists and resources are completely mobile (Young, 1984). Under these conditions, losses in the water selling area are offset by gains in the purchasing area. However, some unemployment and imperfect resource mobility is the rule rather than the exception. Moreover, the distribution of economic activity is of concern to local officials, regardless of aggregate impacts (Nunn, 1985 and 1990; Fort Collins Coloradoan, 1987). A study of the California water bank shows differential secondary impacts by crop type and by region (see Chapters 7 and 8).

Groundwater and Conjunctive Water

Primitive groundwater rights systems tend to be proportionate, based on ownership of overlying land. Under this system, water is typically appurtenant to the land and water trades occur via land transactions. In more sophisticated systems, groundwater rights specify an absolute amount of water that may be pumped each year and water is tradable separate from land (Blumquist, 1995).

In many cases, groundwater is a stock resource rather than a flow resource like a surface stream. Therefore, when water is withdrawn from an aquifer, externalities occur in the form of reductions in the water table, which in turn, increases pumping costs to other water users. The magnitude of these externalities depends, obviously, on the rate of withdrawals in relation to the natural rate of recharge.

Unlike surface water systems, where specific water users are impacted by reductions in return flow, the impact of groundwater withdrawals is spread (although not necessarily evenly) across water users in the aquifer. It is rare for groundwater users to be protected from the externalities associated with declines in the water table. However, in some places (such as Minnesota in the U.S.), a water user must demonstrate that a new well will not interfere with other wells in the area in order to gain a permit. In a few areas, artificial recharge programs have been explored, but they are usually pursued by government agencies for the benefit of all aquifer users, rather than as compensation by and for individual water users.

When water use within an aquifer results in a declining water table, the certainty of water rights is compromised. However, if relatively accurate information is available concerning the rate of draw-down and the impact on the value of rights in the future, then the uncertainty is minimized. As in the case of risky surface water supply, this uncertainty, if it is fairly predictable, does not necessarily stymie markets. It will simply impact prices and the quantities individual water interests must buy to secure a sufficient supply.

Conjunctive water supplies are surface and subsurface water supplies that are physically connected. For example, alluvial aquifers contain groundwater that flows beneath, and sometimes joins, a surface stream and is recharged directly by the stream or via percolation of water applied nearby. Pumping from the alluvial aquifer could mean reductions in surface stream flows.

In order to preserve the integrity of water rights and water markets, institutions must be developed to reflect this connection. In some parts of the U.S., the absence of damage principle has been extended to apply to conjunctive management of surface and groundwater (Young et al., 1986). In practice, this means that in order to pump from the alluvial aquifer, and yet preserve surface flows, sometimes outside water sources must be purchased to supplement the river during pumping or pumping must be limited. Chile also has a groundwater law that allows surface water users to file a protest if someone tries to establish a groundwater right in an area that would affect surface flows (see Chapter 9 devoted to institutions in Chile).

2.3 Organizations and Their Role in Water Markets

Water organizations have a pervasive role in water resource allocation. They may be private in nature, as exemplified by irrigation companies, or public organizations, as characterized by municipalities, regional water districts, or state departments of water resources. These organizations can be instrumental in either fostering water markets or in presenting barriers to market development.

There are highly legitimate economic reasons why water organizations are prevalent throughout the world. Water supplies are naturally concentrated into site-specific common pools or streams. Moreover, water is not perfectly divisible in terms of storage or transportation. As a result, very significant economies of scale often exist in pumping and delivery. For instance, where diversion is necessary, individual irrigators are usually unable to transport small amounts of water in isolation due to costs and conveyance losses. Indivisibility means that water allocation and use must involve group decisions and actions. Therefore, water organizations become a natural management vehicle.

The Philippines exemplifies countries that face serious problems in water administration. Specifically, Cruz et al., (1987) report that the staff charged with implementing the current Water Code lacks the technical ability to estimate true water availability and to evaluate the impact of water allocations on third parties (externalities). Moreover, there is a systematic bias in water allocation in favor of landowners and against irrigators. Finally, due to organizational problems, the process is rife with delays.

There are examples of relatively efficient administrative systems in Asia. Management of the Mahaweli in Sri Lanka and of the Bhakhra Beas Management Board in India come fairly close to meeting the criteria for efficient water allocation (Fredrickson et al., 1993). Both employ a basin-wide focus and use relatively sophisticated evaluation techniques. Real time management is based on a seasonal plan done annually, a monthly policy, and weekly technical meetings.

Ideally, a private or public water organization would facilitate efficient markets by: 1) treating various water interests without bias, 2) serving as a water broker, thereby lowering the transaction costs of water trades, 3) allowing the price of water to vary according to its changing economic value, and 4) ensuring that water transfers do not impose uncompensated externalities on other water users. In reality, water organizations exhibit every combination of policies imaginable but, in general, tend to evolve towards greater recognition of market forces.

Precluding Water Markets via Administrative Rationing

The independence of a government agency potentially involved in orchestrating water transfers is important. The agency must not be biased towards or against a particular user group. Otherwise, water organizations may employ policies that disallow or usurp market functions (Gisser and Johnson, 1983). For example, it is fairly typical for water allocation during drought conditions to be decided through administrative rationing by public water organizations.

Historically in the U.S., agricultural water use typically had first priority, with municipal needs second, and industrial needs third. Given what economists know about relative willingness to pay, this priority is not justified in terms of economic efficiency. Both Mexico and Chile give priority to urban water users. In Mexico, irrigation agencies transfer water from agriculture to cities during droughts without any compensation (see Chapter 11). In this case, markets were not given the chance to generate the transfer from lower to higher valued use.

In response to the 1990-91 drought in Thailand, the Royal Irrigation Department engaged in another kind of rationing. Officially, water for electricity generation carries first priority based on its higher economic return. Consequently, reductions in the acreage of some crops were ordered. Other plantings were delayed while some were increased. Conceivably, this rationing may have mimicked what markets would accomplish but without the compensation that provides some equity through markets.

The Organization as a Water Broker

The organizational skills and leadership ability held by water organizations versus individual water users are a factor in creating water markets. Where individuals lack organizational ability, markets may be unable to function and a private or public organization may be able to supply the integration required for orderly transfers of water. However, if agencies themselves are lacking, market failure may be replaced with organizational failure.

In the U.S., New Mexico employs an administrative system for water reallocations that appears to be quite efficient. Any individual may apply for a water transfer. The application will be approved "if the state engineer determines that the change is not detrimental to existing water rights, is not contrary to the conservation of water within the state, and is not detrimental to the public welfare of the state" (Nunn and BenDavid, 1991). Therefore, great responsibility for determining economic merit rests on the state engineer. Evidence suggests that, in fact, most transfer decisions in recent years have been based on an independent evaluation of aggregate benefits and costs. Given an effective organization and a system-wide approach, this administrative system has kept transaction costs low.

Some public water organizations successfully incorporate market mechanisms to help allocate water more effectively. For example, California has historically relied on reductions in overall use and expanding supply, rather than reallocation between sectors as a means to deal with water shortage. However, in 1992, the California legislature and California Department of Water Resources (CDWR) implemented an emergency drought water "bank" (see Chapter 7 for more details).

The state-wide water bank allowed market-like transfers. However, a significant amount of administrative control was exercised. First, in 1991, all trades were administered by CDWR using fixed prices for water. CDWR bought water at $125 per acre foot and sold it at $175 per acre foot. The difference was estimated to cover transportation, negotiation and administrative costs. The water bank program was highly successful in transferring water from less to more economically sensitive sectors of the state. The big sellers were farmers in the northern part of the Central Valley of

California, while purchases were made by farmers with permanent crops and urban users in Southern California (MaCaulay, 1991).

Organizational Price Fixing and Transfer Approvals

Some private water agencies in the Western U.S. allow transfers, subject to agency-imposed conditions including price restrictions. For example, Kern County Water Agency stipulates that any water sold through the "pool" cannot carry a price in excess of the original cost of the water (Livingston, 1982). Essentially, the policy states that no profit may be achieved through transfers. This policy presumably discourages speculation, a recurrent theme in the public policy debate about water transfers. Unfortunately, in the process of negating profit the vital mechanism that spurs efficiency is rendered inoperable.

In Australia, institutions for water transfer are evolving rapidly (Delforce et al., 1990). There is increasing pressure to develop water institutions that are flexible, because economic development has been impeded by fixed allocations of water. Until recently, transfers in Australia were achieved primarily through changing the conditions attached to permits administered by the state at the time they are renewed. Since the middle 1980s, in response to the problems inflexibility generates, institutional arrangements in Australia have been changing to accommodate transfers. However, they must be pursued through and sanctified by the state bureaucracy. Additional conditions may be attached to the transfer on a case by case basis (Maass, 1990). However, the price attached to the transfer is negotiated without state interference.

Transfer Scale and Assessment of Potential Externalities

Another critical consideration is the scale of the transfer involved. In order for large-scale transfers to be efficient, a system-wide evaluation must be undertaken (Easter et al., 1986). If water transactions are taking place on a small scale and within a given use type, it is more likely that individuals working through markets can function effectively. However, the importance of the interdependency problem grows, in general, with the scale of the proposed water reallocation. Therefore, it is more likely that state agencies must be involved to evaluate, negotiate, and implement a large-scale transfer that entails substantial interdependencies.

The Colorado Big Thompson (CBT) project, administered by the quasi-public Northern Colorado Water Conservancy District (NCWCD), is often cited as an ideal, and certainly one of the very few, water markets which is very wide in scope. The scope of water transfers is made possible largely because the project infrastructure connects three major watersheds with a large centralized stored water supply. Such centralized, stored water supplies are most amenable to water markets because transportation and return flow issues are minimized. Infrastructure is often critical in enabling both simple and sophisticated water markets to operate efficiently (Shah, 1991; NCWCD, 1991). Moreover, the CBT project benefits from the legal technicality specifying that no additional rights could be gained from new return flows created by this "imported" water.

It is interesting to note that while NCWCD facilitates both permanent and rental water markets throughout the district (which extends over roughly 25 percent of the area of Colorado), water transfers outside the district are prohibited. Conceivably, this policy configuration may be due to the fact that potential externalities on other CBT water users become significant only when water leaves the system's infrastructure.

2.4 Changing Water Interests and Their Impact on Water Markets

In any water rights system, governments must be involved, at a minimum, in establishing an initial allocation of water rights and in resolving conflicts when fundamental water interests change (Brown et al., 1982). If governments fail to establish institutional ground rules, devastating impacts on water security and water markets can result. The old saying "not to decide, is to decide," certainly applies to water. Thailand exemplifies the case where there are no formalized institutional rules for water allocation in times of short supply (Johnson et al., 1990).

In the absence of formal rules, the informal rule of right by intrusion and capture applies (Randall, 1983). In this case, a water "transfer" is obtained simply by diverting water from the stream. However, the right is not secure. Additional diversions from the stream are not constrained from damaging existing use. If an entity wishes to increase its water security, the typical approach is to simply increase efforts devoted to water capture, which may negatively impact downstream users. The result is extreme overall resource insecurity. Significant pressure exists to develop institutional arrangements capable of dealing with the inevitable chaos and conflict that result from their absence.

Informal customary rights to water can be somewhat secure but are susceptible to intrusion. In both India and Indonesia, water rights are gained prescriptively, i.e., customary rights are recognized based on historical use of communal systems (Fredrickson et al., 1993). In general, such water users achieve considerable protection from intrusion from outside the basin in question. Typically, they would not be afforded the same guarantees against additional users within the basin itself. For example, the construction of a storage dam on the Chittar River in southern India has resulted in great benefits to upstream users, while downstream users have lost prescriptive rights. Shah, (1991) suggests that water markets are critical to farmers in India who did not share in the initial water allocation. In the Philippines, prescriptive water law has been entirely repealed (Cruz et al., 1987).

Ordinarily, a government establishes what constitutes a legitimate water use and the conditions under which a water right may be obtained and transferred. Even so, over time, the formal definition of a rightful water interest may change fundamentally. Governments are challenged with the task of determining when water legislation should be altered to reflect changes in social concerns. Both Mexico and Chile have recently developed new water laws and rights systems that embody major changes.

In recent decades, environmental concerns have become important in water resource management. In one sense, these concerns can be viewed as part of the change and growth in the demand for water, especially in the developed world. However, they also relate to supply in the sense that human intervention in water supply has increased to the point where environmental goals are compromised. In any case, the challenge has been to modify existing institutional arrangements in order to accommodate environmental concerns within water markets. These new institutional developments are still in their infancy. In general, the institutional arrangements involve establishing instream water rights and water quality levels.

Instream Water Rights

Environmental uses of water usually involve a specific amount of water remaining in the stream at a given location for wildlife or recreational use. Perhaps the most basic institutional issue in this area is to recognize water rights for environmental purposes as legitimate. Where this has been accomplished, two approaches have been taken. In the first case, instream water rights are integrated as "regular" in that they must fulfill the same conditions as other, more traditional rights. In many cases, this has necessitated a change in legal language. For example, in Colorado, until 25 years ago, a potential water right was deemed legitimate only if water was diverted from the stream and applied to some "beneficial" use (which did not include environmental uses). After substantial pressure from interest groups and much controversy, Senate Bill 97 was passed in 1973, expanding the definition of beneficial use and changing the diversion requirement.

In priority systems that incorporate market-based trades, instream water rights can compete directly with other rights. In principle, it would also be possible to provide for the "regular" inclusion of instream water rights in allocation systems based on proportional or administrative rationing. The desire for instream water rights has also been dealt with in a way that "trumps" existing rights. For example, California has implemented a new rule that reduces exports of water from the Delta in order to meet federal water quality standards and to protect endangered species (see Chapter 7).

Some Federal agencies in the U.S. have recently argued that their "reserved" water rights include those necessary to provide instream habitats in the reservation. These proposals have met with fierce opposition because of their potential to supplant existing uses. To date, parties have relied on negotiated compromises rather than "all-or-nothing" outcomes. Instream water rights have also been talked about in the context of "public trust" doctrines. In this case, instream water rights could constitute an up-front constraint on water transfers.

The "consumptive" use transfer rule in priority systems has been complicated by the introduction of instream water rights. If instream water rights are located near the site of a transfer, even the transfer of consumptive use may impact the quantity of water available for an instream right. In this case, the absence of damage rule may disallow any transfer of water (Livingston and Miller, 1986). This issue has been suggested as a problem arising out of the California water bank operations (MaCaulay, 1991).

Water Quality Concerns

The discussion above relates primarily to water quantity but water quality issues have also begun to constrain what water users can and cannot do in production and consumption activities. In addition, water quality concerns may constrain water market activities. In Colorado, Senate Bill 181 requires that the effect of well permits, water transfers, use changes, augmentation plans, and groundwater recharge plans be evaluated in terms of their effects on water quality. (Mexican water law also incorporates consideration of water quality changes.) Market transactions can be voided on the basis of water quality degradation. Therefore, establishing injury, or conversely "absence of damage," with regard to water transactions, now extends to water quality.

The legal requirement (such as Senate Bill 181) that water transfers be evaluated based on their impacts on water quality will raise the transaction costs of water exchanges. It may also mean that the preservation of water quality constitutes a constraint to water markets. Yet markets may also prove to be a vehicle through which water quality can be improved. For example, water rights or options could be purchased to maintain instream uses and improve water quality. Water purchased in order to preserve or recharge wetlands constitutes an example of how markets may enhance water quality and environmental goals.

2.5 Summary and Conclusion

The foregoing discussion rests on the premise that in order to foster efficient water markets, rights to water resources must be sufficiently secure and flexible. Designing institutions to accommodate the physical peculiarities of water in a way that yields accurate incentives and enables efficient water markets is complicated but highly beneficial.

Fundamentally, successful water markets must ensure that water users receive reasonable protection against externalities. Because externalities vary between surface, ground and conjunctive water sources, institutions must reflect these differences. For surface water, it is possible to have water markets in prioritized or proportional water rights, as long as "absence of damage" (i.e., externalities) in return flows to the stream is a legal guarantee.

In many places, groundwater rights need to be strengthened to prevent extensive well interference and aquifer damage due to over pumping. Markets in groundwater can be based on proportional or absolute quantity water rights. The market value of groundwater should reflect the expected decline in the water table and the expected increases in future pumping costs. Conjunctive water resources challenge policy makers to develop water institutions that are integrated in order to enable efficient markets.

Water organizations will likely play a significant role in water markets, especially in relatively sophisticated systems. These private or public water organizations can facilitate the development and operation of water markets by striving to reduce the transaction costs associated with trades and by providing technical expertise regarding the likely (external) impact of trades on other water users.

Over time, water markets must adapt in order to accommodate changing water interests. In particular, instream flows and water quality are quickly emerging as concerns that must be integrated into the institutional framework governing water trades. If carefully crafted, water markets may be useful to environmental interests in achieving their goals, without undue disruption to existing water users.

In order to be efficient and successful, water markets must accommodate the physical nature of the resource and the inevitable interdependence among water users. Institutions establish the foundation for markets and can prevent market failures that would result in the market's demise. Structuring institutions such that the overall incentives that emerge foster sound water markets poses a critical challenge for water resource professionals.

References

Blumquist, W., 1995. "Institutions for Managing Groundwater Basins," in A. Dinar and E. Loehman (eds.), *Water Quantity/Quality Management and Conflict Resolution: Institutions, Processes and Economic Analysis,* Praeger, Westport, Connecticut.

Bromley, D.W., 1982. "Land and Water Problems: an Institutional Perspective," *American Journal of Agricultural Economics*, 64(5):834-844.

Brown, L., B. McDonald, J. Tysseling, and C. DuMars, 1982. "Water Reallocation, Market Proficiency and Conflicting Social Values," in G.D. Weatherford (ed.), *Water and Agriculture in the Western U.S.*, Westview Press, Boulder.

Cestti, R., 1989. "Water Resources: Problems and Issues for the Water Supply and Sanitation Sector," World Bank Discussion paper, INUWS, Washington, D.C.

Chaudhry, M.A. and R.A. Young, 1990. "Economic Impacts of Alternative Irrigation Water Allocation Institutions: Pakistan's Warabandi System" in R.K. Sampath and R.A. Young (eds.), *Social, Economic and Institutional Issues in Third World Irrigation Management*, Westview Press, Boulder.

Ciriacy-Wantrup, S.V., 1956. "Concepts Used as Economic Criteria for a System of Water Rights," *Land Economics,* 4:295-312.

Cruz, C.J., L.B. Cornista, and D.C. Dayan, 1987. Legal and Institutional Issues of Irrigation Water Rights in the Philippines, Agrarian Reform Institute, University of the Philippines, Los Baños.

Delforce, R.J., J.J. Pigram, W. Musgrave, and R.L. Anderson, 1990. "Impediments to Free Market Water Transfers in Australia," in *Transferability of Water Entitlements*, Center for Water Policy Research, Armidale, Australia.

Dudley, N.J., 1991. *Water Allocation by Markets, Common Property and Capacity Sharing: Companions or Competitors*, Centre for Water Policy Research, University of New England, New South Wales.

Easter, K.W. , J.A. Dixon, and M.M. Hufschmidt, (eds.) 1986. *Watershed Resource Management: An Integrated Framework with Studies from Asia and the Pacific*, Westview Press, Boulder.

Fort Collins Coloradoan, 1987. "Increased Demand Creates Controversy: Northern Colorado Fights Back," Dec. 27, Fort Collins, Colorado.

Fredrickson, H., J. Berkoff and W. Barker, 1993. *Water Resources Management in Asia,* WTP 262, World Bank Technical Paper, Washington, D.C.

Gisser, M. and R.N. Johnson, 1983. "Institutional Restrictions on the Transfer of Water Rights and the Survival of an Agency", in T.L. Anderson (ed.), *Water Rights*, Pacific Institute for Public Policy Research, San Francisco.

Howe, C.W. and K.W. Easter, 1971. *Interbasin Transfers of Water: Economic Issues and Impacts*, John Hopkins Press, Baltimore.

Howe, C.W., D.R. Schurmeier, and W. Douglas, 1986. "Innovative Approaches to Water Allocation: the Potential for Water Markets," *Water Resources Research,* 22: 439-445.

Ingram, H. and C.R. Oggins, 1990. *Water, the Community and Markets in the West*, Udall Center for Studies in Public Policy Working Paper, University of Arizona, Tucson.

Johnson, S.H., S. Patamatamkul, T. Onchan, and U. Tan-Kim-Yong, 1990. *River Basin Water Management Proposal Preliminary Study Design* (Thailand), ISPAN Report No. 36.

Livingston, M.L., 1995. "Designing Water Markets: Market Failure and Institutional Response," *Water Resources Management*, 9: 203-220.

Livingston, M.L., 1982. *Water Allocation Agencies: Transfer Policy and Economic Efficiency,* NRE Staff Report AGES820609, U.S.D.A., Washington, D. C.

Livingston, M.L. and T.A. Miller, 1986. "A Framework for Analyzing the Impact of Western Instream Water Rights on Choice Domains: Transferability, Externalities and Consumptive Use," *Land Economics*, 62(3): 269-277.

Maass, A., 1990. "Water Law and Institutions in the Western U.S.: Comparisons with early developments in California and Australia, contemporary developments in Australia and recent legislation worldwide," Presentation at the Natural Resources Law Center, Sept., University of Colorado, Boulder.

Maass, A. and R.L. Anderson, 1978. *"...And The Desert Shall Rejoice":Conflict, Growth and Justice in Arid Environments*, The MIT Press, Cambridge.

MaCaulay, S., 1991. "Water Marketing in California as a Strategy to Meet Future Urban and Irrigation Demands," presentation to the U.S. Committee on Irrigation and Drainage 12th Technical Conference on Irrigation, Drainage and Flood Control, Nov., San Francisco.

Michelsen, A.M. and R.A. Young, 1990. "Drought-year Options on Agricultural Water Rights for Urban Water Supplies," in *Transferability of Water Entitlements,* The Center for Water Policy Research, Armidale, Australia.

Northern Colorado Water Conservancy District (NCWCD), 1991. "District Releases Regional Water Supply Study," *Water News*, Fall, Colorado.

North, D.C., 1987. "Institutions, Transactions Cost and Economic Growth," *Economic Inquiry,* 25: 419-428.

Nunn, S.C., 1990. "Alternative Water Transfer Forums: Social and Community Impacts," in *Transferability of Water Entitlements*, The Center for Water Policy Research, Armidale, Australia.

Nunn, S.C., 1985. "The Political Economy of Institutional Change: A Distributional Criterion for Acceptance of Groundwater Rules," *Natural Resources Journal,* 25(4): 867-892.

Nunn, S.C. and S. BenDavid,, 1991. *Expediting Change in Water Use: Hydrologic Criteria and Market Transactions*, Technical Completion Report Project Numbers 14-08-001-G1646, 1423623, 1345685, New Mexico Water Resources Research Institute, New Mexico.

Randall, A., 1983. "The Problem of Market Failure," *Natural Resources Journal,* 23(1): 131-148.

Saliba, B.C., 1987. "Do Water Markets 'Work'? Market Transfer and Trade-offs in the Southwestern States," *Water Resources Research*, 23: 1113-1122.

Schmid, A.A., 1972. "Analytical Institutional Economics: Changing Problems in the Economics of Resources for a New Environment," *American Journal of Agricultural Economics*, 54(5): 893-901.

Shah, T., 1991. "Water Markets and Irrigation Development in India," *Indian Journal of Agricultural Economics*, 46: 335-348.

Vaux, H.J., 1986. "Economic Factors Shaping Western Water Allocation," *American Journal of Agricultural Economics,* 68(5): 1135-1142.

Young, R.A., 1986. "Why Are There So Few Transactions Between Water Users?," *American Journal of Agricultural Economics*, 68(5): 1143-1151.

Young, R.A., 1984. "Direct and Regional Economic Impacts of Competition for Irrigation Water," in Englebert and Scheuring (eds.), *Water Scarcity: Impacts on Western Agriculture,* University of California Press, Berkeley.

Young, R.A., J.T. Daubert,, and H.J. Morel-Seytoux, 1986. "Evaluating Institutional Alternatives for Managing an Interrelated Stream-aquifer System," *American Journal of Agricultural Economics*, 68(4): 787-797.

3 Meeting Water Needs in Developing Countries: Resolving Issues in Establishing Tradable Water Rights

Mateen Thobani[1]

3.1 Introduction to Market-based Water Management

Recent approaches to meet the growing water demand have focused on voluntary mechanisms to conserve water and to reallocate it among competing uses rather than on developing new sources. They have also focused on alternative ways to finance the operations and management of existing infrastructure and the construction of new infrastructure. These approaches, referred to here as "market-based instruments for water management," may be divided into three categories: pricing policies, informal water markets, and formal water markets. While proponents recognize that such approaches will not resolve all water problems, they believe that they have many advantages over administrative mechanisms.

Pricing Approach

This approach recommends that public authorities raise the user price of water to reflect its true scarcity or opportunity cost, thereby inducing water conservation and making more water available for higher value uses (World Bank, 1993).[2] In principle, if irrigation water near a city could be priced at what a water company would be willing to pay for the crude water (adjusting for conveyance costs), some farmers would give up farming and others would switch to more efficient irrigation or grow less water-intensive crops. The higher water charges would free up water that could be transferred to the water company for treatment and subsequent sale. These charges would also generate fiscal resources that could be used to improve operations and management of the existing infrastructure or for the construction of new infrastructure.

However, because of serious practical and political problems, no country has been able to successfully implement opportunity cost pricing of water administratively. Even if governments could find an inexpensive way to measure and monitor water flow, there are tremendous difficulties in measuring the opportunity cost of water since it varies according to location, water quality, season, year, reliability, and use. Reliable and high quality groundwater near a city that is rationing water will have a high opportunity cost (even after adjusting for pumping costs), while more polluted water along a more distant river with a variable flow will have a low opportunity cost. Even this will vary according to the year and season. In a normal year the opportunity cost will be far lower than in a drought year just as it will be lower in the rainy season than in the dry season. Water used for hydropower generation that is returned substantially intact to the river, in both quantity and quality, will have a very low opportunity cost as compared to water used for agriculture or mining. However, if water for hydropower is stored during periods when it is needed for other activities such as agriculture, its opportunity cost could be significantly higher.

The political problems are even more intractable. It is politically difficult to charge a farmer for water from one river that serves a town (and therefore has a high opportunity cost) a much higher price than a farmer served by an adjacent river. Similarly, it is difficult to charge profitable uses such as hydropower a lower fee than poor farmers. Strong farmer lobbies are typically able to successfully pressure politicians into keeping water charges well below their opportunity cost.

Another political problem in pricing irrigation water at its opportunity cost derives from the fact that the land price already embodies the price of water rights. In areas of low rainfall, irrigated land may sell for ten times the price of unirrigated land, reflecting the expectation that the owner of irrigated land will receive water at a low charge. If charges are later raised ten or twenty-fold to reflect the opportunity cost of water, this land will drop in value to that of unirrigated land, resulting in an effective expropriation of the farmer's assets. While government actions frequently alter the value of private assets, the sheer magnitude of asset expropriation implied, the number of people affected, and the socially disruptive aspects (widespread agricultural unemployment) of such a policy make it highly unlikely that opportunity cost pricing can be introduced within a reasonable time frame.

A unique problem in pricing water has to do with return flows. When a farmer irrigates crops, only part of the water is absorbed by the plant. Depending on the efficiency of irrigation, a significant share of the water – the return flow – will seep underground, flow into a drainage ditch, or into a neighbor's fields. This water may enter an underground aquifer and may be pumped up by another user or it could even rejoin the river and be diverted into a canal. If water were priced volumetrically according to what was received rather than what was actually consumed, farmers using inefficient irrigation (thereby inadvertently helping downstream users) would pay too high a price.

While this does not mean that water should be provided without charge, it does suggest that setting water prices administratively to ensure rational water use will be difficult in practice and could be disruptive socially and politically. Therefore, alternative ways of improving water use should be explored.

Informal Water Markets

Where governments have failed to respond to rapidly changing water demands, spontaneous local (informal) water markets have developed in several water-scarce countries. Such markets are widespread in South Asia (Pakistan Water and Power Development Authority, 1990; Shah, 1991; Saleth, 1996). The author found them to have been widespread in Mexico even before the introduction of formal tradable water rights. Although many of these trades are technically illegal, governments usually turn a blind eye to them, perhaps recognizing that they are politically popular and manage to reallocate water quickly and voluntarily.

In a typical trade farmers will sell a specified volume of their surplus groundwater or surface water for a season or a specified period of time to a neighboring farmer, or a group of farmers will sell part of their water to a neighboring town.[3] In this way water is reallocated to more valuable uses without penalizing existing water rights holders. At the same time, the ability to sell provides an incentive for conserving water and using it more rationally. Meinzen-Dick (1996) finds that informal water markets in South Asia operating without government intervention are able to increase the access to water for some of the poorest farmers.

However, these unregulated trades have sometimes resulted in an economically inefficient allocation of water. In some parts of South Asia, wealthier farmers who own deep tubewells charge smaller-scale neighboring farmers a monopoly price, which results in lower crop output than if the water were priced at its opportunity cost. This also exacerbates income inequality. The opportunity to sell also leads to increased exploitation of groundwater, which can deplete underground aquifers in areas with inadequate recharge (Saleth, 1996). Moreover, it is difficult to monitor such transactions and protect the aquifers. Informal water trading has contributed to depletion of aquifers in parts of South Asia and Mexico.

Such markets also do not compensate for changes in return flows. Thus upstream users may sell more than they actually consume, thereby decreasing return flows to downstream users. Furthermore, since many of these trades are illegal, they cannot be easily enforced, regulated, or taxed. This has tended to limit water trades to spot sales or to sales for a single season, often between neighbors. Longer-term trades are non-existent, depriving potential investors or water companies of secure long-term access to water. Finally, compared with opportunity-cost pricing, informal or spot markets do not generate fiscal revenues nor do they provide sufficient means for the creation of new infrastructure.

Formal Water Markets

Several governments have passed laws establishing tradable property rights for water in an attempt to retain and extend the advantages of informal water markets while reducing some of the negative costs stemming from their illegal and unregulated nature. As with informal markets, formal markets result in water use that reflects the opportunity cost of water. They also allow rentals and "spot" sales of water; in fact, legally tradable water rights facilitate such transactions. Finally, by allocating initial water rights to existing users or holders of water rights without charge, formal water markets can circumvent the political problems associated with raising water charges and setting non-uniform prices.

Legally tradable water rights can be better monitored and enforced. They can also be more effectively subject to laws and regulations aimed at preventing the abuse of monopoly power, at ensuring that sales do not negatively affect the water available to third parties (i.e. at addressing return flow issues), and at protecting the environment.

Chile and Mexico are the only countries that have established formal regimes of tradable water rights at the national level, but many of the western states of the U.S. and some states in Australia have such systems.[4] However, for political and environmental reasons and in order to protect the rights of third parties, most Australian and American states have imposed many restrictions on water trading that have added to the cost of transactions. Depending on the state in question (each state has its own laws and regulations) and on the type of transaction, this can effectively block many potentially beneficial trades. For example, water markets in some states in the U.S., such as Colorado and New Mexico, have functioned quite well for more than 100 years, but restrictions on trading have limited the usefulness of water markets in California. As a result, farmers in California continue to grow low-value, water-intensive crops even when neighboring cities face water shortages and rationing. In California, as in some other regions of the world, negotiations among stakeholders complement other water allocation mechanisms. However, the Chilean and Mexican water market regimes are probably better models for developing countries facing water shortages.

Chile's Water Code. Under Chile's 1981 water code, the state grants existing water users property rights to both surface and groundwater without charge.[5] These rights are separate from land and, except for a few restrictions, owners may sell them to anyone for any purpose at freely negotiated prices. These water-use rights may also be leased, used as collateral, and inherited. New and unallocated water rights may be obtained through a petition from an interested party. If others are interested in the same rights, they will be sold at an auction. If not, the petitioner will receive the rights without charge.

Rights are obtained by being recorded in a public registry as either consumptive or non-consumptive, permanent or temporary (contingent). Non-consumptive rights oblige the holder to return the same volume of water to a specific location (this right is useful mainly for hydropower generation). Temporary rights can only be exercised if all permanent rights have been met. The rights are defined volumetrically (either in liters per second or in cubic meters) but revert to proportional rights (share of the streamflow or of the volume in a reservoir) if the available water does not permit all volumetric rights to be honored. This last feature is particularly important in Chile since few rivers have significant storage capacity.

Water user associations monitor, distribute, and enforce water rights at the level of the river basin, the primary canal, and secondary or tertiary canals. They own and operate the bulk of the hydraulic infrastructure and set the water fees. The government continues to manage the headworks for some large dams and reservoirs, and to charge users for its services.

Mexico's Water Law. Under Mexico's 1992 water law, users may convert their existing non-tradable water rights to more secure long-term concessions (with a typical maturity of 30 years). These concessions may be leased or sold as long as the sale does not negatively affect the water rights of other users. For non-agricultural users, farmer user associations, and groundwater users, the rights are recorded in a public registry and are defined volumetrically. However, since any deficits or surpluses are allocated proportionately, the rights are effectively proportional. For individual farmers using surface water, the rights are defined only in terms of area to be irrigated and are registered by the water user association.

Because the law mentions the possibility of forfeiture if the water is not used efficiently or if it has not been employed for three years, the rights are less secure than in Chile. However, Mexico's National Water Commission does not define what constitutes inefficient water use and the author found no evidence of users having forfeited their rights as a result of inefficient water use. Also, while the rights are for a fixed period, the Water Commission plans to renew the concessions for only a nominal administrative charge. Therefore, in effect, the rights are longer-term and more secure than they appear.

3.2 How Tradable Water Rights Resolve Water Shortages

Once the water rights for existing users are registered, they can be sold or leased to anyone for any purpose.[6] This provides an incentive to the owners of the right to conserve water and sell the surplus to those willing to pay a satisfactory price, thereby allowing water to be reallocated to higher-valued uses. Some users may even decide to sell or lease their entire holdings of water rights. They can thereby extricate themselves from an unprofitable business while being compensated for giving up their water rights. Potential buyers of water also have a strong incentive to conserve water in order to keep their costs to a minimum. Therefore, when considering purchasing new rights, farmers are more likely to use efficient irrigation techniques and water companies are more likely to seek ways to reduce water losses.

The author found many instances of voluntary and flexible water transfers to more productive uses in the water-scarce areas of Mexico and Chile. Some small farmers in Mexico were engaged in unprofitable farming activities and found themselves having to borrow to stay in business. In the past they would have gone deeper into debt or have illegally sold their water rights to more efficient farmers. Now they are selling their rights legally, thereby obtaining a better price and using the proceeds to pay off their debts. Some even work on farms that bought the water rights. Similarly, in the past, firms that needed reliable supplies of water often resorted to extracting groundwater illegally in areas where groundwater rights were exhausted. This led to increasing over-exploitation of aquifers. Now these firms buy groundwater rights legally from farmers.

In Chile, many farmers sold or leased their surplus water rights to more efficient neighboring farmers or to industrial and water companies. The sales and leases have allowed some water companies and industrial users to obtain reliable access to water without expensive alternatives such as building a new dam. For example, the city of La Serena was able to purchase 28 percent of its water rights from neighboring farmers. This allowed the government to postpone the construction of a proposed dam. Similarly, the city of Arica in the arid north of Chile has been able to meet its crude water needs by leasing groundwater from farmers, thereby allowing city residents access to water (Hearne and Easter, 1995). Such measures have contributed to Chile's success in achieving virtually universal urban coverage of water even in very arid regions.

In evaluating the Chilean water market experience, Hearne and Easter (1995) found that buyers and sellers of water in the Elqui Valley received an economic rent of \$3,047 and \$1,156 per share, respectively, after transaction costs (A share is supposed to represent 1 liter per second in a good year; in recent years, a share has averaged about 0.5 liters per second or about 15,800 m^3 per year). After adjusting for the fact that much of the water being sold had not been used by sellers in recent years but was being used by others, the net economic gain-from-trade ranged from \$658 to \$1,139, roughly equal to the purchase price per share. In the Limarí Valley, which enjoys a more reliable water supply because of three inter-connected large reservoirs, the net gains from trading water in the Cogotí Reservoir were estimated to be three times the recent transaction price of \$3,000 per share (a share represents 4,250 m^3 per year). Thus, even after deducting the cost of water transfer, water was worth three times more in its new use. Without markets, it would have been difficult to bring about such transfers without generating serious conflicts.

The author's interviews of water users in Chile and Mexico also showed that water markets led to water conservation. When Santiago's municipal water company, EMOS, was notified that it could no longer receive new water rights without charge, it initially sought to purchase additional water rights. When potential sellers demanded too high a price, EMOS decided instead to rehabilitate its aging pipe structure, thereby reducing water leakage. Similarly, farmers who must pay for water rights to expand production have an additional reason to install efficient irrigation, to use better soil management techniques, or to grow less water-intensive crops. Some Chilean farmers have even used options contracts as a way to save on buying water rights that they may use only in drought conditions. Under such an arrangement, a fruit farmer who needs to ensure that his trees do not die if there is a drought pays an annual fee to a neighboring farmer growing an annual crop for the option of buying the water for a season at a pre-negotiated price in case of a drought.

The conservation occurred even though water charges were not raised. In fact, several water user associations in Chile reported that water charges fell after the new water law was passed in 1981. Even before the introduction of the current water law, the Chilean government was charging users the full cost of system operations and maintenance (O&M), which was being done by public authorities. Under the new law, the government turned over to users the responsibility for setting water fees and for O&M. Since users were able to provide the services at a lower cost, water fees have fallen. The situation was different in Mexico. Prior to issuance of the new water law, the government provided subsidies for O&M that amounted to 0.5 percent of GDP. Just before the introduction of tradable water rights, the government began a gradual move toward charging for the full O&M costs and to turn over responsibility for O&M to users. Although users were able to provide O&M at a lower cost, the savings were not enough to compensate for the reduction in subsidies. Consequently, water users in Mexico are generally paying higher water fees.

Secure and tradable water rights reduce poverty in several ways. First, by allowing scarce resources to be redeployed for more productive purposes, they lead to increases in output and employment. This occurred, for example, when farmers in Chile and Mexico sold their water rights to more productive farmers or cities. Second, they can encourage new investment in activities that require large quantities of water. Businesses considering investment in fruit farms are more inclined to do so if they are assured that their water will not be transferred to a neighboring city in times of scarcity and that they are able to purchase additional water from farmers with lower-value uses in times of shortage. In Mexico, an investor was able to start a water bottling plant using deep tubewells by buying water rights in the same aquifer from a farmer. Not only was the farmer better off, but the increased investment also generated additional employment. A World Bank study on Peru (World Bank, 1994) found that tradable water rights have the potential to increase private investment in hydraulic infrastructure, freeing up public resources for other activities, while allowing rapid and cost-effective development of hydraulic projects.

Third, by empowering user groups to have a say on the issuance or transfer of water rights, they are better able to protect the water rights of the poor. When rights are granted without charge by public authorities it is typically the rich and politically influential who have easier access to them, often at the expense of the poor. This is the case, for example, in Peru. In several of Peru's canals and rivers, influential persons managed to obtain water rights from public authorities even after all the water in the canal or river was fully allocated. This occurred at the expense of water rights of existing farmers. Consequently, some of Peru's rivers and canals have two to three times more water rights than there is water available. Fourth, secure and tradable water rights have a higher value, and are often the most precious assets of poor farmers. In Mexico, many small farmers in the Durango/Coahila area were able to take advantage of their ability to sell their water rights while remaining on the land.

Fifth, by making it is easier for cities to obtain water, such markets benefit the poor since the poor are the most likely urban residents to have been excluded from piped service. A contributing factor in Chile's extensive coverage of urban areas with piped water has been the ability of water companies to obtain "raw" water at a reasonable price (as in Arica and La Serena). In cities such as Lima or Karachi, where municipal water availability is often limited to certain hours of the day, improved availability of raw water could allow households already connected to the municipal water company to receive water 24 hours a day (although improved access to raw water would not, in itself, lead to new connections for those who have no access to piped water). Finally, since the transfer of water to higher-value uses occurs without having to confiscate water from less productive users (farmers) and without having to build new infrastructure, it is cheaper and fairer than alternatives such as confiscating water or substantially raising water charges.

No studies in Chile and Mexico have been done to measure the effect of tradable water rights on economic growth. However, based on anecdotal evidence and on studies showing the gains from trading water, it is likely that secure and tradable water rights facilitated increased economic growth. Agriculture in Chile grew by 6 percent annually in the decade following issuance of the water law. In Mexico, more efficient farmers were able to expand their output substantially by buying surface water rights from small unprofitable farmers. Similarly, by buying groundwater rights from Mexican farmers, industry was able to expand production and employment. Without the opportunity to sell water rights legally, the adjustments following the 1995 changes in relative prices and the fall in domestic demand would have been more difficult.

In principle, water sales, if inadequately regulated, could threaten the environment if minimum flows at the lower sections of rivers are not maintained. In practice, neither the author nor Hearne and Easter (1995) found any evidence of such phenomena in Chile, mainly because Chilean farmers rarely sold all their water rights, rivers in Chile are short, and some water user associations forbid trades that might reduce water availability downstream. In both Chile and Mexico, the approval of water user associations and the public water authorities are required for trades to be legally registered. Although the Chilean water authorities recently reported instances of environmental degradation caused by water sales, such transactions do not appear to have caused serious environmental problems (Box 3.1).

In fact, by inducing conservation, water markets have postponed the need to build additional dams such as the proposed dam near La Serena in Chile, thereby averting potential environmental problems. In addition, by inducing water conservation, formal water markets may well reduce soil salinization, a phenomenon which often stems from over-irrigation upstream. While soil salinization is not an important problem in Chile, it is an important issue in countries such as Pakistan or Mexico, given their propensity for salinization in the lower reaches of their long rivers.

Box 3.1. Chile's Misunderstood Water Problems

Despite its success in improving water use, Chile still suffers from conflicts between hydropower companies and farmers, from water quality problems, and from anti-competitive behavior in electricity generation. The conflicts between farmers and hydropower companies developed when the privatized companies were granted non-consumptive water rights without an obligation to release specified volumes from their reservoirs when farmers needed them. The problem was exacerbated when companies requested additional non-consumptive rights in rivers where consumptive rights had not yet been assigned (Rios and Quiroz, 1995). This experience underscores the need to specify weekly releases for non-consumptive rights – whether or not the rights are tradable is irrelevant.

Similarly, the water quality problems arose because regulations to enforce the high standards set in the environmental law have not yet been enacted. The security or tradability of the rights was not the issue.

While not directly related to water markets, a shortcoming in Chile's water code has allowed one power company to obtain non-consumptive rights to most rivers in a bid to keep out competition. Moreover, they were obtained without charge since auction rules specify that the rights are granted to the petitioner without charge, if no other parties approach. This points to the need to have a minimum reservation price and to impose taxes on holdings of water rights (analogous to land taxes) rather than to require holders to exercise the rights within a prescribed period.

Despite these problems, Chile has far fewer conflicts and makes better use of its water than its neighbors. However, water markets clearly do not solve all water problems; nor do they work everywhere.

3.3 Issues in Establishing Tradable Water

If water markets have all these advantages and have worked well in Chile,[7] why have most countries failed to adopt them? There are many possible reasons, with the relative importance varying by country. First, for cultural or religious reasons, some countries oppose the idea that life-sustaining water should be bought and sold like any other good. Second, some fear that rich individuals or companies will buy up all the rights, thereby excluding the poor from access to water and raising equity and monopoly concerns. Third, some believe that small-scale farmers, either in desperation or ignorance, will sell their rights for a pittance and lose their livelihood. Fourth, some are concerned that water transfers will damage the environment by depleting aquifers, increasing water pollution, or changing ecosystems. Fifth, the few that stand to gain from the current system may effectively oppose changing it. Sixth, because water use often has social benefits that exceed private ones, many believe that public control of water is necessary to ensure adequate investment and low prices.

Perhaps the most important reasons for why water markets have not evolved relate to the costs stemming from setting up a new legal, regulatory, and institutional framework; from defining, measuring, and enforcing water rights; and from making necessary changes in water intakes and conveyance infrastructure to effect the transfers. Closely related to these costs are the difficulties of implementing the initial allocation of water rights, of ensuring that sales of water by one user does not affect the water rights of others (the return flow problem), and of establishing or strengthening public and private institutions to permit a well-functioning market. Given these costs, the potential benefits from trading water must be sufficiently large for governments to consider establishing water markets.

Many of the problems described above exist even under publicly-administered systems of water allocation. Water rights are difficult to define, measure, and enforce even when they are nontradable. Institutions to manage the allocation and distribution of water must still be present. Although a water market may have higher infrastructure demands to allow individual transfers of water rights, the conveyance infrastructure to transfer water to more desirable uses needs to exist or be built even under an administered system. Environmental problems and issues of access to water by the poor exist even when water is nontradable and publicly-administered. Most publicly-administered systems have had a poor record in protecting the poor. Efforts to keep water charges low have often resulted in poor service and excluded the poor even while better-off farmers and urban residents obtained water at highly subsidized prices. To protect against potential monopolies, experience has shown that a regulated private monopoly often results in a better outcome than government management (e.g. power, telephone, and gas utilities in the U.S.). Similarly, where social benefits exceed private benefits, a subsidy may be preferable to government ownership and control.

Formal water markets are likely to reduce the extent of the water problems. For example, by increasing the implicit value of water rights and by empowering users, water markets provide better incentives to define, measure, and enforce rights and to strengthen private institutions such as water user associations. The infrastructure needed to implement the transfers is likely to be built cheaper by water users than by public authorities. Moreover, the users will only build when it is economical to do so. Often governments, succumbing to vested interests, build expensive tunnels and other conveyance infrastructure even when the benefits are much smaller than the costs.

The analysis above suggests that there are certain minimum conditions that must be met before governments should consider establishing a costly, new legal and institutional framework. *First*, because of the costs of identifying potential trades and of making, recording, and enforcing changes in water intakes and conveyance infrastructure, water must be quite scarce and, therefore, of high value. In addition, the infrastructure must be flexible enough to allow trades (e.g. gates and not fixed flow dividers). *Second*, society must be willing to respect private property rights for water and enact legislation that is credible and recognizes such rights for a reasonably long period. Where cultural, constitutional, or religious factors preclude this commitment, effective formal markets are unlikely to develop or to stimulate private investment and improved water use.[8] *Third*, minimum institutional capacity in the public and private sector must exist or be developed prior to establishing tradable water rights. Private institutions, such as water user associations that operate at the level of a ditch, canal, and river basin, are needed to help establish the initial allocation of water rights and to operate the system. Public institutions are needed to initially establish the legal and regulatory framework, to register the rights, to continue to operate parts of the system that users cannot, and to resolve any disputes that cannot be resolved by user associations. *Finally,* because of the likely opposition from those with a vested interest in maintaining the status quo, the political leadership must be prepared to withstand the opposition.

If these conditions are met for a substantial part of the country, governments should consider enacting national or state[9] legislation to establish tradable water rights. The following section offers suggestions on the design and implementation of the legislation in order to address some of the concerns raised above and to ensure the effective functioning of formal water markets.

3.4 Designing Legislation for a Water Market

Once key policy makers are convinced of the advantages of tradable water rights over administered systems of water allocation, they need to address several issues to ensure the smooth implementation of a tradable rights regime. In addition, they must be strongly committed and patient. Because water allocation causes emotional responses, because there are technical and legal issues to be handled, and because there may be opposition from those that stand to gain from the status quo, the process of changing to a new regime may take several years.[10] Although the design and implementation of tradable water rights needs to be tailored to specific country circumstances, the following guidelines, which build upon the Chilean success but avoid its mistakes, may be useful.

Conduct an Information Campaign. Before a law establishing tradable water rights can be issued, it is important to have an information campaign to explain to users how formal property rights for water can help make their rights more secure and how water trading can result in mutual gains. Such a campaign would help overcome the opposition from those that have a vested interest in maintaining the status quo. The mechanics of trading could be explained, perhaps by representatives from countries with effective water markets. Subsequently, water users and other stakeholders should be invited to participate in the design and implementation of the legal framework. Discussing draft versions of the law and demonstrating willingness to accommodate the concerns of farmers and other users is key to successful implementation.

Register Rights of Existing Users Without Charge. The best way to ensure support for the law from users is to assign them rights without charge based on their historic usage. While this may give a windfall gain to some farmers, it acknowledges that the land price already reflects access to water at low prices and that the government is unlikely to directly recover the capital costs of its investment in infrastructure. Since this also unfairly rewards those users that may be taking more than their fair share, there may be merit in trying to partially rectify some of the most egregious wrongs. However, if the government were to try to use this opportunity to correct all such mistakes or to confiscate all illegally obtained rights, there is a good chance that the legislation will be blocked and the injustices will continue.

Once the rules are established, the individual registration process should be coordinated by water user associations using "block titling" methods (Holden and Thobani, 1995). Under this system, public authorities allocate rights in blocks to water user associations based on historical use. The user associations then assign them to individual users, at which point they are recorded in a public registry. The government should establish the registry and staff it with officials who are perceived to be honest and are given an adequate budget. One way to ensure fiscal autonomy for the registry is to have a small registration charge to cover operational expenses. The public media should be used extensively to enhance compliance with registration and to make sure that farmers are fully aware of the consequences of their failure to register or to sell their rights. Where there are large quantities of non-consumptive rights (for hydropower), the government should ensure that the initial allocation process does not prejudice consumptive rights. This may require specifying the minimum volume of water that will be released each week (based on the historic use of consumptive rights holders).

Assign New Rights via Auction. New and unallocated water rights should be sold via auctions conducted in an open and transparent manner, making active use of the public media. Prior to any auction, the government should verify that the water is not being used by others and that it is not needed for environmental or recreational purposes. The government needs to reserve water rights for such instream uses. It should also establish a minimum reservation price prior to the auction. Information on prices and volumes of past auctions and on transactions in the area should be made public. Any costs to enter the auction should be kept as low as possible.

Protect Water Rights of Third Parties. Even if the initial allocation procedure ensures that the water rights of existing users are protected, subsequent sales could infringe upon the water rights of third parties. This is because of the return flow problem described earlier. In countries where reuse of return flow is substantial, it is essential that procedures to protect third parties be instituted prior to authorizing trades. One way to do this would be to specify that all water rights have both a consumptive and a non-consumptive portion. While the consumptive portion can be sold without restriction, the non-consumptive part can be sold only if it does not deprive other water users. Thus for most transfers within the same river basin for the same use, owners would be free to sell 100 percent of their water rights. However, if farmers were to sell their rights to a water company whose return flows do not return to the same aquifer or river, they could sell only the purely consumptive part of the right.[11]

Because of the technical difficulties in calculating the purely consumptive portion on a case-by-case basis, this approach may not be appropriate for developing countries. However, as a basis for trade, it may be suitable to calculate averages that specify the volume of water consumed by a certain crop or activity. These published volumes are all that owners would be free to sell to those whose return flows do not return to the same aquifer or river. This procedure would reduce the need for each seller to justify the consumptive portion of the water while giving sufficient protection to downstream users (Holden and Thobani, 1995). Moreover, it would work equally well for surface and groundwater. Even though the system still has shortcomings, it would be a vast improvement over prohibiting all transfers or having no controls, as occurs in informal water markets.

In addition to the third-party hydrologic effects, there may be important third-party economic effects. While it is neither feasible nor desirable to protect against all third-party economic effects, two effects need particular attention since ignoring them could block potentially desirable transactions. First, when sales of water from one canal system to another result in the loss of revenue to operate the system, it may be desirable to have a procedure to partially compensate the water user association losing the income. For instance, some water user groups in the La Lagunera region of Mexico have developed a system whereby the buyers must pay 70 percent of the water charges to the original association and 30 percent to the new association. Second, where municipalities lose significant revenues when water rights are transferred to other regions, arrangements such as having to pay property taxes on water rights to the original municipality or introducing an exit lump-sum tax may be worth considering. This is an important issue in some western states of the U.S. where rural municipalities lose their revenue base when irrigation water is transferred to cities.

Protect Against Monopolies through Taxes and Regulations. Because of the large number of consumptive water rights holders and the high prices they will demand to give them up, there is little danger that establishing tradable water rights will lead to monopolies in consumptive water use, with two possible exceptions. First, a monopoly could occur when assigning new non-consumptive water rights, as happened in Chile (Box 3.1). Ensuring an appropriate minimum reservation price in auctions should help protect against this outcome. This could be accompanied by establishing taxes on the holdings of both consumptive and non-consumptive water rights. The tax should be imposed on the holdings of the rights and not on the purpose for which the water is used or for the quantity of water used. Such a tax has some desirable properties similar to those for land taxes: it does not distort production decisions and it helps recover public investment costs in infrastructure. The level of the tax could, at a minimum, be set at the difference in land taxes between irrigated and unirrigated land. For equity and administrative ease, small holdings of water rights should be exempt from this tax.

Second, monopolies could occur when large volumes of new water rights (consumptive and non-consumptive) are awarded to private entities when privatizing large public hydraulic projects. To protect against this, an appropriate regulatory framework for each hydraulic project being privatized should be developed prior to privatization (World Bank, 1994).

Some have suggested that water rights legislation should protect against monopolies and inefficient water use by specifying that water rights would be forfeited if they are not used for a beneficial purpose or if they are used inefficiently. However this is not usually advisable. Since the definition of beneficial or efficient use is very subjective, this could give public authorities the power to threaten arbitrary forfeiture, making the rights less secure. Especially in unregulated rivers (rivers without storage reservoirs), farmers may well choose to hold substantially more water rights than they need in a normal year. Similarly, a mining company may legitimately wish to hold large quantities of water rights because it is hoping to strike gold or silver in the future. The ability to sell water is a sufficient incentive to use it efficiently and the minimum reservation price and water taxes mentioned above should deter monopolies. If necessary, these measures can be complemented by anti-monopoly legislation.

The experience in the western U.S. and Mexico with the beneficial use criterion, while fairly effective against monopolies, has led to additional problems. In the western U.S., speculators get around the restriction by buying marginal farmland that they have no intention of farming. This needlessly raises the cost of doing business and results in higher prices for consumers. In Mexico, small unprofitable farmers who wished to sell their rights sometimes found that the only buyer was a large farmer offering them a low monopsonistic price. In contrast, in Chile, speculators have helped create more competition by adding to the number of agents in the market.

Protect Against Water Pollution and Aquifer Depletion. There is no need to change water quality standards or their enforcement when establishing tradable water rights. If the standards need revision or their enforcement needs to be improved, it could be done independently of the law establishing tradable water rights. Based on the experience of many countries, laws that threaten forfeiture of rights for water pollution activities have been difficult to enforce. It is politically and socially difficult to withhold water from a water company or even an industrial plant or farmer. The alternative of imposing escalating fines is likely to be more effective as it is more credible.

But in one area, water markets could exacerbate environmental degradation through aquifer depletion. Water markets can provide an additional incentive to pump more water from the ground. However, if the water law requires that groundwater rights be registered, formal water markets provide a feasible way to protect against aquifer depletion. In regions where aquifer depletion is a concern, this could be done by requiring groundwater users to form an association with the aim of monitoring, with public assistance, the level of the water table and each others' use. If the aquifer were not recharging adequately, the user association would decrease the extraction limits of its users proportionately. In addition, if exploitation of groundwater by a user is detrimental to others who are legally entitled to the water, public authorities, at the request of the affected parties, could establish proportional reductions in volumetric rights and bar new exploitation. Such a system is preferable to most existing regimes, whereby owners of the land above an aquifer have full rights to its water, even when their use will deplete the aquifer.

3.5 Conclusion

Three market-based approaches to water management show potential for meeting water needs: opportunity cost pricing, informal water markets, and formal water markets. Because of some practical and political difficulties, publicly-administered opportunity cost pricing is usually unworkable. Informal markets, which evolve spontaneously in response to inflexible methods of water allocation, can quickly lead to improved water use. They are also politically easy to implement. However, their illegal and unregulated nature often results in problems. Formal water markets have greater potential for success.

Economic principles and the Chilean experience suggest that formal enactment of tradable water rights permits rapid and voluntary changes in water allocation in response to changing water demands, thereby improving water use. These formal water markets can also increase user participation in water allocation and investment decisions and stimulate investment as investors are assured of their access to water. Formal water markets also help the poor by increasing employment opportunities resulting from more productive water use and increased investment in water-intensive activities. In addition, by making rights more secure and giving users increased say in the granting of new rights, the rights of the poor can often be better protected than when government officials allocate them. Thus formal water markets retain the advantages of other market-based approaches to water allocation while reducing some of their negative aspects. However, formal water markets are not a panacea and effective tradable water rights are not easy to introduce. But administrative solutions to water allocation are also difficult. Chile's experience and the demonstrated superiority of markets over publicly administered means of resource allocation suggest that markets are preferable when society is willing to respect tradable water rights, when water is scarce, when the infrastructure to effect transfers exists or can be cheaply developed, when there is a minimum institutional capacity to implement trades, and when there is political will to establish appropriate legislation.

Once these minimum conditions for the potential success of water markets exist, the challenge is to enact suitable laws to establish tradable water rights. The design and implementation of the legislation should involve all those that have a stake in how water is used. It should also pay particular attention to the initial allocation of water rights, to the methods of allocating new water rights, to the creation and maintenance of a water rights registry, and to the protection of the rights of third parties. As with any system, public authorities should ensure that adequate environmental laws exist and are enforced, and that steps are taken to protect against potential aquifer depletion.

Endnotes

[1] Another version of this article appeared in the 1997 *World Bank Research Observer,* 12(2):161-79. The author would like to acknowledge useful comments sent by Lorena Alcazar, Ariel Dinar, K. William Easter, Karin Kemper, Larry Simpson and Ashok Subramanian.

[2] Note that a policy of pricing water to cover the full cost of building and managing the infrastructure (the long-run marginal cost) is not optimal if the infrastructure was ill-conceived and built at high cost. If full cost pricing could be enforced, most farmers, who typically account for the bulk of water use, would find irrigation farming unprofitable and would be forced out of irrigated farming.

[3] It is useful to distinguish between the water charge paid by a user for system operation and maintenance and the price paid to a seller for the use of a given volume of water or for leasing a certain volume of water for a given period. To draw an analogy from the condominium market, the former is the condominium fee while the latter is the fee paid for renting the condominium from its owner.

[4] See Rosegrant and Gazmuri Schleyer (1995) and Pigram et al. (1993) for descriptions of the California and Australian systems respectively, as well as Chapters 7 and 8 in this book. There are also pockets of informal water markets in some countries. In the town of Reus in Spain (Tarrech et al., 1994) and in the Cariri region in the Brazilian state of Ceara (Kemper et al., 1996), well-regulated water markets have existed for over 100 years, even though they are not consistent with the respective national and state water laws.

[5] However, the registration process itself is costly because it includes the placing of announcements in major newspapers. This has discouraged most small farmers from registering their rights. Even so, these farmers' rights continue to be honored; however, they are unable to sell such rights independently of land. See Chapter 9 in this book for more discussion of Chile's water institutions and Chapter 10 for a detailed discussion of the economic gains from water trading in Chile.

[6] Continuing with our condominium analogy, the sale price is the cost of buying the condominium while the lease price, as in informal water markets, is the rent that the condominium owner may charge a tenant.

[7] Given limited sales of water rights in Chile as well as some serious water-related problems, some observers question this statement. In fact, while there are few trades in the high rainfall southern regions and in canals that use fixed flow dividers, there are many beneficial sales and leases of water in the water-scarce north. Also, Chile's water problems are largely unrelated to water markets (Box 3.1).
[8] It may be possible to get around the constitution as was done in Mexico by calling water rights long-term concessions but treating them in a similar manner to property rights for land.
[9] Having a national law as in the Chilean and Mexican cases is preferable; however, if political factors make this very difficult, or where constitutionally the states have responsibility for water, then one should consider separate state laws such as in Australia and the U.S..
[10] In Peru, despite more than three years of debate and technical assistance, passage of a law establishing tradable water rights is still uncertain.
[11] In Chile, where rivers are short and there are few return flow issues in most rivers, water user associations allocate return flow losses proportionately among all users. In rivers with sizeable return flows, user associations prohibit trades between certain sections of rivers.

References

Hearne, Robert R., and K. William Easter. 1995. *Water Allocation and Water Markets: An Analysis of Gains-from-Trade in Chile.* World Bank Technical Paper No. 315. World Bank, Washington, D.C.

Holden, Paul, and Mateen Thobani. 1995. "Tradable Water Rights: A Property Rights Approach to Improving Water Use and Promoting Investment," *Cuadernos de Economía.* No. 97. (December), p. 263-289.

Kemper, Karin, Yarley Goncalves and William Brito. 1996. *Water Resources Allocation and Use in the Cariri Valley – Ceará, Northeast Brazil.* Linköping University, Linköping, Sweden.

Meinzen-Dick, Ruth. 1996. *Groundwater Markets in Pakistan: Participation and Productivity*, IFPRI Research Report 105. International Food Policy Research Institute, Washington, D.C.

Pakistan Water and Power Development Authority. 1990. "Trading of Canal and Tubewell Water for Irrigation Purposes." P&I Publications 358 (December).

Pigram, John J., Robert J. Delforce, Michelle L. Coelli, Vol Norris, George Antony, Raymond L. Anderson, and Warren Musgrave. 1993. "Transferable Water Entitlements in Australia," Report to the Land and Water Resources Research and Development Corp, Center for Water Policy Research, Armidale, Australia.

Rios, Monica, and Jorge Quiroz. 1995. "The Market for Water Rights in Chile: Major Issues," *Cuadernos de Economía*, No. 97. (December), p. 317-345.

Rosegrant, Mark W., and Renato Gazmuri Schleyer. 1995. "Reforming Water Allocation Policy through Markets in Tradable Water Rights: Lessons from Chile, Mexico and California," *Cuadernos de Economía.* No. 97. (December), p. 291-315.

Saleth, R. Maria. 1996. *Water Institutions in India: Economics, Law and Policy Institute of Economic Growth.* Commonwealth Publishers, New Delhi, India.

Shah, Tushaar. 1991. "Managing Conjunctive Water Use in Canal Commands: Analysis for Mahi Right Bank Canal, Gujarat." In Ruth Meinzen-Dick and Mark Svendsen (eds.), *Future Directions for Indian Irrigation*. International Food Policy Research Institute, Washington, D.C.

Tarrech, Rámon, Manuel Mariño, and Glen Zwicker. 1994. "The Siurana-Riudecanyes Irrigation Subscribers Association and Water Market System." Draft Mimeo.

World Bank. 1994. *Peru: A User-Based Approach to Water Management and Irrigation Development.* Report Number 13642-PE. World Bank, Washington, D.C.

World Bank. 1993. *Water Resources Management, A World Bank Policy Paper.* World Bank, Washington, D.C.

4 The Application of Water Market Doctrines in Texas

Ronald C. Griffin

4.1 Introduction

The State of Texas is the U.S.'s second largest state in terms of land area and it has recently become the second largest state in terms of population (U.S. Department of Commerce, 1995). It is hydrologically diverse and, at over 692,000 km^2, it is larger than some countries. As a coastal state, some river basins lie fully within the State while other rivers originate in upstream states. Precipitation at Texas's arid western tip averages 20 cm annually (Texas Department of Water Resources, 1984). There, large expanses of land may be too parched to provide forage for livestock, groundwater availability is meager, and the few available surface water supplies are heavily used. Contrarily, at the humid southeastern edge of the State, precipitation averages 142 cm annually (Texas Department of Water Resources, 1984). There, water is abundant, rice is an important crop, and flooding is a periodic problem.

Between these two extremes lies a full spectrum of water availability issues, but Texas's recent population and commercial growth has put pressure on its water supply in all but the easternmost part of the State. Consequently, intersectoral competition for water, either by market, political, or judicial means, has become fierce in some areas. These tensions sometimes extend beyond the State's borders. For example, water-related litigation between Texas and the State of New Mexico has occurred in recent times. In some situations legally enforceable river basin compacts specify inflows that states upstream of Texas must provide. Similar agreements involving treaties between the U.S. and Mexico have substantial implications for Texas, which has a very long river border with Mexico, but these agreements lack the same enforceability of interstate compacts. This fact became pronounced during the 1996 drought that accentuated institutional failures concerning water allocation.

Reservoirs are important components of the current water supply infrastructure in Texas, but the federal government's withdrawal from dam building has shifted all costs to local authorities, effectively halting reservoir construction. The protection of streamflows for the maintenance of water quality, recreation, and endangered species has become an important matter during the past decade. Likewise, wetland loss is regarded as a social problem in Texas. Groundwater exhaustion is an issue for most Texas areas with groundwater. Saltwater intrusion and subsidence have become problematic in some coastal aquifers.

Texas circumstances indicate ripe opportunities for water market institutions of various kinds. In some noteworthy situations, Texas has harnessed the positive incentives created by water markets. This has been accomplished where a properly designed system of property rights to water has been established. In other ways, however, Texas has failed to employ water rights that support efficient water allocation. Lessons learned here can be as educational as the success stories. Both kinds of experiences are discussed following an overview of important Texas water institutions.

4.2 Surface Water

Surface Water Institutions[1]

"Texas, one of [the] 'dual-doctrine states', first adopted the riparian doctrine and later superimposed on it the prior appropriation system" (Templer, 1981). Prior to statehood, Texas water administration was influenced by the Spanish settlement that commenced during the 1600s. Spanish explorers and missionaries brought their legal principles relating to water. In this system, land was apportioned by governmental grant with or without specific rights for water access (Kaiser, 1987). Spanish land grants law had aspects of both prior appropriation and riparian doctrines (Dobkins, 1959). It was similar to the prior appropriation system insofar as the right to divert water had to be explicitly obtained from the government. It was similar to the riparian system in that only riparians could have rights, and rights were not limited to specific water quantities.

In 1840, the Republic of Texas[2] embraced English common law with its riparian water law doctrine. Unlike the Spanish system, the English riparian doctrine gave all landowners adjacent to streams the right to use surface water as long as their diversions were reasonable. However, the drought in the late 1880s and early 1890s demonstrated the need for a new system. The Texas Legislature looked to other western states for a model law and adopted the prior appropriation doctrine. A series of statutes, specifically the Irrigation Acts of 1889 and 1895, progressively required individuals to obtain water rights from the state instead of exercising riparian claims. The unappropriated waters in the arid regions were all declared state property. Riparian rights were not removed though and new appropriations were to be allowed only when they could not harm preexisting riparian rights (Skillern, 1988). In 1913, a new law required application to a state agency to obtain a permit to take water from Texas streams (Willatt, 1995). Still, the combination of riparian and appropriation doctrines in the system led to many conflicts.

When claimed water rights greatly exceeded water available during the drought of the 1950s in the Rio Grande Valley, the multiple systems of water rights collided noisily. In 1956, the State of Texas filed a lawsuit against 40 water districts and some hundreds of corporations and individuals to have a state court adjudicate the water rights in the lower part of the Rio Grande (Caroom and Elliott, 1981; Holbert, 1984). Fifteen years later, adjudication of Valley water rights was completed by the courts. This established the first conducive environment for Texas water marketing. The enormous effort and expense[3] spent upon this litigation, however, indicated that another approach to defining surface water property rights was needed.

In 1967, the Legislature adopted a statute to merge all surface water right claims into a uniform permit system. This statute initiated the administrative adjudication of all surface water rights. Under this Act, individuals claiming riparian water rights and unrecorded users of state water were requested to file claims by September 1, 1969. As a result of subsequent proceedings, all the river basins of Texas are now completely adjudicated – meaning that the administrative process of resolving competing claims has been finished for surface water. Except for the Rio Grande Valley, Texas surface water rights have been quantified and each has an assigned seniority based on the date diversion began. In the Rio Grande Valley, all surface water rights are somewhat correlative because the court chose to manage it that way – there are no assigned seniorities except that municipal rights have greater seniority than irrigation rights (Characklis et al., 1997). In periods when water supply is insufficient to satisfy all right holders, the shortages are shared equally by irrigators. In periods of unusually high water supply, all right holders are allowed to use water in excess of their entitlements.

Adjudication was completed for the Rio Grande Valley about 20 years prior to the rest of the State, so the earliest water marketing pertains to this region. In fact, surface water marketing remains rare in all areas of Texas except for the Rio Grande Valley. Reasons for this phenomena are likely many, but a few can be readily identified. In part, surface water markets are "thin" in much of the State because either (i) water right enforcement is lax, (ii) groundwater is an available, loosely controlled option, (iii) a water supply organization dominates the provision of water in the region, or (iv) water scarcity is low.

Surface Water Organizations

In addition to the unique body of water law used in Texas, the Legislature has authorized over 1,000 water districts of various types (Smerdon and Gronouski, 1986). Many of these districts have been influential in the regional management of surface water. The most powerful of the surface water management districts are generally labeled "river authorities" although their jurisdictions may not include an entire river basin. Texas river authorities often hold title to some surface water rights, but they do not own or control all the water in any basin. Authority and ownership of water may be a consequence of reservoir construction in which the authority participated or of acquisitions made by the river authority. Across Texas, 13 river authorities are thought to possess 19-25 percent of the consumptively used surface water of the State (Harper and Griffin, 1988; Kaiser, 1996).

The sizable water right holdings of river authorities and other water districts exert some influence on the role of water marketing in Texas. In some regions, authorities or districts are very dominant water supply organizations with the implication that these organizations possess market power. In such areas, growth in water demand tends to be accommodated by using long-term contracts with the authority/district. Managers of these organizations are adverse to the sale of water rights because these rights are viewed as essential to the security of the organization. They do favor the financial stability of long-term leases, and, as organizations which do not receive any state funding, they are generally interested in serving higher-valued water demands (Harper and Griffin, 1988).

Surface Water Transfer

Every sale of surface water rights requires approval of a state agency, the Texas Natural Resource Conservation Commission. Short-term leases may be separately regarded by the agency if the exchange is local and intrasectoral – thus minimizing third-party considerations (such as from one irrigator to a neighboring irrigator). Procedurally, water sales obtain approval by *amending* the water right in a four-step process. The first step is the filing of an application for amendment with the Executive Director of the Commission. Once an application is filed, it is reviewed for completeness with regard to informational requirements and for technical information.

After satisfactory review by the Executive Director, action by the Commission begins. Public notice is unnecessary if the amendment application does not contemplate an additional consumptive use or an increased rate or period of diversion and if, in the judgment of the Commission, the amendment application has no potential for harming other water rights. However, typical water market transfers may adversely affect other water rights and therefore require that the Commission give public notice by mail and publication. If, within the 30-day period following publication of the notice, no request for a public hearing has been submitted by the Executive Director, the applicant, or an affected person, then the Commission may take action on an application without holding a public hearing. However, if a request for public hearing is made, a hearing will be held within 30 days after the notice of application has been published or mailed before the Commission can take action.

After the hearing proceedings, a final determination shall be made by the Commission for resolving the application, following the hearing examiner's proposal. In the event that someone wants to appeal the final decision, a motion for rehearing can be filed. If the motion for rehearing is granted, the original decision is nullified and a rehearing is conducted by the Commission. Further review in the judicial system is necessary in the case of continued appeal.

Because of its earlier adjudication by suit and its unique hydrological character, special rules have been adopted for water resource administration in the lower segments of the Rio Grande Valley. The general amendment procedure applies to the Valley with two exceptions. First of all, water rights cannot be transferred from the lower and middle segments to points upstream. Second, mailed and published notices are not required as long as transfers are made within either the lower or middle segment. The rationale for public notice is to inform any party whose rights may be adversely affected, so that they may contest the application. Because of the aridity and drainage of the region, there is little possibility of return flow to the Rio Grande and, hence, diverted water is considered to be completely consumed. Water withdrawn from the river either evaporates or drains away from the river, thereby minimizing third-party effects. Therefore, no public notice concerning potential water market transfers is deemed necessary. For these same reasons, the leasing of water receives summary approval in the region as long as the exchange is intrasectoral.

The Economics of Surface Water Marketing

An earlier study by Chang and Griffin (1992) found 152 water market sales involving Rio Grande water rights. Transfers from agriculture to municipalities dominated the amount of actual water transferred. Transfers of 94,312,569 m^3 (74,966 acre-ft) from agricultural to municipal use constituted nearly 94 percent of total transfers in terms of water volume.[4] More broadly, 99 percent of all water transferred was from agriculture to nonagriculture. Furthermore, and crucially, 45 percent of all Valley municipal water rights existing in 1990 had been obtained by transfer during the preceding two decades. This clearly suggests how important water markets can be for an arid or semiarid region experiencing substantial population and economic growth. There are no broadly empowered river authorities operating in the region but in addition to the city utilities, there are several large irrigation districts. These districts hold title to the majority of agriculturally employed water in the Valley. This is an important fact because past agriculture-to-urban water sales have transferred privately-held water rights, not district-held water rights. As noted earlier, water districts view water sales negatively, opting instead for leasing and delivery contracts. As the supply of privately-held agricultural water rights dwindle, irrigation districts are likely to become active in water marketing, though perhaps through leases and delivery contracts. In order for leasing to offer a practical approach for irrigation district participation, the underlying water rights must first be converted to nonagricultural water rights.

Although recent prices have been higher[5], Rio Grande water rights have generally traded at prices ranging roughly from \$350 to \$475 per 1,000 m^3, and water leases for one season have ranged from \$12 to \$14 per 1,000 m^3. Chang and Griffin (1992) selected representative 1983-84 sales to estimate the agricultural value of traded water as well as its new value in municipal use. The difference between these values is an indication of the net benefits of water marketing.

Consultations with transactors indicated that much of the agricultural water sold would have otherwise been unused by its owners[6]. Agricultural water values can be bounded from above (that is, overestimated) by assuming that irrigation water would have been devoted to cotton production, the primary irrigated crop in the region. A budgeting methodology detailed in Chang and Griffin (1992) was employed to obtain the value of water in cotton irrigation. The procedure involved a careful comparison of production budgets for irrigated and dryland cotton. For a range of cotton yield and price scenarios, the value of irrigation water was determined to be –\$3.50 to \$77.41 per 1,000 m^3 per year. The slight negative values occur for yield-price situations in which it would have been more profitable to pursue dryland cotton production. When transformed to a present value using a 6 percent discount rate and a 50-year horizon, as is required to estimate the market value of a permanent water right transfer, this range becomes – \$59 to \$1,300 per 1,000 m^3. These values seem consistent with regional water market prices, because the assessment technique tends to overestimate the value of agriculturally employed water, and, since water market transactions presumably create some surplus or rent, it is reasonable to expect that sellers would receive a portion of this surplus.

Similarly, it is anticipated that the buyers' assessment of water value would exceed sales prices. The net present value of urban water supply increments (such as water right purchases) can be calculated from a measure of consumer surplus change. A modeling procedure outlined by Griffin (1990) employs scenarios composed of community water demand functions, population growth, water rate growth (the price paid by households), and the temporal distribution of the increased supply during the year to compute consumer surplus changes. Knowledge of a point on a water demand function and an estimated elasticity at that point permits local approximation of the demand function. After demand has been specified in this manner, expected population growth can be used to shift demand over time. If supply is increased by water right purchase, the value of the increment in each time period is obtained as a consumer surplus change. With real water charges increasing over time and a downward-sloping demand curve, as typifies the Texas experience, the per capita quantity demanded is progressively curtailed and consumer surplus per capita is reduced for the water right purchase. It is therefore important to consider the influence of water charges on the municipal value of a water right. Working in the opposite direction, population growth expands total consumer surplus.

Using Texas urban water demand elasticities provided by an earlier study and specific water and population information for two Rio Grande Valley cities, Chang and Griffin have estimated a range of potential urban assessments for water right purchases (Chang and Griffin, 1992; Griffin and Chang, 1991). To produce comparable values to the agricultural water evaluations calculated previously, a 6 percent discount rate and a 50-year planning horizon was also used to compute the net present value of water in urban use. Using actual conditions across both cities, the net present value of acquiring 1,000 m^3 of water ranged from \$7,700 to \$17,100.

These findings indicate that the net value of these transfers can be rather large. If the average transfer has indeed produced \$12,000 of net benefits per 1,000 m^3 of water, there is a sizable aggregate value for the 94,000,000 m^3 transferred from agricultural to municipal use prior to 1991. However, the extension of these results to the rest of Texas or other areas of the world is ill-advised for several reasons. The return flow externality and secondary economic effects issues are largely absent in this region. Return flows are negligible because of proximity to the Gulf of Mexico and water transfers are generally local. Moreover, instream flow values are relatively unaffected by marketing in the Rio Grande Valley.[7] Finally, this is a region of very high population growth which causes high municipal benefits.

These results provide some testimony, albeit a likely best-case scenario, regarding the potential merits of market-based policy. In the absence of transferable water rights, what would have transpired during the twenty-five years since adjudication in the Rio Grande Valley? After all, marketing operations have approximately doubled municipal water supply during this period. In all likelihood, there would have been a substantial amount of political haranguing, calls for legislative action, formation of special interests groups, and excruciating interdisciplinary studies of the water problems of the Rio Grande Valley basin. Instead, reallocation in the Rio Grande Valley has occurred steadily and without fanfare over many years.

All is not well, however, with surface water allocation in the Rio Grande Valley. The recent summer drought of 1996 exposed some design flaws in the institutional makeup of the Valley's water market which is based on a different framework than that employed across the rest of Texas and most other western states. Two such failures seem identifiable.

First, the absence of a seniority system for Valley water rights may have exacerbated drought unpreparedness. Because Valley water rights are correlative within the distinct municipal and agricultural classes, shortages are shared equally across all water rights in each class. Under this water rights structure, more risk-averse agricultural water users have had incomplete opportunities to acquire senior water rights as a means of protecting their valued interests during drought. Fortunately, the degree of this problem has been dampened by turning to short-term water leasing, but such alternatives can be incomplete remedies in some situations. For example, the prohibition on intersectoral water leasing is arguably a source of potential inefficiency even though intrasectoral water leasing is allowed (Characklis et al., 1997).

Second, the drought has underscored the problems of water management along an international border. To support a workable market, a water rights system requires considerable resolve on the part of government authorities to enforce and protect property rights to water. In an international setting, both governments must possess this resolve, since either one can fracture the functionality of water markets.

4.3 Groundwater

Groundwater Institutions[8]

The legal doctrine for managing Texas groundwater resources evolved differently than that for surface water. Groundwater law is incongruent with surface water law in Texas, as is often the case in other state jurisdictions. The unique feature found in Texas is that groundwater is administered according to the principle of *absolute ownership* – another doctrine derived from English common law. Texas landowners may extract groundwater from beneath their land as long as their use is not wasteful. The condition of no waste is a very light burden since Texas interpretation considers only the landowner's water-use practices, not in relation to alternative, and perhaps more valuable, uses by other individuals. The landowner does not hold title to specific units of water beneath his land and, consequently, groundwater marketing has been limited to two forms, as described below.

The right to use available groundwater, not the groundwater itself, is transferred with the land, so one rough method of groundwater transfer is to transfer the land. There are difficulties with this method. Parties to the transfer may desire water exchange, not land exchange. The necessity of land exchange may be economically burdensome to the parties and may easily dissuade an otherwise advantageous water trade. Additionally, the water underlying a given land tract is commonly a mobile resource and there are no institutional remedies to prevent its extraction by neighbors. Therefore, the transfer of land resources offers no guarantee that underlying groundwater will be possessed by the land purchaser.

A second available form of groundwater transfer is that the landowner may "reduce groundwater to ownership" by pumping it, after which it can be sold and transported. In this case, the site of withdrawal must be sufficiently proximate to the water user(s) to allow economic transfer of water by pipeline. Hence, the groundwater aquifer is not a legal conduit for groundwater transfer. In Texas, it is not possible to employ aquifers as water conveyances even when it would be economically efficient to do so.

Despite occasional use, these two market forms are severely limited in their ability to address growing water scarcity. A fundamental problem with the absolute ownership doctrine is that a groundwater user cannot contract with another groundwater user for a reduction in one's pumping so that the other may increase pumping. This is a serious issue that erects a great barrier to achieving economically efficient water allocation. A second problem is that owners of small land tracts are not encouraged to defer water pumping to the future, because conserved water may be taken by neighbors. Economists have argued in favor of legal reform to achieve more comprehensive groundwater marketing (Gisser, 1983). The basic reform would be to adjudicate groundwater rights and assign quantitative permits to users. By adjudicating both the existing stock of groundwater and future recharge, users can bargain regarding reallocation, use the aquifer as a conduit, and defer water use to the future by refraining from pumping and storing water in the aquifer.

The extreme freedom extended by absolute ownership to landowners in making pumping decisions infers that Texas groundwater is "managed" more as an open-access resource than a common-property resource. Open access resembles a first-come, first-take, free-for-all that is barren of restrictions, whereas common property arrangements employ social expectations and rules of conduct governing resource use (Ciriacy-Wantrup and Bishop, 1975).

To some degree, however, the adjudication of groundwater has begun to take shape in Texas and transferable groundwater rights appear to be in Texas's future. A very remarkable and important groundwater body in south central Texas, the Edwards Aquifer, has been the focus of extremely intense debate, litigation, and legislative deliberations during recent years. Part of the contentious nature of the Edwards debate stems from the consideration of two major springs which are fully dependent on aquifer outflow. These constant-temperature springs support four endangered or threatened species[9] and they also form a very important recreation resource for water-based sports. In addition to concern for springflow protection, the capacity of the aquifer to support its many well owners has been jeopardized by rapidly growing pumping. The city of San Antonio, the nation's ninth largest city, obtains all of its water supply from the Edwards Aquifer. Due to the geology of the aquifer, its level changes very quickly in response to either recharge or withdrawals. This implies that the aquifer is a stochastic resource, much like a river.

Although the smoke has yet to fully clear from this political and legal battleground, new legislation has initiated a new path for allocating Edwards Aquifer water. Many experts believe that these changes may eventually serve as a model for groundwater legal reform across the State.

Groundwater Transfer

The new approach for the Edwards Aquifer parallels Texas's painful abandonment of the riparian doctrine for surface water. As noted earlier, the transition from riparian principles began more than a century ago, but it took a recent adjudication process to fully lay the riparian doctrine to rest. It may not take so long for the Edwards Aquifer.

Recent legislation for the Edwards Aquifer includes directives and procedures for adjudicating groundwater rights. These rights are to possess a high degree of transferability. Since the passage of this legislation in 1993, it has been beset by legal challenges contesting its constitutionality. The Texas Legislature acted to refine this legislation in 1995, but challenges continued to thwart its actual application. As these legal struggles are completed, the process of assigning groundwater rights is beginning – thereby initiating the severance of land and water rights. Subsequently, the exchange of water rights by sale or lease will be possible and the absolute ownership doctrine will be extinguished for the Edwards Aquifer.

The primary features of the new legislation for the Edwards Aquifer are to:

- establish an oversight Water Authority;
- require mandatory well metering at the Authority's expense;
- adjudicate groundwater rights based on proven water use in recent times;

- establish declining target levels for total aquifer water rights to improve springflow protection over time;[10]
- charge the Authority with drought management responsibilities to be exercised using a discretionary/nondiscretionary classification of all water uses;
- allow the sale of water rights by water right owners installing water conservation equipment; and
- allow the lease of water rights, limited to 50 percent of an irrigation water right.

Close inspection of the terms of this new water doctrine reveals additional parallels with the evolution of surface water law and some economic flaws as well. For example, in spite of the climate-driven variability of Edwards groundwater, this law establishes correlative groundwater rights (as with Texas's first experience with surface water rights in the Rio Grande basin). A seniority system is not employed. The primary mechanism provided for addressing dry-year water allocation issues is the suspension of "discretionary" water use hinging on the Authority's chosen distinctions. This places the Authority in an unenviable position and, unless the Authority opts for an explicitly economic criterion, water allocation during drought cycles will likely exhibit economic inefficiency.

A seemingly unfortunate provision of this law is the required prerequisite for the sale of groundwater rights. The transferred water must be from water freed by the installation of "water conservation equipment." For example, this clause does not allow water transfers made possible by the simple cessation of irrigation by a water-right holder. It may become possible to circumvent this economically inefficient oversight by regarding the installation of a well cap as "water conservation equipment." Ultimately, however, this judgment falls squarely to the new Authority.

In establishing the "conservation equipment" clause in the new law, the Texas Legislature demonstrated sensitivity to the politically visible desires of the agriculturally-oriented region overlying the upgradient portion of the aquifer. Many people of this region have a disdainful attitude for the prospect of water marketing. They see water sales to San Antonio as the export of their economic base which is somewhat dependent upon irrigated agriculture. The same concern is the likely origin for the 50 percent limitation on water leasing by irrigation water right holders.

The Economics of Groundwater Marketing

Although the Edwards groundwater market has yet to get underway, evidence regarding its potential economic rewards is obtainable through the economic modeling of alternative water uses. A recent dissertation undertakes both a static and a dynamic analysis of water market potential for the aquifer (Williams, 1996). Whereas the static work involved an economic model of agricultural, municipal, and industrial demand under alternative water supply and demand conditions, the dynamic analysis was based on a multi-year model in which each year's decisions influence the stock of groundwater available for the following period. The competing demand for springflow is not incorporated in the surplus-maximizing model that is central to these analyses, so springflow concerns are addressed by imposing limits on aggregate pumping from the aquifer. Using the first of the pumping limits being phased in by the new legislation noted previously, the static analysis finds that a water market can raise regional net benefits (surplus) by $3,000,000 annually. For the second, more severe pumping limit, the net benefits of water marketing increase to \$7,400,000 due to the heightened scarcity. These results do not incorporate the transaction costs consequent to the necessary exchanges of water rights. The dynamic analysis reveals similar results.

There is fresh evidence that water marketing will soon commence in earnest for the Edwards Aquifer. The aquifer's newly created Authority has initiated a pilot effort in which agricultural producers will be paid not to irrigate during the 1997 season. Present plans call for some of the pumping reductions to be used for spring-flow enhancements. Bids have been received from 123 farmers concerning over 66,000 irrigated hectares and municipal interests have earmarked \$2,000,000 thus far for entering into one-year leases with the most advantageous bids (Needham, 1997). Interestingly, initial information about this pilot program indicate one-year lease prices which average \$93 per 1,000 m^3 – more than five times the values judged necessary by a recent research project focused upon dry-year lease opportunities for the Edwards (McCarl et al., 1996). One must be mindful, however, that water right holders who are leasing their water property expect to receive some of the economic surplus attributable to such transactions.

4.4 Texas Water Banking

In addition to the surface water and groundwater markets discussed above, 1993 Texas legislation established a complementary marketing institution known as the Texas Water Bank. Water banks are a relatively new tool for water reallocation in the western U.S. (see Chapters 7 and 8 that discuss the California Water Bank). While the available features of a water bank are highly variable, they have been broadly characterized as "an institutionalized mechanism specifically designed to facilitate the transfer of water use entitlements" (MacDonnell et al., 1994). Rather than relying fully on the rational behavior of potential water buyers and sellers to motivate trade, a water bank provides a more active role for the government as an intermediary for and facilitator of trade. The government's water bank accepts "deposits" of water rights from right holders and allows for the "withdrawal" of these permits by others. Naturally, deposits are monetarily compensated and withdrawals are similarly charged. While some theorists see water banks as an efficiency-enhancing policy tool, capable of moderating transaction costs and streamlining third-party checks (MacDonnell et al., 1994), others view water banks suspiciously. From the latter perspective, water banks can be viewed as a potential source of non-competitive market power or, alternatively, as a clumsy mechanism unable to advance the specialized desires of water market participants (Smith and Vaughan, 1994).

The Texas Water Bank has been widely empowered as a very active partner in water marketing (Texas Water Development Board, 1994). The Bank can purchase or sell surface water or groundwater rights. It can negotiate transfers or serve as a "bulletin board" for posting offerings and requests. Water transfers can still occur without Bank involvement. Indeed, this appears to be the norm thus far, as the youthful Texas Water Bank has not, as of last report, enabled any water transfers (Texas Water Development Board, 1994). Interestingly, only 50 percent of any particular water right may be exchanged through the Bank although non-Bank transfers are not subject to any such limitation.[11] Even though the Texas Water Bank has been idle, the establishment of this institution may find importance in future years and its presence does provide evidence of a political commitment to achieving self-determination and economic efficiency in water allocation.

4.5 Conclusion and Lessons Learned

Throughout many decades of trials with various water doctrines, Texas water law has progressively evolved toward doctrines which allow water transfer. For surface water, this process is highly mature, though one dare not say it is finished, and water transfers are still uncommon in much of Texas. In the case of groundwater, much of the legal evolution to transferability remains, but the process does appear inexorable. Growing scarcity has exposed the burden of inefficient rules and laws, suggesting the need for better institutions.

The general evidence of the Texas experience is that water marketing has provided large net benefits in water allocation. This is especially apparent under conditions of rapid population growth and water scarcity, although many other factors also influence the economic returns available from water markets. It is interesting that the most solid Texas data in support of the economic efficiency of water markets comes from an institutionally distinct river basin where correlative water rights are emphasized. All other basins use the prior appropriations doctrine. But the net benefits achieved in this region appear attributable to high water scarcity in this basin and the long-standing presence of transferability well in advance of transferable water rights in other basins. Still, it should be recognized that correlative rights are easily comprehended and that their uniformity creates a simple basis for water marketing. In addition, social fairness principles are sometimes supportive of a water doctrine that places all water users on equal footing. It is not surprising then that the initially adopted practice of transferable rights tends to employ correlative systems. But when water availability is subject to annual variations and drought, correlative water rights are economically inferior to transferable appropriative rights. Seniority is an important element of transferable water rights when supply is stochastic. Where the region possesses water storage facilities sufficient to ameliorate year-to-year supply variations, the absence of seniority is less problematic. Also, water leasing can offer a significant avenue for achieving risk-responsive allocations of water even within correlative systems.

Finally, Texas experience demonstrates that the development of a system of transferable water rights is a path-dependent process that is inseparable from the institutional steps taken along the way. Each new revision of the water law must acknowledge and accommodate earlier laws no matter how antiquated they may be. Transferability can therefore be difficult to achieve. Inevitably, the existence of a previous water doctrine that prohibits water transfer can cause long delays in the reformation of water law to take advantage of efficiency-enhancing institutional change.

Endnotes

[1] This section and the next two draw heavily on Chang and Griffin (1992).
[2] After gaining independence from Mexico in 1836, Texas was a distinct country until it joined the U.S. in 1845.
[3] The cost of this litigation was reportedly $10 million in attorney fees and court costs (nominal).
[4] 1,000 m^3 is equal to 0.811 acre-ft.
[5] Recent prices have been drought influenced and approach $650 per 1,000 m^3.
[6] In some cases, the unused water that was purchased appeared to be the result of land urbanization but no formal study of this issue is available.
[7] Rio Grande flows are slight except for releases from upstream reservoirs. Consequently, water administrators view the lowermost reservoir as the diversion point for all downstream diversions. Because reservoir releases travel across essentially the same stretch (diverters are concentrated close to the Gulf), it is felt that transfers are inconsequential for either return flows or instream flows.
[8] This section draws heavily on Griffin and Boadu (1992).
[9] The four are the San Marcos gambusia, the Fountain darter, the San Marcos salamander, and Texas wildrice. The first two are small fish.
[10] The legislation introduces initial total water rights of 450,000 acre-ft per year (555 million m^3) to be reduced to 400,000 acre-ft per year (493 million m^3) by 2008. The Authority is also authorized to modify these amounts if necessary to insure springflow protection or if possible to permit greater water use.
[11] The 50 percent limitation is imposed by the Bank's administrative agency rather than by the Bank's enabling legislation. This agency ruling stems from the interpretation of the intent of Texas legislators in establishing the Bank (Schwartz, 1994).

References

Caroom, Doug and Paul Elliott, 1981. "Water Rights Adjudication-Texas Style." *Texas Bar Journal,* 44 (November): 1183-90.

Chang, Chan and Ronald C. Griffin, 1992. "Water Marketing as a Reallocative Institution in Texas." *Water Resources Research,* 28 (March): 879-90.

Characklis, Gregory W., Ronald C. Griffin, and Philip B. Bedient, 1997. "Improving the Ability of a Water Market to More Efficiently Manage Drought in the Lower Rio Grande." In *Sustainable Development: Managing the Transition*, De Lange Woodlands Conference, pp. 317-36. Energy & Environmental Systems Institute, Rice University, Houston, March.

Ciriacy-Wantrup, S. V. and Richard C. Bishop, 1975. "'Common Property' as a Concept in Natural Resources Policy." *Natural Resources Journal,* 15 (October): 713-27.

Dobkins, B., 1959. *The Spanish Element in Texas Water Law.* Austin: University of Texas Press.

Gisser, Micha, 1983. "Groundwater: Focusing on the Real Issue." *Journal of Political Economy,* 91: 1001-27.

Griffin, Ronald C., 1990. "Valuing Urban Water Acquisitions." *Water Resources Bulletin,* 26 (October): 219-25.

Griffin, Ronald C. and Fred O. Boadu, 1992. "Water Marketing in Texas: Opportunities for Reform." *Natural Resources Journal,* 32 (Spring): 265-88.

Griffin, Ronald C. and Chan Chang, 1991. "Seasonality in Community Water Demand." *Western Journal of Agricultural Economics,* 16 (December): 207-17.

Harper, Jayson K. and Ronald C. Griffin, 1988. "The Structure and Role of Authorities in Texas." *Water Resources Bulletin,* 24 (December): 1317-27.

Holbert, J. E., 1984. "Rural/Urban Conflict over Water Control in the Lower Rio Grande Valley of Texas." Ph.D. dissertation, Texas A&M University.

Kaiser, Ronald A., 1987. *Handbook of Texas Water Law: Problems and Needs.* Texas Water Resources Institute, Texas A&M University.

Kaiser, Ronald A., 1996. "Texas Water Marketing in the Next Millennium: A Conceptual and Legal Analysis." *Texas Tech Law Review,* 27: 181-261.

MacDonnell, Lawrence J., Charles W. Howe, Kathleen A. Miller, Teresa A. Rice, and Sarah F. Bates, 1994. *Water Banks in the West,* Natural Resources Law Center, University of Colorado School of Law. Research Report No. 12.

McCarl, B.A., L.L. Jones, R.D. Lacewell, K. Keplinger, M. Chowdry, and K. Yu, 1996. *Final Report on TWDB Project: Evaluation of 'Dry-Year Option' Water Transfers from Agricultural to Urban Use.* Texas Agricultural Experiment Station, Texas A&M University. Unpublished report.

Needham, Jerry, 1997. "SAWS Trustees Give Nod to Aquifer Program Pledge." *San Antonio Express-News*, January 22, pp. 1B, 7B.

Schwartz, Suzanne, 1994. "The Future of Water Transfers and Marketing: Current and Potential Problems; Role of the Texas Water Bank." Presented at *The Fourth Annual Conference on Texas Water Law*, November 3.

Skillern, Frank F., 1988. *Texas Water Law.* Vol. I. San Antonio: Sterling Press.

Smerdon, Ernest T. and John A. Gronouski, 1986. *Texas Management Issues.* Bureau of Engineering, The University of Texas at Austin. CRWR 221.

Smith, Rodney T. and Roger Vaughan, 1994. "Water Banking: Facilitator of Trade or Mechanism of Control." *Water Strategist,* 8 (July): 1-2, 12-3.

Templer, Otis W., 1981. "The Evolution of Texas Water Law and the Impact of Adjudication. " *Water Resources Bulletin,* 17 (October): 789-98.

Texas Department of Water Resources, 1984. *Water For Texas: A Comprehensive Plan for the Future.* Texas Department of Water Resources. GP-4-1.

Texas Water Development Board, 1994. *Texas Water Bank: A Report to the Governor and Members of the Texas Legislature Concerning the Texas Water Bank.* Texas Water Development Board.

U.S. Department of Commerce, 1995. *Statistical Abstract of the United States, 1995.* 115th ed. Washington: U.S. Government Printing Office.

Willatt, Mike, 1995. "Buying and Selling Water Rights in Texas." Presented at *Semi-Annual Conference of the Texas City Attorneys Association*, June, South Padre Island, Texas.

Williams, R. Lynn, 1996. "Economic and Hydrologic Implications of Edwards Aquifer Management Plans." Dissertation, Texas A&M University.

5 Water Markets in Colorado: Past Performance and Needed Changes

Charles W. Howe

5.1 Water Law and Historical Market Performance

The evolution of water law provides a fascinating example of the responses of law to changing social and economic conditions. Scott and Coustalin (1995) and Miller (1996) have provided the details of water law change from Roman times to the 20th century. The law of prior occupancy (an early version of the priority doctrine), wherein the earliest water users had first call on available water, was adopted in England from Roman Law. At the start of the industrial revolution, it became clear that these historical uses were preventing water access for the newer, more technical industries. To accommodate these needs, a "reasonable use" doctrine evolved, allowing new activities access as long as they did not "unreasonably" reduce availability for others or "unreasonably" degrade water quality. This was the "riparian doctrine" of English Common Law that was brought to the eastern U.S. by the English settlers. Until the post World War II period, the doctrine served these well-watered regions well enough, although "reasonable use" remained vague and ultimately had to be defined in court.

The riparian doctrine was ill-suited to the water allocation needs of semiarid and arid regions where water was much scarcer, much more variable in supply, and where uses were often removed from the streambank. The evolution of the "priority doctrine" in this setting has been well documented (Scott and Coustalin, 1995; Miller, 1996).

Water rights under the priority (or prior appropriations) doctrine have the following characteristics:

1. they are considered personal property of the owner and can be sold or otherwise transferred to other owners and uses;
2. the priority of the rights is established by the date of first use and is transferred with the right;
3. the initial appropriation and subsequent transfers must prove that the water is being put to a "beneficial use" - a concept whose meaning has changed slowly over time;
4. transfers of water rights are not allowed to damage other water users, and
5. rights are defined in terms of: (a) diversion rate, (b) dates of diversion, (c) use of the water, (d) whether the right is a direct flow right or a storage right, and (sometimes) (e) total volume of water allowed annually.

While one might conclude that such a system is inefficient (see Falkenmark and Lindh, 1993, for such a view), the priority system has allowed water to move from older, lower-valued uses to emerging, higher-valued use and has allowed those water users who highly value reliability of supply to obtain that reliability by buying senior (high priority) rights. This stands in contrast with the eastern states of the U.S. and many countries where water is not tradable and where reliability can be guaranteed only by building excess supply capacity.

Colorado has long had active water markets. The data in Table 5.1, while now dated, show that Colorado, Utah, and New Mexico have had large numbers of permanent water transfers.

Table 5.1. Total Number of Change of Water Right Applications by State, 1975-1984

State	Number of Applications Filed
Arizona[1]	30
California[2]	3
Colorado[3]	858
New Mexico	1,133
Utah	3,853
Wyoming	40

Source: Lawrence J. MacDonnell, 1990.

[1] These are server and transfer proceedings involving a change of purpose and/or place of use.

[2] No applications were filed in California until 1982.

[3] Applications filed in Colorado, New Mexico, and Wyoming may include more than one water right.

In addition to permanent sales of water rights, there are many temporary transfers (or "rentals") to accommodate short-term needs, especially during drought. Howe et al., (1982) documented the many types of short-term transfers used (especially in the agricultural sector) during the 1976-77 drought in Colorado. More recently, "water banks" of various types have been organized to facilitate short term transfers during drought. The California (Chapter 7 and 8 in this volume) and Idaho water banks have been widely publicized (Israel and Lund, 1995; MacDonnell et al., 1994).

The direction of permanent water rights transfers has been primarily from irrigated agriculture to municipal and industrial uses. This is not surprising since nearly 85 percent of all water diversions in the 17 western states and close to 90 percent of all consumptive uses are in the agricultural sector (U.S. Geological Survey, 1993). The composition of permanent transfers is shown in Figure 5.1.

While the western states all follow variations of the priority doctrine, the institutional mechanisms for effecting transfers differ from state to state. In Colorado's closest neighbor states, Utah and New Mexico, applications for new appropriations and for permanent water right transfers are made through the Office of the State Engineer. That office has the technical skills to determine whether or not the new appropriation or the requested transfer will adversely affect other water users. The Engineer Office can then approve or modify the request. While appropriators and transferors have recourse to court appeals, 99 percent of the time the Engineer Office's decision is accepted.

Colorado, by contrast, long ago adopted a water court system under which proposed transfers must be advertised and can be challenged by parties which perceive themselves to be injured by the transfer. Figure 5.2 shows that a majority of transfer applications in Colorado are protested, in contrast with much lower percentages in the other five states.

Figure 5.1. Approved Changes of Water Rights by Purpose of Use, by State (Percent of Total Approvals)

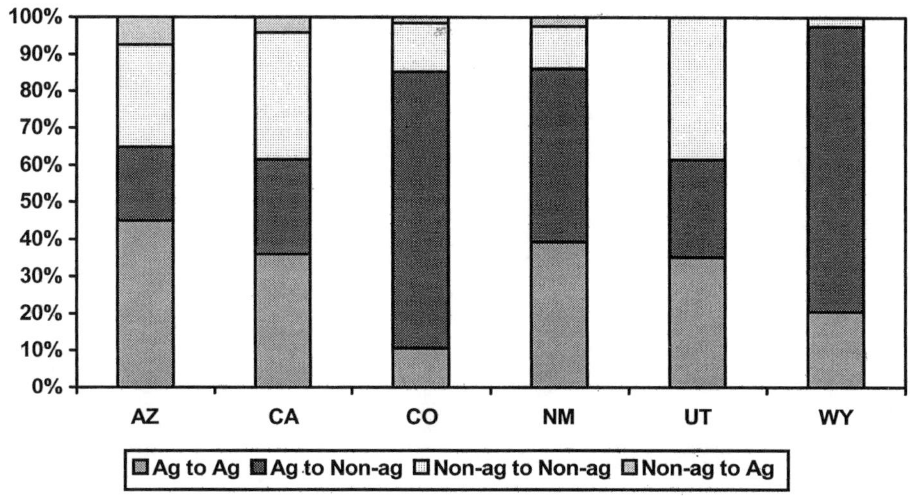

Source: Adapted from MacDonnell, 1990

The larger Colorado cities make a practice of protesting all proposed transfers of water rights on the rivers in which the cities own rights. The protest provides time for the cities to carry out studies of possible injury. If no protest is entered, there is no recourse once a transfer is approved. Since a protest is likely to lead to water court hearings that require all parties (transferor, transferee, and protestors) to have legal and engineering representation, the process is costly and precludes legitimate protests by small water users. In contrast to the New Mexico and Utah systems, no one looks out for the small water user in Colorado. The results are injustices to small users and much higher transaction costs to all parties. This is reflected in Figure 5.3 that shows the average time it takes to effect water right transfers in the six states.

Another interesting contrast exists between the Colorado and New Mexico systems: the New Mexico State Engineer *closes* the appropriations on streams when the "reliable" supply has been appropriated. "Reliable" is not legally defined but is in the neighborhood of the 80th percentile flow, i.e. the flow that is present or exceeded 80 percent of the time. Chile also uses the 80 percent rule. In Colorado, the State Constitution forbids closure of appropriations, with the result that many streams have more water rights outstanding than they ever have water. Thus, Colorado must continuously "administer" these streams (i.e. determine by priority which rights get cut off) while New Mexico seldom has to administer its streams. New Mexico may lose some water to other states on interstate streams as a result but the costs of administration are greatly reduced.

One of the most interesting and active water markets in Colorado is that of the Northern Colorado Water Conservancy District (NCWCD) that collects and distributes the water from the (federal) Colorado Big Thompson Project (CBT) (for fuller descriptions see Howe, et al., 1986; Michelsen, 1994). The District covers a large part of northeastern Colorado and has developed an elaborate distribution system for the water imported from the Colorado River headwaters on the western slopes of the Rocky Mountains. The distribution system makes it physically easy to transfer CBT water from any owner to other parties in the District.

Figure 5.2. Percentage of Applications Formally Protested or Opposed, by State: 1975-1984.

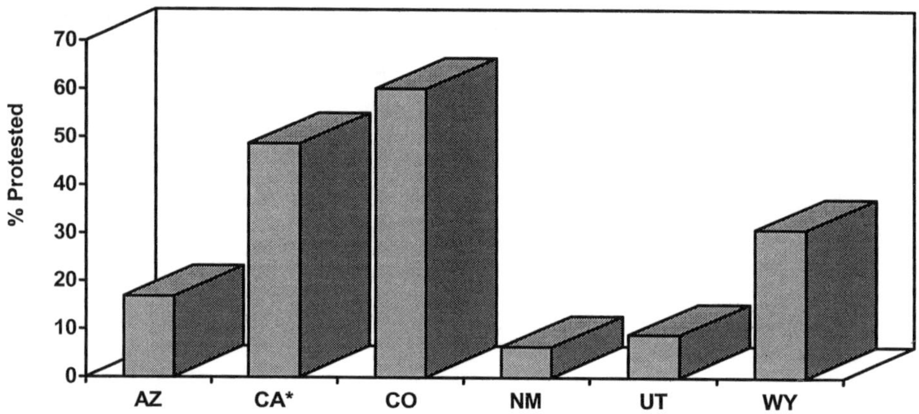

*For applications filed between 1982 and 1989.
Source: Adapted from MacDonnell, 1990.

In 1957 when the CBT Project was completed, the Bureau of Reclamation wanted to require a uniform allocation of water to all agricultural lands in the District, mainly as a way of guaranteeing project cost repayment. Irrigators objected strongly since different lands had quite different local water supplies and thus quite different needs for CBT water. In the District, 310,000 shares ("allotments") had been created, representing proportional shares in the Project's maximum annual delivery capacity of 382.23 million m^3 (310,000 acre-ft). The Bureau relented, allowing the District to make the shares available to any party in the District who was willing to pay the $2.25 annual fee (the largest part of the District's revenues come from real estate taxes on all property within the District). It was several years before all the shares were taken by water users.

At first it was unclear whether or not these shares would be tradable but, faced with the desires of many water users to adjust their water holdings, the District Board ruled the shares to be tradable. Initially, very few shares were traded but, as urban areas and industry in the region expanded, there was increased trading activity. The volume of trades over the 1970-1993 period is shown in Figure 5.4.

Trading is done by advertising, by posting offers on the District bulletin board and through brokers. Two characteristics of the system greatly facilitate trading: (1) the extensive distribution system and (2) the special status of return flows from users.

Ordinarily in priority doctrine states, the return flow from a water use becomes subject to appropriation. Other water users establish rights to that water. Thus, when a transfer takes place, water equivalent to these return flows must be left in the stream to avoid injury to those water users. This greatly complicates the transfer process since studies are required to quantify the amount and timing of the return flows. Colorado water law, however, specifies that any party importing water from one major drainage basin to another owns the water totally and is entitled to "use it to extinction." Thus, when transfers of imported waters occur, there is no legal responsibility for return flows and the owner can use or exchange the return flows. This creates a bias in favor of interbasin transfers relative to further development of local supplies. The Bureau of Reclamation originally wanted the District to try to sell the return flows (a further guarantee of repayment of project costs), but the District wisely refused. Thus, while transfers in NCWCD may exert both positive and negative externalities on return flow users, there is no legal liability for those externalities.

Figure 5.3. Average Months to Decision for Approved Cases, by State: 1975-1984.

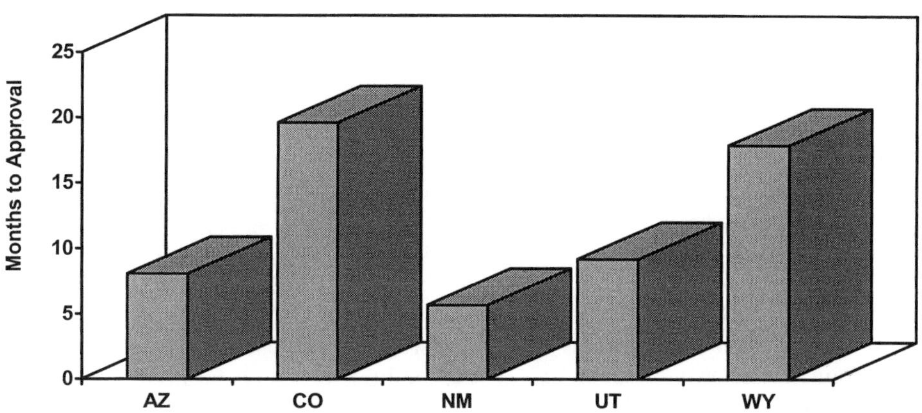

Source: Adapted from MacDonnell, 1990.

This brings out an interesting asymmetry in western water law that was first pointed out by Hartman and Seastone (1970). Transferors are held liable for the return flows at the original point of use but get no credit for the enhanced return flows at the new point of use. From an economic efficiency point of view, this creates a bias against transfers. The magnitude of this bias has not been quantified.

Over time, NCWCD shares have changed hands easily and frequently (see Figure 5.4) with a shift in ownership towards municipal and industrial uses as shown in Figure 5.5. The high quality and reliability of CBT water makes it highly desirable for these purposes.

Figure 5.4. CBT Unit Transfers by Category of Seller

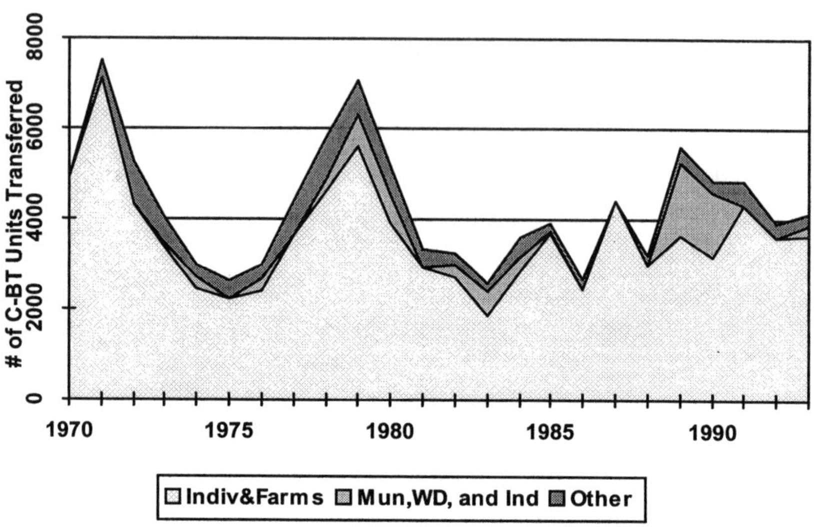

Source: Adapted from Michelsen, 1994.

Since municipalities and industry frequently buy water supplies in anticipation of future needs and to guarantee supplies during droughts, the actual use of water has not shifted to the same extent as its ownership (see Figure 5.6).

Municipalities typically "rent" part of their raw water supplies back to agriculture on a year-by-year basis, calling the water back during severe drought events.

The structure of water rights underlying the CBT Project and the distribution of water by NCWCD bring out several interesting features. The Bureau of Reclamation actually appropriated the water in the Colorado River headwaters and owns the corresponding priority water rights. While the priority date is relatively recent, the water is nonetheless reliable because of the large volume of water in the Upper Colorado Basin and the slow development of that part of the state.[1] The water yielded by these water rights is then *proportionally divided* among the share owners of NCWCD. Thus, there is a conversion of priority rights to proportional rights under which all owners share risk equally.

In summing up the Colorado water marketing experience, one can say that the appropriations doctrine with its priority and tradable water rights system has, in general, served the state well. It has allowed water to move from lower-valued (primarily agricultural) uses to higher-valued uses (primarily municipal and industrial). However, because of the establishment of the water court system for administering transfers, the transaction costs associated with transfers have been high. Small water users are not protected from injury from transfers. Also, because of the narrow interpretation of "beneficial use" as required for water right ownership, an expanding array of public values is being slighted in the appropriation and transfer processes.

Figure 5.5. Ownership Evolution of CBT Shares, 1962 and 1992

Source: Adapted from Michelsen, 1994.

Figure 5.6. CBT Water Deliveries for Agricultural Use and Municipal and Industrial Use.

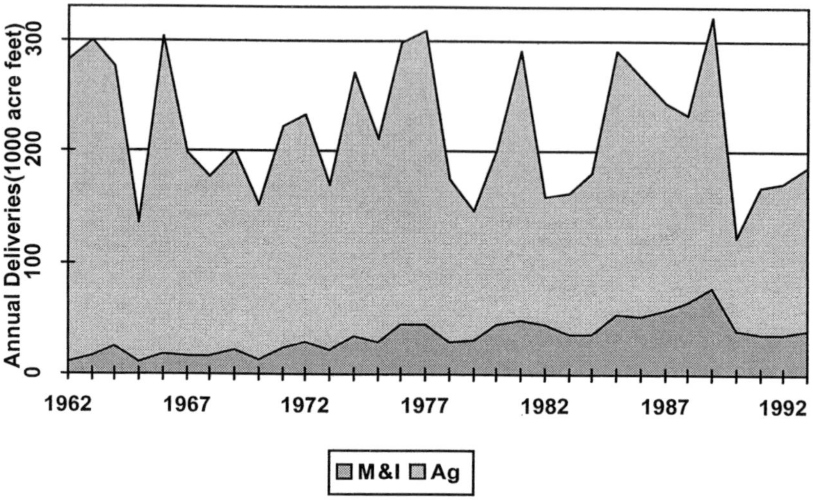

Source: Adapted from Michelsen, 1994.

5.2 Needed Changes in Colorado Water Administration

One major change has been suggested above: shifting the administration of appropriations and transfers from the water court system to the State Engineer Office as is done in New Mexico and Utah. This would greatly reduce transaction costs by allowing disinterested expertise to evaluate appropriations and transfers, including the determination of all third-party injuries. The State Engineer Office can develop rules of thumb for subregions of the state concerning consumptive use by crop, return flow, etc. Parties would have recourse to the courts if they disagree with these determinations, but in New Mexico, most of the determinations are accepted without appeal. Naturally, this move will be resisted by the lawyers and engineers who make their living preparing court cases and will probably be resisted by some of the traditional water organizations that would like to see water transfers slowed.

The resistance to water transfers has been growing on a wider basis. As large transfers have dried up large tracts of formerly irrigated lands, people of the affected region as well as people statewide who place a value on maintaining agriculture have increasingly voiced objections to out-of-basin transfers. An example of such a transfer would be the sale of the water of the Colorado Canal in Crowley County in southeastern Colorado to the cities of Denver and Aurora. This transferred 98.64 million m^3 (80,000 acre-feet) of Arkansas River water from 40,000 acres of irrigated crops. The multiplier effects on agriculturally-related activities and the reduction of the county's tax base were devastating (Howe et al., 1990). There have also been threats of large diversions from the recreationally rich Gunnison Valley for municipal use in the Denver area and of diversions for minor hydropower from the unique Black Canyon of the Gunnison National Monument.

As the regional population contemplates these large-scale impacts, they turn against water transfers generally. There have been several attempts in the state legislature to prohibit or severely restrict out-of-basin transfers. Some of these attempts may eventually succeed.

Continuing transfers of water from agriculture to expanding urban uses will, however, be necessary. Even larger transfers will be needed for environmental and recreational purposes. Arbitrary, severe restrictions on transfers will doubly harm the environment by requiring new reservoirs and by making it difficult to keep water in the streams.

The cause of the resistance to transfers is that the current market system, in connection with existing water law, fails to take into account important public values generated by the various uses of water. The basic problem is the narrow definition of the concept of "beneficial use."

The Constitution of Colorado states that "the waters of the State belong to the people, subject to private appropriation which shall not be denied." Private appropriations are to be established through "diversions from the streams to beneficial uses." It was not until 1967 that a legislative provision was made to consider instream uses as a "beneficial" use. While a major tenet of the priority doctrine is the "no injury" rule relating to appropriations and transfers, only "beneficial uses" are protected from injury. Thus, when this term is narrowly defined, some public values remain unprotected.

What are some of these unprotected yet increasingly important public values? Many of them are instream environmental values: ecosystem protection, riparian habitat and general stream aesthetics. Even recreational uses remain unprotected and unrecognized unless instream water rights have actually been acquired. Thus the city of Aurora with its proposal to export large quantities of water from the Upper Gunnison River Basin is allowed to do so (there is lots of unappropriated water) with no recognition of the large recreational and environmental benefits that would be lost. That water, if left in place, also has a very high value for hydropower and water quality maintenance. Howe and Ahrens (1988) estimated the hydropower value at \$0.06 per m^3 (\$72 per acre-ft) and water quality benefits at \$0.03 per m^3 (\$38 per acre-ft), for a total value of \$0.09 per m^3 (\$140 per acre-ft), not counting any recreation benefits or opportunity costs in the lower Colorado River Basin. Only if a downstream value is represented by an established water right in the same state is it protected against out-of-basin transfers. The Bureau of Reclamation power dams of the Aspinal Unit on the Gunnison River have minimum flow rights, but flows always exceed those minimums. Every acre-foot taken out represents a loss of power. The large recreational benefits are not protected at all. Thus major appropriations and transfers take place with no recognition of very large opportunity costs.

Other public values that are not taken into account (i.e. not counted as "beneficial uses") are important social and cultural values. A major issue arises from the forward and backward linkages to agriculture (input and processing industries) when irrigation is phased out. Suppliers find the demand for their projects reduced. Agricultural processors suffer shortages and have to turn to higher cost sources. Unemployment of human and mobile capital result. If we assume (as most economic analysis does) that these resources move quickly and costlessly to uses where their value-productivity will be nearly as high, then these secondary impacts ("pecuniary externalities" to economists) can be ignored. That is just the market at work efficiently reallocating resources.

However, the processes of reallocating these human and capital resources are not costless in real economic terms. Resources often remain unemployed for long periods of time. Job search, retraining, and moving costs are high in economic terms and even higher in community terms. Tax bases are eroded and social services suffer (Howe and Easter, 1971).

We do not really want to stop resource transfers, but two types of policy modification are warranted: (1) such secondary costs should be compensated by some form of payment to the area of origin; and (2) the speed of large transfers should be moderated to allow more time for human adjustments. Water courts frequently slow transfers to allow for revegetation of the land being dried up. We can do the same to allow for human adjustment. The issue of compensating areas of origin has been treated by MacDonnell and Howe (1986) and MacDonnell et al. (1994).

It might be argued that areas of origin could protect their social, environmental, and recreational values by applying for water rights or by purchasing existing water rights. This should be encouraged by counting instream uses as "beneficial uses." There are, however, some obvious problems with that approach. The first is the income and wealth discrepancy between rural riparian areas and wealthy metropolitan areas that would compete for the water rights. The second is the "public goods" nature of most of the benefits generated by instream flows. These benefits stem from recreational activities largely on public lands. They do not generate revenues. It would be difficult to charge and/or exclude people from these activities. Thus these "public goods" must be protected by public agencies that have the authority to acquire or reserve water for these purposes. Hopefully, such reservations would be sized in accordance with careful assessments of the benefits generated and the opportunity costs incurred. Montana, still a water plentiful state, has made water right(s) reservations to protect streamflows (McNally and Matthews, 1995).

Cultural values represent a case where reservations of water rights by public agencies are needed. A famous case in point is the "sleeper" case[2] in New Mexico wherein some of the members of an old Spanish community decided to sell their share of water from the old "*acequia*" (ditch) system to a new ski area. The State District Court enjoined the transfer on the grounds that it would disrupt the culture of the community that centered around the mutually-maintained ditch system. That decision was later reversed by the Court of Appeals, but the case reflects the increasing social concern with the maintenance of traditional cultures.

Another mechanism that can slow the permanent transfer of water rights from agriculture while assisting cities to meet their needs for reliable supplies is "water banking," i.e. any set of arrangements to facilitate the *temporary* transfer between irrigation districts and towns (MacDonnell et al., 1994). The 1991-92 water bank in California (Chapters 7 and 8) and the long-standing water banks in Idaho have received attention (see MacDonnell et al., 1994). Water rentals from Bureau of Reclamation reservoirs on the Snake River system have existed since the 1930s. Their problem is lack of price flexibility in that the Bureau requires a low uniform price that is independent of supply and demand conditions.

A similar but long-term contractual arrangement is the "agricultural lease-out" or "dry year lease-out," wherein towns contract with irrigation districts to take part of their water under certain drought conditions. In return, annual payments are made to the district, with much larger payments in years when water is actually taken. The

Metropolitan Water District of Southern California has entered into such contracts with the Palo Verde and Desert Irrigation Districts.

5.3. Summary and Conclusion

It has been argued that the prior appropriations system of water rights has, overall and over time, served Colorado well. Water has moved slowly to higher-valued uses and instream flows are now recognized as "beneficial" uses. Important public values, however, are not protected nor reflected in the water allocation process. These include many environmental values, "public goods" types of recreation, secondary income losses in areas of origin, and cultural values.

My recommendations for modifications of the Colorado water rights system are: (1) to shift from the water courts system to one of administration of appropriations and transfers by the Office of the State Engineer to avoid the costs of the current litigious environment; (2) to modify the "beneficial use" doctrine to include almost any use and, in particular, to consider nearly all instream uses to be beneficial; (3) to require payment to basins of origin for the opportunity costs incurred from out-of-basin transfers, such payments should also cover secondary and community costs; (4) to authorize appropriate state agencies to reserve or acquire water for "public goods" types of recreational and aesthetic values after doing appropriate benefit-cost sizing of the reservations; (5) to facilitate the reservation of water in protection of cultural values, perhaps by vesting water rights in the community rather than in individuals; (6) to encourage local governments and special districts to acquire water rights in protection of instream, social, and aesthetic values; and (7) to encourage water banking and agricultural lease-outs to preserve agriculture while meeting the risk-mitigation needs of towns.

If Colorado adopts these recommendations it will be moving in directions already taken by neighboring states, bringing the priority system up to the challenges of the twenty-first century. As developing countries contemplate the use of water markets under encouragement from the World Bank and other aid organizations, consideration should be given to the same points.

Endnotes

[1] Recent agreements with the City of Denver and the Colorado River Water Conservancy District resulting in the construction of the Wolford Mountain Project have increased this reliability.

[2] 760 P. 2nd 787 (NM APP 1988). Cert quashed 759 P 2nd 200 (NM, 1988).

References

Falkenmark, Malin and Gunna Lindh, 1993. "Water and Economic Development," in Peter H. Gleick (ed) *Water in Crisis: A Guide to the World's Fresh Water Resources,* Oxford University Press, New York.

Hartman, L. M. and Don Seastone, 1970. *Water Transfers: Economic Efficiency and Alternative Institutions*, The Johns Hopkins Press for Resources for the Future, Inc., Baltimore.

Howe, Charles W. and K. William Easter, 1971. *Interbasin Transfers of Water: Economic Issues and Impacts*, The John Hopkins Press, Baltimore.

Howe, Charles W., Paul K. Alexander, and Raphael J. Moses, 1982. "The Performance of Appropriative Water Rights Systems During Drought," *Natural Resources Journal*, 22(2):379-389.

Howe, Charles W., Dennis R. Schurmeier, and William D. Shaw, Jr., 1986. "Innovations in Water Management: Lessons from the Colorado-Big Thompson Project and Northern Colorado Water Conservancy District" in Kenneth D. Frederick (ed), *Scarce Water and Institutional Change*, Resources for the Future, Inc., Washington, D.C.

Howe, Charles W. and W. Ashley Ahrens, 1988. "Water Resources of the Upper Colorado River Basin: Problems and Policy Alternatives," Chapter 5 in Mohamed T. El-Ashry and Diana C. Gibbons (eds.) *Water and Arid Lands of the Western United States*, Cambridge University Press, New York.

Howe, Charles W., Jeffery K. Lazo, and Kenneth R. Weber, 1990. "The Economic Impacts of Agriculture-to-Urban Water Transfers on the Area of Origin: A Case Study of the Arkansas River Valley in Colorado, "*American Journal of Agricultural Economics,* 72(5):1200-1204.

Israel, Morris and Jay R. Lund, 1995. "Recent California Water Transfers: Implications for Water Management," *Natural Resources Journal*, 35(1): 1-33.

MacDonnell, Lawrence J. and Charles W. Howe, 1986. "Area-of-Origin Protection in Transbasin Water Diversions: An Evaluation of Alternative Approaches," *University of Colorado Law Review*, 57(3):527-548.

MacDonnell, Lawrence J., 1990. "The Water Transfer Process as a Management Option for Meeting Changing Water Demands," Natural Resources Law Center, School of Law, University of Colorado, Boulder, Colorado.

MacDonnell, Lawrence J., Charles W. Howe, Kathleen A. Miller, Teresa A. Rice, and Sarah F. Bates, 1994. "Water Banks in the West," Natural Resources Law Center, School of Law, University of Colorado, Boulder, Colorado.

McNally, Mary and Olen Paul Matthews, 1995. "Changing the Balance in Western Water Law? Montana's Reservation System," *Natural Resources Journal*, 35(3): 671-694.

Michelsen, Ari M., 1994. "Administrative, Institutional and Structural Characteristics of an Active Water Market," *Water Resources Bulletin*, 30(6):971-982.

Miller, Kathleen A., 1996. "Property and Water: A Historical Analysis," paper presented at the 7th Annual Meeting of the European Association of Environmental and Resource Economists, Lisbon, Portugal, June 27-29.

Scott, Anthony and Georgina Coustalin, 1995. "The Evolution of Water Rights," *Natural Resources Journal*, 35(4):821-980.

U. S. Geological Survey, 1993. "Estimated Use of Water in the United States in 1990," U. S. Geological Survey Circular 1081, U. S. Government Printing Office.

6 Negotiated Transactions as Conflict Resolution Mechanisms: Water Bargaining in the U.S. West

Bonnie G. Colby

6.1 Water and Conflict in the American West

Disputes prompted by competition for scarce water supplies have proliferated worldwide, generating enormous costs and uncertainties for affected parties. Economists have extolled the virtues of markets and voluntary transactions for well over a century as a means to achieve efficient resource allocation. This chapter examines voluntary water transfers from a somewhat different perspective than that of allocative efficiency. It explores negotiated transactions as a tool for resolving water disputes through analysis of conflicts in the American West. The study of mechanisms to successfully resolve water disputes is a field of inquiry in its own right, one in which economics is just beginning to play a role (Dinar and Loehman, 1995). Interesting challenges arise in applying economic paradigms to evaluate dispute resolution processes. Conflicts over water often are difficult to evaluate using economic efficiency constructs because the baseline entitlements from which costs and benefits should be measured are themselves the subject of contention. While allocative efficiency is problematic in such situations, other economic criteria can be used to compare the strengths and weaknesses of different approaches to resolving water disputes.

The chapter begins by providing a brief overview of water conflicts, of the development of water resources and water markets in the western U.S., and of policies that govern water rights. Case studies of disputes involving tribal/non-tribal water users and environmental water needs highlight the role that voluntary transactions have played in resolving western water disputes. They illustrate the advantages and disadvantages of voluntary negotiated water transfers as a means to address conflicts, as compared to litigation and administrative water reallocations.

Characteristics of Water Conflicts

Water conflicts have a number of characteristics which affect the costs and likely success of dispute resolution attempts. First, disputes over water inevitably consist of multiple rounds of interaction because hydrologic systems are dynamic and the disputants periodically must regroup to deal with drought, flood, and other unanticipated changes. Moreover, scientific understanding and social values evolve overtime and the parties involved must grapple with new technical knowledge about rivers, lakes, and aquifers and with new concerns regarding water use and the environment. The ongoing nature of water disputes has several implications. The parties have incentives for different behavior than in a one-time dispute. They know they will meet again and must deal with one another over time. They have the opportunity to learn cumulatively how to better resolve conflicts and they have incentives to build trust and to gather improved data on their watershed so that the next round of problems can be solved more cost-effectively.

Water conflicts involve multiple parties with different capabilities to reach internal consensus and to "bind" their constituents to an agreement. For example, governments (cities, counties, states, tribes, and irrigation districts) usually have a formal decision and budget allocation process, while citizen groups, environmentalists, and an angry-but-diffuse public typically do not. Consequently, it may be difficult for the latter groups to define their positions and to commit to a proposed solution, such as a purchase and transfer of water. Parties to water disputes also have diverse financial capabilities that affect the sharing of costs for resolving disputes (including the costs of acquiring additional water supplies) and their ability to hire experts and to obtain authoritative legal and technical information. The widespread use of "dueling" experts and attorneys to gain bargaining power in negotiations implies that the financial status affects a party's staying power in the dispute resolution process, lending credence to the adage "the one with the deepest pockets wins." The ongoing, multi-party nature of water disputes presents a challenge for voluntary transactions, as well as for other dispute resolution approaches.

Approaches to Resolving Conflicts

Approaches to dealing with conflicts over water in the American West have evolved over the past century and a half. In the early years of European incursion into the region, bloody battles erupted over the most productive and well-watered areas and conflicts were resolved by military force, with Native Americans eventually being subdued and resettled on reservation lands. Conflicts over water for mining and farming arose during the gold-rush years and led to the development of rules for determining who had rights to use how much water and for allocating water during dry years. These rules became formalized, as states incorporated the prior appropriation doctrine into their water laws. The prior appropriation doctrine gives first priority to the most senior water rights and lower priority to the rights established at later points in time ("first in time, first in right"). The prior appropriation doctrine facilitated rapid mining and agricultural development consistent with the national imperative to settle the West and to solidify U.S. control over territories recently acquired by treaty and conquest.

During the first half of the 1900s, western states litigated and negotiated over the division of interstate rivers and developed compacts to divide rivers among themselves. Once a basic property rights structure was in place among the states and for individual water users, development of water proceeded rapidly, with large infusions of Federal money. During the first seven decades of the 1900s, disputes arising as local supplies became inadequate to meet competing needs were addressed by the parties mounting a campaign (generally successful) to obtain Federal money for new water projects. When Federal largesse began to diminish in the late 1970s and environmental opposition to new water projects became more adamant, conflicts over scarcity had to be addressed in new ways. Litigation over water for Native Americans and the environment became more prevalent and markets for water began to emerge in some areas as a means to transfer water out of agriculture to other uses. In the modern West, there are three primary approaches to addressing conflicts over inadequate water supplies: (1) litigation to wrest water away from current users, (2) political maneuvering to get public water project operations altered and their supplies reallocated, and (3) voluntary purchases or leases of water. This chapter compares and evaluates these three approaches to resolving conflicts over scarcity.

6.2 Water Development and Emergence of Markets

In the western U.S., water has been reallocated from one use to another throughout the 1900s. Through the first half of the century, water development projects reallocated vast amounts of water from streams (to the detriment of recreation, fish, wildlife, and water quality) for offstream irrigation, mining, and urban uses. At that time, these water development projects were not regarded as reallocations because the benefits associated with instream flows were largely unacknowledged, as were Native American claims to water. Indeed, water remaining in streams was widely regarded as wasted, water development was viewed as a new claim on an unused resource, and irrigation districts, mining companies, ranchers, and towns were accorded property rights in the resource.

The West's economic transition from ranching, irrigated farming and mining to urban growth, services, tourism, and industry brought strong pressure to transfer water out of agriculture, which still accounts for over 80 percent of water use in most western states. Western cities pioneered the era of water marketing by purchasing irrigated land, sometimes entire irrigation districts, to acquire water rights for urban development (Saliba and Bush, 1987). The City of Los Angeles' acquisitions of irrigated lands in the Owens Valley, beginning in the 1920s, stimulated decades of lawsuits and political backlash. Denver area cities began acquiring and transferring water from the western slope of the Rocky Mountains in the 1930s to accommodate new industry and urban growth. Cities and developers in Arizona bought tens of thousands of acres of irrigated land in the 1980s to acquire water for urban growth. In the 1990s, water transfers to support wildlife, fisheries, and recreation have become more common (Colby, 1993). Transfers have become progressively more complex and innovative in order to respond to drought and to environmental and community concerns.

Voluntary transfers are by no means the only mechanism used to reallocate water. Complex legal proceedings, termed adjudications, are taking place in many areas of the western U.S. to quantify and prioritize the competing claims of Native Americans, wilderness areas, cities, and farms. For example, the Arizona General Stream Adjudication seeks to clarify water claims for most of Arizona and involves over 100,000 claimants, including 20 tribes and several wilderness areas. As described later, a decades-long adjudication process and a series of court rulings in the Wind River Basin have reallocated water from irrigation districts and farmers to tribes seeking to use the water instream for fisheries and recreation. Adjudications are only one of the ways in which involuntary reallocations occur. Litigation based on the Endangered Species Act, the Clean Water Act, federally reserved rights, and the public trust doctrine has forced reallocation of water to enhance streamflows for recreation, fish, wildlife, and water quality. In addition, tribes and environmental interests have been able to influence the political process in order to achieve administrative changes in the operations of water projects that favor their needs over those of historic project beneficiaries (hydropower users and irrigators).

Water Rights and Policies Governing Transfers

States in the U.S. have the primary responsibility to govern property rights for water and to regulate their transfer. In the western states, the party seeking to transfer a water right (the applicant) to a new use must file an application with the state water agency and often hires attorneys and engineers to help prepare the application. Applicants typically face uncertainty regarding the quantity of water that may be transferred and the conditions and costs that may be imposed. Most western states require public notice to alert parties who may be affected by the transfer. Parties who object to the proposed transfer (called objectors or protestants) may file a protest with the state. However, not everyone who may be adversely affected can file a protest. The earliest, and still the most common, legal basis for protesting a proposed transfer is impairment of existing water rights. When protests are filed, transfer approval is delayed as the applicant and protestants argue over the magnitude of transfer impacts and the extent of mitigation or compensation. If the applicant and objectors can negotiate a compromise, their negotiated modifications of the transfer proposal are submitted to the state agency and transfer approval normally follows. When unresolved protests remain, the state agency will hold hearings and rule on the transfer application.

The threat of a costly, protracted hearing, followed by an unpredictable ruling on the transfer application, provides incentives for the parties to negotiate a settlement. However, those interests that have no legal standing to file a protest generally do not have a seat at the bargaining table. Recreationists, wildlife advocates, and rafting outfitters all have an economic interest in maintaining streamflows but generally do not own water rights. Consequently, in most western states, they cannot file a protest to protect their interests. States limit the externalities that are considered when a transfer is proposed because their policies limit the purposes for which water rights may be held. They may disallow protests by those who do not hold water rights and otherwise constrain the basis on which a valid protest may be filed. Despite the costs and uncertainties of the regulatory process, MacDonnell (1990) finds that over 80 percent of the applications to transfer water are eventually approved in Colorado, New Mexico, and Utah (three western states with a high level of water transfer activity).

Water Transfers and Transaction Costs

Transaction costs arise for various reasons. The complex nature of the resource and the conflicts associated with transferring water have generated layers of public policies that apply when water transfers are proposed. Further, the "thinness" of most regional markets results in the absence of low-cost market information and brokering services. In addition, measurement costs are non-trivial in establishing and enforcing property rights in water. While many western states require that large diversions be measured, groundwater use and small surface water diversions frequently are not metered, so more precise measurement is needed when transfers are proposed, requiring investment in measuring devices.

Numerous public concerns come into play when transfers are proposed, and so conflicts are common. These include concerns over public access to water for recreational, environmental, and cultural purposes, concerns over impacts on rural economies whose water is being transferred, and concerns regarding community stability and cohesion, especially where traditional Hispanic, Mormon, or Native American communities are involved. Regulatory processes attempting to account for these concerns impose costs on those seeking to transfer water.

Transaction costs associated with voluntary transfers are an important policy issue. If the costs of implementing water transfers become too high, some beneficial transfers will not take place and society will lose the efficiency gains that those transfers would have generated. On the other hand, the ability to impose transaction costs on those proposing to transfer water represents bargaining power in the water allocation process. Parties engage in market transactions for economic gain and the power to erode this expected gain by imposing transaction costs gives third parties leverage with transfer proponents. This leverage can be used to induce transfer proponents to internalize some of the external costs of transfers. Transaction costs necessarily are incurred when public agencies evaluate proposed transfers, seeking to account for externalities and public values. Transaction costs also reflect the absence of "free" information and the need for hydrologic, legal, and economic data to address externalities in an efficient manner. If voluntary transfers are to be an effective dispute resolution tool, then transaction costs associated with policies that govern transfers must be balanced—neither so high as to discourage desirable transfers nor so low as to allow unmitigated externalities.

6.3 Recent Conflicts and the Role of Voluntary Transactions

Voluntary transfers, litigation, and political maneuvering to achieve reallocation have been instrumental in addressing conflicts involving tribal water claims and disputes over water for streams, wetlands, and endangered species. Four specific conflicts are discussed below to highlight the role of, and to illustrate interactions between, different conflict resolution mechanisms in resolving disputes.

Controversies raging across the American West over Native American water rights highlight the classic features of water disputes. When tribal communities were placed on reservations as the West was settled, early court rulings affirmed that water was reserved for tribes, along with the reservation land (*Winters v. U.S.*, 1908). However, the Federal government proceeded to dam and divert western rivers for irrigation projects, settlers gained water rights under state laws and most water users never considered that there might be a competing property claim. Tribal water claims garnered the attention of western water users only after a dispute over the waters of the Colorado River culminated in a U.S. Supreme Court decree (*Arizona v. California*, 1963). The court ruled that the tribes had rights to as much water as needed to grow crops on every irrigable acre of their reservation (the Practicably Irrigable Acreage (PIA) standard), even though fulfilling those rights would deprive others of water they were accustomed to using and to which they believed they had firm rights. In a similar manner, court rulings regarding the public trust, endangered species, and water quality have cast a cloud of uncertainty over the ability of water users to continue to divert water they have been using for decades.

Those attempting to resolve competing claims to water confront two very different legal perspectives on property rights, which the opposing sides use as their baseline in negotiating. Each side attempts to convince the other that a particular outcome is "fair" given their own starting point regarding the presumption of water rights and to convince opponents of their legal strengths, if litigation were to proceed. To date, negotiated settlements over tribal claims have protected non-tribal access to water through new water development projects, the importation of additional water supplies, water conservation measures, and water leasing provisions. Voluntary transactions have played a role in several negotiated settlements of tribal claims. While some settlements ban the marketing of tribal water altogether, recent Arizona settlements authorize the leasing of tribal water in designated areas within Arizona and include complicated water transfers, exchanges, lease-backs, and water contract modifications.

Voluntary transactions for environmental restoration also illustrate the challenges of using a voluntary, negotiated approach to satisfy unmet water needs. During the 1980s and early 1990s, a number of voluntary transactions resulted in water being transferred from agricultural uses to streamflow and wetland enhancement in the American West. Purchasers included State and Federal agencies, environmental organizations and local recreation clubs. The transactions included an innovative variety of purchases, leases, exchanges, conservation improvements, and option arrangements (Colby, 1993; MacDonnell, 1993). Nevertheless, a number of obstacles prevent environmental interests from being better served by market transactions. First, environmental organizations do not have legal access to water rights on the same terms as farmers, cities, and industry do, though some states have provided the means to dedicate water to instream needs. Furthermore, transaction costs are generally higher for acquiring and transferring water rights to instream uses. Organizations wishing to purchase water rights to maintain streamflows often face opposition by neighboring water users and have to incur high costs in overcoming objections. In addition, many State agencies have little experience in handling applications to change the use of a water right from irrigation to instream flow maintenance. New procedures and criteria often have to be developed first, creating delays and additional costs for the instream use applicant. Consequently, market purchases are seldom used to acquire water for environmental needs, as compared to litigation and political lobbying to acquire water.

Ak-Chin Negotiated Settlement of Tribal Rights

The Ak-Chin Water Settlement in 1978 was the first Native American water right dispute resolved through a negotiated settlement. The settlement was amended in 1984, and again in 1992, illustrating the difficulty of achieving finality in resolving complex conflicts (Checchio and Colby, 1993). The Ak-Chin Reservation was established in 1912 and is located in the midst of a productive agricultural area in central Arizona. Neighboring irrigators as well as the tribal farming enterprise were wholly dependent on groundwater. However, extensive off-reservation pumping caused sharp declines (approximately 20 feet per year) in groundwater levels beneath the reservation and the Native American community's irrigated acreage declined from 10,000 to less than 5,000 acres in less than a decade.

The Ak-Chin tribal community requested that the U.S. file suit against the non-tribal pumpers and the Federal government entered into negotiations to settle the dispute over declining groundwater. An agreement was reached after two years which provided a replacement water supply to meet the needs of the Native American community and obligated the Federal government to meet specified water delivery obligations (P.L. No. 95-328, 92 Stat. 409). The 1984 amendment was needed because the anticipated water supplies proved insufficient and too costly. The revised agreement provided that the Federal government pay damages for failure to make timely delivery of the water and modified the delivery schedule. Under the amended settlement, the bulk of the Ak-Chin's water entitlement was procured through the transfer of a Colorado River water entitlement from an irrigation district several hundred miles away. The imported water was an unused portion of the irrigation district's Federal water project supplies. The Federal government provided the irrigation district with a $9.4 million grant for improvements and conservation measures and forgave Federal loan obligations worth approximately $17.8 million in exchange for the water transferred to the Ak-Chin.

The original settlement did not provide for off-reservation leasing of Ak-Chin water because non-tribal interests strongly opposed this concept. It was not until 1992, 14 years after the original legislation and after the passage of several other Arizona agreements which permitted off-reservation leasing, that the settlement was again amended to allow off-reservation leasing of tribal water in specific areas of the State. The Ak-Chin community negotiated an option to lease some of its water to one of the southwest's largest real estate developers for use in new developments around the Phoenix area. Voluntary water transfers proved important at two junctures in the Ak-Chin water dispute: first, in procuring water to satisfy the agreement with the tribe and, second, allowing the tribe to earn revenues from leasing its water assets for urban growth. It is important to note that the ever-present threat of litigation was essential in motivating these voluntary solutions.

Litigation Over Water Rights in the Wind River Basin

"Courts provide narrow answers to narrow questions, yielding a crop of new questions." Jeff Fasset, Wyoming State Engineer, 1992.

Decades of conflict over water resources in Wyoming's Wind River Basin highlight some of the advantages and disadvantages inherent in resolving complex water disputes through litigation as contrasted with negotiation. The Wind River litigation has been lengthy, bitter, and expensive. Its legacy includes continuing uncertainties over water supplies and strained intergovernmental relations. However, the rulings have clarified specific issues, including the large quantity and high seniority of tribal water claims. The case, carefully observed throughout the West, serves as a critical reference point for other tribes and states and provides a graphic portrayal of the benefits and hazards of litigation as a means to settle water disputes. Only the most basic aspects of the litigation are summarized here (Checchio and Colby, 1993).

In the early 1800s, settlers began to enter the traditional homelands of the Shoshone Tribe in northwestern Wyoming. In an effort to avert the bloody conflicts that erupted elsewhere as Native American homelands were encroached upon, the U.S. and the Shoshone Tribe negotiated a treaty that recognized vast portions of the present-day states of Colorado, Utah, and Wyoming as Shoshone territory. Five years later, Shoshone lands were reduced substantially under a subsequent treaty which created the Wind River Indian Reservation. In 1878, the U.S. moved the Arapaho Tribe onto the Wind River Reservation, despite objections from both tribes. The Shoshone and Arapaho Tribes have jointly occupied and governed the reservation since that time.

The reservation, which includes the most productive agricultural region of Wyoming, attracted farmers who homesteaded there during the allotment era (1887 to 1934) when the Federal government opened "underutilized" reservation lands to anyone for settlement. The new settlers developed extensive irrigation on the reservation under Federal reclamation programs, building Wyoming's largest irrigation canal. This canal diverts the entire flow of the Wind River during periods of intensive irrigation, leaving a dry riverbed below the diversion dam.

In the mid-1970s, the State of Wyoming initiated the Bighorn Adjudication in State court to clarify the rights of non-tribal water users with respect to the Wind River Tribes. The Wyoming District Court eventually quantified the rights, after six years of hearings, involving over 100 attorneys, 15,000 pages of hearings transcripts, and "unprecedented expense to parties on all sides" (Roncalia, 1982). The tribes' reserved water rights were 617.38 million m^3 (500,717 acre-ft) per year, with an 1868 priority date, making them the most senior water rights in the basin. The parties appealed the decision to the Wyoming Supreme Court in 1988 and then to the U.S. Supreme Court in 1989 (Rights to Use Water in the Bighorn River, 1989).

Following the U.S. Supreme Court ruling, the State and the tribes negotiated a one-year agreement in which all parties accepted "equally shared surpluses and shortages" of water in the basin. In return for this protection of non-tribal water users, the State paid the tribes $2 million in a compensation fund, granted $1 million for repair and expansion of the reservation water delivery systems, and forgave 75 percent of the oil and gas severance taxes it collects from reservation oil and gas production. The agreement expired in 1990 and was not renewed (WIM, 1990). Further voluntary negotiations to deal with the conflict were thwarted by the high level of bitterness and distrust among the parties, as well as by irrigators' hopes that renewed litigation would protect their customary diversions by preventing the tribes from using their water rights to maintain instream flows for fish, wildlife, and recreation.

The parties have gone back to court many times. Court rulings have failed to effectively address questions of water management and jurisdiction, and further litigation is expected, leaving the region clouded by uncertain access to water and the lack of coordinated water management. The Wind River case highlights a litigious approach to resolving water disputes and illustrates the difficulty of switching to voluntary negotiations after years of litigation have produced bitterness and mistrust. The case also illustrates how the hope of victory through litigation can undermine collaborative water transfer arrangements.

Litigation and Negotiation in the Pyramid Lake Basin

The Truckee-Carson Water Settlement of 1990 involved the Pyramid Lake Paiute Tribe, the State of Nevada, Federal interests, local water users, and environmental advocates. The case highlights the role of bargaining power in motivating conflict resolution. The Truckee and Carson Rivers flow into the western Nevada desert from their headwaters in the Sierra Nevada Mountains of California and provide water for tribes, cities, fisheries, wetlands, and a large Federally developed irrigation project. The Pyramid Lake Indian Reservation, ancestral home of the Pyramid Lake Paiute Tribe, surrounds Pyramid Lake, residence of the cui-ui (an endangered fish species) and the threatened Lahontan cutthroat trout. This high desert lake is the terminus of the Truckee River and depends solely on the river for inflows. Conflict in the Truckee-Carson Basins began in the early 1900s when the Truckee Canal (completed in 1905) began diverting approximately half the flow of the Truckee River into the neighboring Carson Basin for irrigation deliveries. Reductions in Truckee River flows led to precipitous declines in Pyramid Lake, depriving the Pyramid Lake Paiute Tribe of fish and water upon which they had relied for centuries. The situation was further complicated by the fact that important wetlands in the Carson Basin depend on return flows from irrigation, and thus on Truckee River diversions.

The Pyramid Lake Paiute Tribe pursued negotiations and litigation for decades to restore the stream and lake levels for fisheries (Checchio and Colby, 1993). Following ten years of litigation, the U.S. Supreme Court dealt the tribe a serious blow by rejecting its water rights claims to Truckee River flows (*Nevada vs. U.S.*, 1983). However, the tribe persevered and turned to the Endangered Species Act as its new source of bargaining power. The tribe resumed litigation based on the Endangered Species Act and the courts found that uncommitted Truckee River water stored behind a Federal dam could only be used to protect endangered fish in Pyramid Lake, and not (as was planned) to satisfy the growing urban water needs of the Reno-Sparks area (*Carson-Truckee Water Conservancy District vs. Clark,* 1984). Finally, other water users began to recognize the tribe's commitment to litigation as a serious threat to their own water supplies. Negotiations began in earnest to satisfy the competing demands of Native American tribes, irrigators, endangered species, wetlands protection, and urban growth. The negotiated agreement, passed by Congress in 1990, provides protection for Pyramid Lake and Truckee River fisheries, continued water deliveries for irrigators and cities, and Federal money to buy out irrigation rights and to promote agricultural water conservation in order to assure water for wetlands. The voluntary transfer program endorsed by the settlement is playing a crucial role in diffusing conflict and reconciling endangered species and wetland concerns. As of 1996, numerous water rights acquisitions have been made by Federal and State wildlife agencies, local conservation groups, and national environmental organizations seeking to restore fisheries and wetlands.

Several aspects of this case are relevant to the discussion of the role of voluntary transactions in resolving water conflicts. No voluntary solutions were forthcoming until the tribe had enough bargaining power to command the attention of the other parties. The tribe's negotiating position was fortified not only by the Endangered Species Act, but also by the tribe's insistence that urban water users make more efficient use of their existing water supplies. Specifically, the tribe demanded that cities install water meters and charge customers for the amount of water used, instead of continuing to charge a flat water fee. This demand was met with strong initial resistance by urban water providers but it gave the tribe a "moral advantage" in the national media coverage of this dispute. Few could argue that this request was unreasonable given water conservation programs in other western cities. Local water utilities bowed under the pressure of public opinion and agreed to install water meters and begin charging water users according to the quantities used.

Political Pressure to Alter Operations of Glen Canyon Dam

The Glen Canyon Dam case illustrates how reliance on political pressure, combined with threats of litigation and some voluntary reallocation, can achieve changes in water allocation. Unlike the earlier cases, this case does not involve a change in the ownership of water but rather a reallocation of stored water and a change in the way a Federal water facility (the dam) is operated. Glen Canyon Dam is located on the Colorado River at Lake Powell in northern Arizona. The dam was constructed in the late 1950s to store water and to generate large quantities of low-cost hydropower for the Southwest. Controversies over the operations of Glen Canyon Dam focus on tradeoffs between water released to generate valuable peak hydropower and the effect of these releases on recreation and ecosystems along the river below the dam. Disputants include economically dominant western hydropower users, environmental advocates, river recreation businesses, several Native American tribes, and an array of Federal and State agencies with jurisdiction over water, wildlife, recreation, and electric power.

Dam operations designed to maximize peak hydropower production can raise and lower the river level in the Grand Canyon by several feet a day, creating hazards for recreationists, destabilizing beaches and vegetation, and impairing habitat for fish and wildlife (National Research Council, 1996). These operations went largely unquestioned until the early 1980s when the Bureau of Reclamation proposed to increase generator capacity and output. The Bureau wanted to avoid an Environmental Impact Statement (EIS) (none had been conducted for the construction of the dam, since it preceded passage of the National Environmental Policy Act in 1970) and therefore initiated the Glen Canyon Environment Studies (GCES) to collect data, in lieu of an EIS process. The GCES initially were confined to examining the environmental resources near the dam and the operating regimes to maintain or enhance hydropower production. However, environmental, recreational, and tribal interests effectively used their political muscle and threats of litigation to force initiation of an EIS process that considered operating regimes that actually would curtail peak hydropower production in order to protect downstream ecosystems. In the course of the multi-year EIS process, it became clear that the traditional way of operating the dam to maximize peak hydropower production was no longer politically viable. Proponents of the new emphasis on environmental values and recreation obtained an act of Congress (The Grand Canyon Protection Act of 1992) modifying dam operations to restore the riverine environment and endangered species below the dam, with specific limits on how much river flows can be raised or lowered for power production. This legislation critically altered the bargaining power among the disputants, giving further leverage to environmental and recreational interests. Though no change in ownership of water occurred, the Act reallocated resources by revising dam operations and shifting water releases from peak power generation to river flow stabilization.

The Glen Canyon Dam conflict illustrates the use of political maneuvering, culminating in an act of Congress, to reallocate water releases and modify dam operations. The bargaining power of those pushing for modifications rested on the Environmental Policy and the Endangered Species Acts and was bolstered by Native American cultural concerns and recreation interests. This bargaining power became enough of a threat to established hydropower interests that they agreed to give up some peak hydropower generation in order to reach a compromise and settle the conflict.

Markets as a Component in Conflict Resolution

The preceding four cases illustrate that voluntary water transfers can play a key role in addressing disputes over water scarcity. Acquisitions of water, primarily from farmers, have satisfied the demands of tribes, endangered species, and recreationists. Often several rounds of transfers are negotiated, first to address the immediate water needs of tribes or endangered species, and then to provide alternative supplies for those who gave up water in the initial transfer. Such staged transfers of water were a condition of several agreements to settle tribal claims in Arizona. Ownership of water previously held by others was transferred to the tribe, which then leased the water back to the original water users in order to guarantee their continued water supplies while also generating revenues for tribal economic development. In the case of Glen Canyon Dam, the parties agreed that water temporarily could be redirected back to peak hydropower production (in violation of the new limits on dam operations) for a specific number of hours each month and during electrical power emergencies. This is an example of a temporary and conditional transfer embedded within the larger agreement, which shifted water away from peak power generation toward ecosystem restoration (National Research Council, 1996).

The case studies also illustrate the interrelationships between voluntary transactions, litigation, and political lobbying. In all four cases, voluntary transfers were negotiated in a broader context of viable threats to involuntarily reallocate water. "Voluntary" water transfers as a means of resolving conflict are not free of coercion and are typically motivated by the threat of high transaction costs and involuntary transfers through court or administrative actions. Disputants pursue water reallocation using the approach that appears to yield the highest net payoffs, weighing the likelihood of achieving their goals and their expected costs. Those that can issue bonds or raise taxes to acquire water (such as growing cities) favor market acquisition. Those with strong legal arguments and long-term goals of revising water policies (tribes for instance) tend toward litigation. Those with political clout and public support (wildlife groups) pursue political lobbying to achieve their objectives. Each form of reallocation favors somewhat different types of disputants, depending on the mix of economic, legal, and political power they command. Since water conflicts involve multiple disputants, each of whom may pursue a different reallocation strategy, water disputes are characterized by a combination of voluntary transfers, litigation, and political pressure.

Markets, as a negotiating framework, assume that property rights are well-defined among the potential trading partners so that the nature of the property rights themselves is not being negotiated in a market setting. However, in many water conflicts, rights to use water are fraught with ambiguity. Disagreements over the characteristics of the rights must be resolved while the parties are also discussing price and other terms of a transaction. The term "bargaining" refers to this more general situation in which individuals and organizations are exploring the possibility of a mutually beneficial voluntary agreement, in the face of conflicts regarding the nature of the rights to be exchanged (Nash, 1950). Complex water conflicts typically involve disputes over who has the rights to how much water, and for what purposes.

Well-functioning water markets with common agreement on the nature of the rights being exchanged are the exception, not the norm, in the American West.

6.4 Bargaining Over Water

Regardless of the mix of litigation, market transfers and politics that characterizes a particular dispute, the parties will engage in bargaining with one another. A common theme in such negotiations is disagreement over the baseline rights which each side claims as its starting point for negotiations. Many different types of baseline entitlements are invoked. The Helsinki Rules, developed in 1966 as principles to guide international water allocation, state that each jurisdiction in a river basin is entitled to an "equitable apportionment" of basin waters – a vague criteria that parties interpret variously and to their own benefit. Some parties use baselines that refer to geography – calculating their reference entitlements based on where the river originates, how much of the watershed falls within their jurisdiction, or on who owns the land alongside the watercourse. Other parties base their claims on the chronology and history of water use, with the earliest diverters and dam builders making the claim for primacy. Still another type of baseline refers to the relative water "needs" of the respective parties. One of the few international water agreements ever negotiated in the Middle East (though not ratified) was a needs-based apportionment of the Jordan River between Israel and Jordan, taking into account current and projected water uses in the two nations (Wolf, 1996).

In general, each party will select an approach to define its baseline that results in the biggest claim. There are no objective criteria for choosing *among* the different baseline positions in multi-party bargaining since each baseline can be justified from the particular perspective of the party advocating that baseline. Generally, rights-based claims lead to legal wrangling and the development of competing legal positions, while needs-based claims generate differing assessments of the parties' "legitimate" water needs and also incite behaviors that amplify needs-based claims, such as high population growth and development of new irrigated areas.

Public policies play an essential role in defining property rights, the parameters within which negotiations and conflict resolution occur, and thus bargaining power. The property rights structure determines the distribution of costs among the disputants, for example, which party must bear the costs of initiating action to protect their interests. The ability to impose costs on others represents bargaining power. It is one thing for wildlife advocates to complain in the press about the impact of irrigation diversions on a particular species. However, it is quite another level of influence entirely for those same advocates to be able to initiate litigation that will diminish the irrigators' diversions. In this case, the irrigators have a genuine incentive to bargain with wildlife interests. The Endangered Species Act, the Clean Water Act, the National Environmental Protection Act and many other laws have shifted the distribution of bargaining power, providing new incentives for negotiations between environmental interests and water users. A parallel set of laws and court rulings has given tribes new bargaining power. Public policies can grant disputants a seat at the bargaining table and provide the "chips" with which to threaten and bribe other parties in order to win concessions from them.

The policy framework determines, to a large degree, the effectiveness of voluntary transactions in resolving conflicts. Voluntary transactions require facilitating institutions (law of contracts, property ownership, registration of transfers) and these institutions require the political system to decide whose interests will be protected. In western water disputes, typically there is legal and political ambiguity when urban, agricultural, tribal, recreational, and environmental needs clash. Water rights are defined based on collective consensus which shifts as collective values and norms change regarding the environment, justice for tribes, and the importance of instream recreation. These shifts generate considerable conflict and make voluntary transfers more difficult. The ambiguity of water rights and water policies also aggravate disputes as the parties each interpret ambiguous provisions of regulations and court rulings in their own favor.

6.5 Evaluating Conflict Resolution Processes

Little has been written regarding rigorous evaluation of dispute resolution processes from an economic perspective (Raab, 1994). Economic efficiency concepts can be difficult to apply in analyzing water disputes since costs and benefits must be measured from a specific baseline. The selection of that baseline determines the analytic results but, in disputes over water entitlements, the appropriate baseline itself is in dispute. Even the use of the status quo as the baseline is arbitrary because, in a full-swing dispute, some interest groups clearly believe the status quo is unacceptable.

A comprehensive examination of the social costs and benefits of dispute resolution should focus on both the outcome achieved *and* the processes used to achieve the outcome. Efficiency analyses can assess the desirability of the resource allocation achieved as compared to other possible outcomes – inquiring whether all potential gains have been realized, which parties are better off and which are worse off, and whether gains outweigh losses. While such analytic efforts must confront the choice of a baseline from which to measure gains and losses (a choice likely to please some disputants and infuriate others), the analysis of benefits and costs can provide information and insights to assist the parties in crafting a solution.

Other economic constructs can be invoked to evaluate the dispute resolution processes themselves. Raab (1994) suggests "process-related resource savings" as one criterion to examine whether some processes are less expensive in terms of time, money, and other resources, as compared to a baseline case of litigating the dispute. Resource savings, or cost effectiveness, encompass both long- and short-term aspects. Short-term considerations include the up-front benefits and costs to disputants who participate in the process and the benefits and costs to others who are not direct participants – such as taxpayers, the environment, resource users not at the table, and the "public interest." Long-term cumulative costs relate to the stability and flexibility of the agreement, the frequency with which the parties need to revisit its provisions and negotiate adjustments, and the effectiveness of the initial process in building capacity among the parties to resolve future disputes. A cost-effective process facilitates the development of a good working relationship among the parties and a shared information base regarding the resource, so that future disputes can be resolved more readily.

The stability of negotiated agreements is another criterion for evaluating success, but this can only be assessed over time as the implementation of an agreement proceeds and the parties face new challenges that test the strength of the original agreement. One would like to be able to predict the stability of an agreement ex-ante, as it is being formulated, so that appropriate adjustments to enhance stability can be built in during negotiations. Features to enhance stability include incentives built into the agreement to reinforce the parties' commitment to its provisions, such as penalties for non-compliance. Stable agreements need the capacity to withstand new environmental constraints and changing values, and should include mechanisms for dealing with the unexpected (droughts, floods, new court rulings) as well as establish a process for negotiating modifications as needed. The relationship that evolves among disputants over the course of the conflict affects the stability of the agreement. Development of mutual assurance, reciprocity in "give and take," and trust that other disputants will not use their influence to undermine implementation of the agreement all contribute to stability (Easter, 1993; Runge, 1984).

Perceived fairness is an emotionally charged criterion that can make or break a dispute resolution process and yet it is not amenable to precise definition. Disputants have different beliefs regarding their baseline entitlements and these determine their perceptions of the fairness of a particular outcome. One economic component of fairness involves apportioning the costs of a particular solution among the parties. To what extent should differing financial capabilities play a role in allocating costs? For instance, should both wealthy and poor disputants bear equal shares of the costs of obtaining additional water to resolve their conflict? Should differences in historic access to and benefits from the disputed resource affect how benefits are distributed under the new agreement? In the Glen Canyon Dam conflict, for instance, the electric power users who had benefited from low-cost access to hydropower for decades had the most to lose under the new agreement. Does their prior favored status make their losses *more* fair or *less* fair? The power users argued that they had made substantial investments relying on low-cost power and ought to be compensated for losses, while other parties argued that power users had been benefiting from subsidized power for years at the expense of taxpayers and the environment and merited neither sympathy nor compensation.

Perceived fairness is a criterion that applies not only to the particulars of a proposed agreement (cost sharing, water allocation) but also to the processes used to reach agreement. Voluntary negotiations in which all disputants have a voice and the parties build up trust in one another's integrity and ability to abide by agreements are more likely to be perceived as fair than are solutions imposed by courts and bureaucrats.

The three criteria discussed are interrelated. Perceived fairness and stability affect the cost effectiveness of dispute resolution processes. A process that is deemed unfair by key participants is likely to be less stable and more readily undermined, leading to longer and more costly periods of unresolved conflict.

6.6 Strengths and Weaknesses of Alternative Reallocation Processes

Three alternative (but often complementary) approaches to resolving water disputes have been discussed: voluntary reallocation of water, litigation to achieve reallocation, and political pressure to change public water management and allocation.

Voluntary transactions are just one option to address conflicts, but one commonly advocated by economists. Stimulated by Pigou's seminal work (1932), economists long have advocated using market incentives to further environmental protection and to improve resource allocation. The U.S. Department of Interior has released a series of policies advocating market transfers of Federally supplied water and a number of western states have enacted specific legislation designed to encourage water markets (U.S. Department of Interior, 1988 and 1994). Economists' advocacy for voluntary transfers to improve efficiency has been a vital policy contribution. However, public officials typically resist the initial creation of a market framework. Their ambivalence stems from a matter often overlooked by economists, but of great importance to stakeholders: policies that encourage market transactions imply endorsement of transferable rights (the items of value to be traded in the market place). The initial assignment of these rights confers wealth and bargaining power, and determines which parties reap benefits and incur costs from the new policies. While it may be easy for policy makers to adopt the economists' enthusiasm for markets in the abstract, it can be difficult to assign transferable rights to affected parties, a policy task with very tangible consequences. The assignment of private, transferable water rights has often been viewed as inconsistent with preserving public environmental and recreational amenities. This initial granting of rights is where the heart of resistance to markets, as a dispute resolution mechanism lies. The assignment of water rights has generally been to those historically diverting water, but others, who benefit from but do not divert water (tribes, recreationists, wildlife advocates), perceive the assignment of rights as unfair. Consequently, many recreational and environmental interests have objected to market processes which they believe reward an unjust initial allocation.

Voluntary transactions acknowledge the existing water right structure and fully compensate right holders who sell their water, reinforcing status quo entitlements so that post-trade allocations of wealth do not depart notably from pre-trade endowments (Griffin, 1991). The costs of acquiring water for streams, wildlife, tribes, and recreationists are borne solely by the organizations and agencies advocating such changes, and are not shared by status quo right holders. For a market transfer to help resolve a water conflict, the disputants must concede that the current resource user holds a right which legitimately can be transferred for economic gain. This endears the market mechanism to those favored under status quo entitlements, but makes it considerably less attractive to those advocating major changes in water allocation. For them, litigation and lobbying may be preferable.

Involuntary transfers appeal to environmental and tribal advocates because they can shift the distribution of property rights and set precedents for the future. The costs of reallocating water through litigation and through revised water regulations are spread among the parties to the conflict and are thus not borne only by environmental or tribal interests. Such interests may be willing to spend more to *litigate* for a favorable ruling than they would pay to purchase the rights being disputed, because a favorable ruling will set a precedent that will make subsequent conflicts less costly by enhancing their bargaining power in future disputes. Many of the voluntary transactions that have reallocated water from consumptive uses to tribes, wetlands, and wildlife were preceded by decades of litigation. For instance, as described earlier, the Pyramid Lake Paiute Tribe could not have influenced water allocation in Nevada's Truckee-Carson Basin to a significant degree without their prior successes in court based on the Endangered Species Act (Colby et al., 1991). The costliness and uncertainty of litigation and the threat of an unfavorable court ruling are powerful incentives for parties to consider voluntary transactions as a lower-cost means to resolve conflicts.

While litigation has desirable features for some disputants, it is costly, protracted and controlled by the courts rather than by the disputants themselves. Litigation evokes a sense of antagonism among those whose interests are at stake – to the detriment of future cooperative water management efforts, once the immediate conflict has been addressed by the courts. Furthermore, litigation is characterized by the use of technical information as a weapon and persuasive device, rather than by a collective search for the best data on the watershed and on competing water needs. The parties generally know the characteristics of their basin and their respective demands for water more thoroughly than a judge can learn this information, even in an extended trial. In addition, courts generally are vague regarding the implementation of their rulings and address only narrow legal questions. This leaves the parties with the task of interpreting and working out the implementation themselves, or going back to court for further clarification. In a consensus-oriented process, the parties can address practical matters of water management, going beyond the narrow jurisdiction and capabilities of the courts.

With regard to the three evaluation criteria discussed earlier, voluntary transfers are likely to be more cost-effective than either litigation or political lobbying. Resources are expended on acquiring water rather than on attorneys and lobbyists and the parties can pool their knowledge of the watershed to design transfers that best meet their needs. Voluntary approaches also have the greatest potential to build trust and stable working relationships among the protagonists. This is a form of building social infrastructure. Just as investments in building dams and canals pay off in terms of improved water storage and conveyance, "investment" in voluntary processes can build conflict resolution capability and pay off in cost savings over the long run. Given the long-term, multi-round nature of water disputes, improved problem-solving capabilities save resources over time as the parties learn to resolve conflicts more quickly and cost-effectively. Thus, in many conflicts, voluntary agreements are likely to be more cost-effective, stable, and fair than involuntary reallocations. Yet voluntary agreements may not work in those conflicts where key disputants bitterly object to status quo water entitlements, view market acquisitions as reinforcing an unfair distribution of wealth, and seek to change the current water allocation structure through setting new policies and legal precedents. In such cases, purchases of water from current users may be unacceptable and litigation or political reallocations will be pursued.

6.7 Conclusions

Voluntary transfers require the disputants to collaborate in gathering information and to reach agreement on price and on baseline entitlements. In the process, the parties build social capital in the form of trust and sound working relationships. Even if the reallocation that is achievable in a voluntary setting is not satisfactory to some disputants, they may choose to stay with the process for the sake of further long-term improvements when the next round of problems prompts further negotiations. Voluntary reallocations, in the context of a water conflict, often are motivated by threats of litigation and other coercion, leading one to question whether the virtue of "voluntariness" is meaningful in this setting. Moreover, voluntary transfers involve full compensation of status quo water users and require that the disputants acknowledge the prevailing distribution of water rights. Consequently, such transfers cannot accomplish significant redistributions of wealth and so may be unacceptable to those who believe a major distributional change is warranted.

The two primary alternatives to voluntary reallocation (litigation and political pressure) also have their strengths and pitfalls. They can alter the rules for water allocation, set precedents for future disputes, and serve as threats to bring disputants into voluntary negotiations. However, litigation and strong-arm political tactics erode rather than build trust and problem-solving capacity among parties who must co-manage shared water resources.

This chapter has explored various processes for resolving water disputes. Economists traditionally focus on the economics of alternative *outcomes* – resource allocations and property right configurations – rather than on the *processes* for achieving those outcomes. Water disputes present economists with the challenge of evaluating processes and not just the outcomes. While economic efficiency cannot neutrally determine whose interests ought to be favored, economic constructs can be used to compare the advantages and weaknesses of various conflict resolution strategies and economic incentives can be built into negotiated agreements to enhance their stability, flexibility, and cost-effectiveness.

References

Arizona v. California, 1963, 373 U.S. 546.

Carson-Truckee Water Conservancy District v. Clark. 1984. 741 F. 2d, 257, cert. denied , 1985, 105 S. Ct. 1842.

Checchio, E. and B.G. Colby, 1993. "Indian Water Rights: Negotiating the Future." The University of Arizona, Water Resources Research Center.

Colby, B.G., 1993. "Benefits, Costs and Water Acquisition Strategies: Economic Considerations in Instream Flow Protection." In *Instream Flow Law and Policy*, ed. L. MacDonnell. Boulder: University of Colorado Press.

Colby, B.G., M. McGinnis and K. Rait, 1991. "Mitigating Environmental Externalities Through Voluntary and Involuntary Water Reallocation." *Natural Resources Journal,* (31) 757–783.

Dinar, A. and E. Loehman, 1995. *Water Quantity/Quality Management and Conflict Resolution*. Praeger Publishers, Westport, Connecticut.

Easter, K. William, 1993. "Economic Failure Plagues Developing Countries and Public Irrigation: An Assurance Problem," *Water Resources Research,* (29)1913–1922.

Griffin, R.C., 1991. "The Welfare Analytics of Transaction Costs, Externalities and Institutional Choice." *American Journal of Agricultural Economics,* (73) 601–614.

Grand Canyon Protection Act of 1992. Public Law 102-575.

Helsinki Rules, 1966. International Law Association, Finland.

MacDonnell, L.J., 1990. "The Water Transfer Process as a Management Option for Meeting Changing Water Demands." Report for the Natural Resources Law Center, University of Colorado, April, Boulder, Colorado.

MacDonnell, L.J., ed., 1993. *Instream Flow Law and Policy.* Publication of the Natural Resource Law Center, University of Colorado, Boulder, Colorado.

Nash, J.F., 1950. "The Bargaining Problem." *Econometrica,* (18) 155–162.

National Research Council, 1996. *River Management in the Grand Canyon*. National Academy Press, Washington, DC.

Nevada v. U.S., 1983. 463 U.S. 110, 113.

Pigou, A.C., 1932. *The Economics of Welfare, Fourth Edition,* Macmillan, London.

Public Law no. 95-328, 92 Stat 409, 95th U.S. Congress, 2nd Session, 1978. Washington, D.C.

Raab, J., 1994. Using Consensus Building to Improve Utility Regulation. American Council for an Energy Efficient Economy. Washington, D.C.

Rights to Use Water in the Bighorn River, 1988, 753 p. 2d, 76, Wyoming.

Rights to Use Water in the Bighorn River, 1989, 488 U.S.1040; 109 S.Ct. 863.

Roncalia, T. , 1982. "Special Master Report Concerning Reserved Rights Claims By and On Behalf of the Wind River Indian Reservation," (Wyoming).

Runge, C. F., 1984. "Institutions and the Free Rider: The Assurance Problem in Collective Action," *Journal of Politics,* (46) 154–181.

Saliba, B. C. and D. B. Bush, 1987. *Water Marketing in Theory and Practice: Market Transfers, Water Values, and Public Policy*. Boulder, CO: Westview Press.

U.S. Department of Interior, 1988. "Water Transfer Policy Statement," Washington, DC.

U.S. Department of Interior, 1994. "Proposed Rules and Regulations for the Colorado River," Washington, DC.

WIM (Water Intelligence Monthly), 1990, March, p.13. Published by Stratecom, Inc., Claremont, CA.

*Winters v. U.S.***,** 1908, 207 U.S. 564.

Wolf, Aaron T., 1996. "Middle East Water Conflicts and Directions for Conflict Resolution," Food, Agriculture and the Environment Discussion Paper 12, International Food Policy Research Institute, Washington, DC.

7 Expected Transaction Costs and Incentives for Water Market Development

Sandra O. Archibald and Mary E. Renwick

7.1 Introduction and Background

California provides the opportunity to analyze both the potential social gains from water transfers and the institutional constraints, including transaction costs, that can limit water market performance. Common law principles, constitutional provisions, State and Federal statutes, court decisions, water agency policy and contracts all govern how water is developed, allocated and used (State of California, 1995). This complex institutional setting creates a heterogeneous system of water-use rights resulting in uncertainty regarding the precise legal entitlements conferred with use rights. State and Federal law and policy governing water transfers and liability for any related economic and environmental impacts have been clarified over the past ten years to facilitate water market development. Despite these clarifications, uncertainty regarding use rights, transfer procedures, and transaction costs remains. As a consequence, water markets are yet to develop as anticipated. Drawing on recent experience, this chapter explores both the potential for social gains-from-trade in water-use rights and the institutional barriers constraining such trade, particularly focusing on the impact of heterogeneous water-use rights and transaction costs.

It is well-documented that significant gains could be realized if a well-functioning water market existed in California to reallocate supplies from lower valued agricultural uses to higher valued agricultural, urban, and industrial uses (Vaux and Howitt, 1984; Howitt, 1994; Zilberman et al., 1993). While water-related institutions and the legal specification of water rights severely constrained the development of water markets in California historically, a number of developments in the late 1980s laid the foundation for significant institutional change. In particular, a six-year statewide drought (1987-92), coupled with chronic regional water scarcity, provided the impetus for State and Federal institutional reforms to facilitate water transfers. In 1991 and 1992, changes in California State law clarified property rights related to both short- and long-term water transfers (California Water Code Sections 1435, 1725-72 and 1735). Concurrently, Federal legislation contained in the 1992 Federal Reclamation Reform Act (PL102-575), solidified water-use rights for Federal contractors, eased transfer restrictions, and gave individual agricultural producers the right to engage in transfers to overcome potential institutional resistance to water transfers by water supply agencies.[1]

The inherent potential for market failure due to water's pervasive non-rival and non-exclusive characteristics was also explicitly addressed with the adoption of institutional rules that place restrictions on transfers and establish transfer approval mechanisms. Federal and State regulations require that only water used consumptively can be transferred, provided that it imposes no injury to other vested rights and results in no unreasonable effect on groundwater or fish and wildlife. Reforms set rules and procedures for identifying and remedying negative externalities, but have yet to clarify liability for third-party impacts.

In addition to these policy changes, government failure to include non-use values in previous water allocation decisions has been addressed by reallocating surface water supplies from current uses to meet critical environmental demands. The 1995 "Bay-Delta Accord" and the subsequent State Water Resources Control Board's "Water Quality Control Plan" reduce the quantity and control the timing of water that can be exported from the Sacramento/San Joaquin Delta to meet Federal water quality standards and protect endangered species habitat.[2] This reallocation of water to meet environmental uses has a significant impact on the availability of municipal and agricultural water, as the Delta is the State's primary water export hub. It provides a further incentive for well-functioning water markets.

While many of the policy reforms were supposed to facilitate the development of water markets, they have yet to achieve their full intent. Since 1982, only 1.7 million acre feet (acre-ft) of water have been transferred and all were spot or short-term (annual lease) transfers (State of California, 1996). No long-term (greater than one year) or permanent transfers have been arranged under revised policies. For the first time in 1995, the State Department of Water Resources purchased water supply options from sellers in the face of an expected "dry" year (Jercich, 1997). None of these options was executed.[3] California has, however, developed some experience with informal intra-seasonal spot markets and annual lease markets. The State-run "Water Bank" that operated in 1991, 1992, 1994 and 1995 has handled 40 percent of all of the water transfers that have occurred in the State since 1982. The Water Bank demonstrated that annual lease markets could profitably reallocate supplies and that buyers and sellers would be willing to participate in an options market.

Despite the success of the Water Bank, it did not develop as economists had anticipated. In all but 1991, trade within the agricultural sector dominated rather than the anticipated intersectoral trade from agricultural to urban uses.[4] In that year, surface supplies were reallocated from lower-valued agricultural cropland, yet concentrated third-party impacts limited sources of supply in subsequent years to groundwater substitution and reservoir stocks.[5] This has raised new questions about impacts on groundwater supplies and the need to identify "real" water. Uncertainty over whether or not producers were eligible to participate in the Water Bank also precluded some profitable trades. Other institutional policies related to water-use rights and water source limited market participation. Producers with less secure rights (e.g., groundwater or water rights settlement contractors) did not participate as sellers, whereas holders of more secure surface riparian rights did. It is likely that uncertainty over property rights affected expected transaction costs. Likewise, expected transaction costs were affected by uncertainty regarding impacts on other vested rights and third-party liability (Archibald and Kuhnle, 1992; Howitt, 1994).

While the basic legal framework for water transfers exists, the performance of California's water markets to date highlights the likely need for further institutional change if transfers are to help meet future needs. However, before policy changes to facilitate the development of water markets can be recommended, factors affecting current market performance need to be better understood. For example, do state and Federal laws designed to protect vested rights and third parties raise expected transaction costs to preclude all but the most economically beneficial transfers? To what extent have institutional reforms failed to address issues related to differential market access among potential participants? How do heterogeneous water-use rights, and any asymmetries in transaction costs among market participants due to the current institutional framework, influence gains-from-trade?

To provide insight into these issues, a gains-from-trade model is developed to evaluate water market potential in California and to assess the impact of transaction costs and other institutional constraints on market performance. The gains-from-trade model is applied to select agricultural water agencies (or "districts") in California's Central Valley to measure potential social gains-from-trade when transaction costs are explicit and heterogeneous property rights prevail. Water market potential is defined as the gains that could be achieved with economically feasible trades in water-use rights. Water market performance is defined as the gains associated with trade that actually occurred relative to the potential gains. The value of water in its current use is calculated as a measure of the value of a water-use right. The willingness of agricultural producers to enter into water market transactions is modeled as a function of bid prices, the marginal value product of water in its current use (conditioned on the type of property right), and the existence and magnitude of transaction costs. Previous studies of water markets in California have not considered heterogeneous water- use rights, market access constraints, and transaction costs (Vaux and Howitt, 1984; Howitt, 1994; Zilberman, et. al., 1993). As a consequence, they overstate water market potential.

The agricultural water agencies selected for study are located in California's Central Valley and represent nearly 20 percent of irrigable acreage in the State and approximately 40 percent of its developed surface water supplies (Archibald et. al, 1992; Archibald and Kuhnle, 1992). These districts were selected for study for a number of reasons, including differences in types of water-use rights, sources of water, and a priori information on the differences in the value of water in agricultural production. Analysis is based on district specific water allocation and use data (crop, yield, and price information) and actual water transfer data. Secondary sources were relied upon for water market prices and transaction costs.

7.2 Institutional Setting and Water Market Performance

Beyond influencing the potential gains-from-trade in water-use rights and protecting against external effects of market transfers, the institutional setting in which water markets operate will influence actual market performance by affecting participation incentives. When the "incidence" of institutional arrangements varies, due to heterogeneous property rights among trading partners for example, potential market participants may face differential water market access constraints or highly asymmetric transaction costs. Failure to account for institutional constraints and their distributional implications may overstate the willingness of buyers and sellers to participate and the potential efficiency gains associated with transfers. In California, the differences in institutional arrangements among potential market participants are likely to play a significant role in explaining the current level of market activity.

Water market access, or the ability of a holder of a water-use right to transfer water, depends on the type of property right. In California, water held pursuant to a riparian right is not transferable from place to place, although downstream appropriators may contract with riparian rights holders to leave water in the stream for potential downstream diversion. Appropriative water-use rights, based on the principle of "first-in-time," may be transferred. Contractual water rights that are based upon an exchange for riparian rights may be transferable subject to the terms of the exchange contract. Transfers of groundwater and groundwater substitution arrangements (i.e. substituting groundwater for transferred surface supplies) are, in some cases, prohibited by statute to avoid overdraft and to maintain local control of the resource. Water held pursuant to a water supply contract, such as typically held by contractors to large State and Federal water supply projects, is potentially transferable (State of California, 1995).

Access to markets and the conditions of participation are also influenced by legal and institutional parameters. Water transfers must comply with complex legal requirements set by the State. Water transferred from the Federal Central Valley Project (CVP) must comply with additional and more stringent access requirements. (See Figure 16.1 on page 265 for a diagram of the California water system.) Table 7.1 summarizes California State water code requirements for different types of water transfers and lists the additional requirements for Federal CVP water transfers. It illustrates the complex procedures as well as the potential sources of transaction costs associated with meeting these requirements.

Market participation is also influenced by other policies. For example, during the first year of operation of the State Water Bank contractors in the Federal CVP were prohibited by Federal policy from participating as sellers. Buyers were originally limited to those with "critical" water needs as defined by the State Department of Water Resources. Despite the reforms in the 1992 Reclamation Reform Act which allow CVP water to be transferred out of its service area, CVP contractors retain the right of first refusal for any planned out-of-project water transfers. Water districts may set additional requirements for transfers within the CVP service area.

Transaction costs can be aggregated into two broad categories: administratively-induced transaction costs (AICs) and policy-induced transaction costs (PICs). Table 7.2 describes common transaction costs associated with water transfers by these two categories and indicates their incidence by market position under present policy. AICs include the costs of searching for trading partners and negotiating the terms of exchange and other contract provisions. These costs are typical for any transfer of property. PICs are the costs of obtaining approval for water transfers. Colby (1990) initially defined them to include the costs of identifying the legal and hydrological characteristics of the water-use right, complying with state or Federal laws and other institutional requirements, and assessing and mitigating potential third-party impacts. The specific PICs related to water transfers in California can be derived from the transfer procedures outlined in Table 7.1. Clearly, AICs and PICs are a function of transaction type as well as property right. Table 7.2 makes it clear that the incidence of transaction costs falls asymmetrically between buyers and sellers. Heterogeneous property rights and differential rates of market access work to accentuate these asymmetries.

Table 7.1. California State and Federal Water Market Requirements (continued on next page)

California Water Code Requirements for Water Markets

Transfer Type	Water Code Section	Requirements	Environmental Actions	Comments
Temporary Urgency Change (one year or less)	1435	• Urgent need. • No injury to vest rights. • No unreasonable effect on fish and wildlife. • Use in public interest. • Show diligence in seeking the permit or long-term change.	Normal CEQA process	• Petition must be filed with SWRCB. • Change good for up to 180 days • Can be renewed. • Board notice and action.
Short-Term Transfer (one year or less)	1725 - 1732	• If applicable, petitioner must have been diligent in petitioning for a permanent change. • Involves only water used consumptively or stored. • No injury to vested rights. • No unreasonable effect on fish or wildlife.	Exempt from CEQA	• Permittee notifies SWRCB of proposed change. • SWRCB must make findings. • Hearing may be required. • Effective 5 days after SWRCB approval. • Good for one year or less.
Long-Term Transfer (more than one year)	1735	• No injury to vest rights. • No unreasonable effect on fish or wildlife.	Normal CEQA Process	• Petition must be filed with SWRCB. • SWRCB provides notice and opportunity for hearing. • Good for any period in excess of one year.

Table 7.1. California State and Federal Water Market Requirements (continued from previous page)

Federal Requirements for CVP Water Transfers [1]

- Secretary of Agriculture reviews and approves all transactions.
- Transactions involving more than 20 percent of a district's total are also reviewed by the contracting district.
- Transferred water is limited to that which would have been used consumptively or irretrievably lost to beneficial use during the year or years of the transfer and amount transferred cannot exceed quantity of water under contract annually.
- Water must be repaid at the greater of full-cost or cost-of-service for agriculture or cost-of-service or the rates for municipal and industrial users. The difference between the contract rate and the full-cost or cost-of-service rate is allocated to the Fish Restoration Fund.
- All transfers must be for beneficial use.
- Parties within CVP have right of first refusal to all out-of-project transfers.
- Transfers must comply with Federal law and not interfere with the ability to meet CVP contractual obligations or obligations to fish and wildlife because of conveyance or pumping capacity.
- Must have no significant long-term impact on groundwater conditions.
- No unreasonable impact on the water supply, operations, or financial conditions of the contracting district.
- Cannot result in a significant reduction in the quantity or quality of water used for fish and wildlife unless other benefits offset adverse effects; requires mitigation to provide fish and wildlife benefits equivalent to those lost.

[1]Conditions on Water Transfers Under Section 3405 of the 1992 Federal Reclamation and Reform Act.

Table 7.2. Distribution of Transactions Costs between Buyers and Sellers under Current Policy by Type of Market Participation

Type of Transaction Cost	Incidence	
	Seller	Buyer
Administratively-Induced (AICs).		
Searching for Trading Partners.	X	X
Negotiating Terms:		
establishing price, quantity, and quality;	X	X
negotiating payment terms;	X	X
establishing delivery dates;	X	X
negotiating physical transfer.	X	X
Policy-Induced (PICs).		
Identifying Legal Characteristics of Water-Use:		
ability to transfer;	X	
return flow obligations;	X	
timing of transfers.	X	
Identifying Hydrological Characteristics of Rights.	X	
Complying with State and Federal Law regarding transfer application and approval process.	X	
Complying with Other Institutional Requirements:		
project approval process;	X	
water district approval process.	X	
Adjustment Costs of Changing Resource Base:		
third-party impacts;	X	X
litigation for damages;	X	X
litigation/risk.	X	

7.3 Modeling Gains from Water Trades

The gains-from-trade model measures potential social gains from the exchange of water-use rights. Before gains can be measured, the value of water-use rights must be known. The model is developed initially under the assumption that agricultural producers are the likely participants in developing water markets. A producer-level profit maximization model is employed to calculate the value of water in its current uses for four agricultural water districts in California's Central Valley which represent the region's range of property rights and water sources. The resulting marginal value product of water in its current use provides a measure of the short-run (annual lease) value of the water-use right, conditioned on the type of right and water source. The profit maximization model explicitly incorporates water allotments and purchase price based on water-use right and source, water market buy and sell decisions, and associated transaction costs. It can incorporate other types of water market transactions and intersectoral trade with simple modifications.

Each agricultural producer (i) maximizes profits, π_i, by choosing the volume of irrigation water, x_{ij}, for each crop (j) given the simplifying assumption of optimal levels of land, l_i*, and other non-water inputs, z_{ij}*. Producers face a water allocation constraint, A_i, reflecting their appropriative or riparian water-use right or contractual entitlement. Producers pay a per unit price for this allocation of w_i, which can vary from the market price of water to reflect implicit subsidies. Producers can choose to use their water-use right allocation, A_i, entirely for agricultural productive purposes; augment it with water market purchases, m^B_{ij}; or sell some or all of their allocation, m^S_{ij}, where m_{ij} is the acre-ft of water bought or sold by producer i for crop j.

In the market, water is bought and sold at an exogenously set market price of P^m.[6] However, the net price paid in the case of a buyer or received in the case of a seller depends on the existence and magnitude of transaction costs. Transaction costs, δ^K_{ij} (where k = B for buyers and S for sellers), are modeled as a function of the types of transaction cost, τ^K_{ij}, associated with buying or selling water (i.e. administratively-induced transaction costs or policy-induced transaction costs). Since the nature and magnitude of transaction costs may vary among buyers and sellers and cropping patterns, $\delta^K_{ij} = f(\tau^K_{ij}) \geq 0$.[7] Given the above, the problem takes the following form:[8]

$$\text{Max } \pi_i = \sum_j \left[P_j f_{ij}(x_{ij} \mid l_i^*, z_{ij}^*) - m^B_{ij}(P^m + \delta^B_{ij}) + m^S_{ij}(P^m - \delta^S_{ij}) - E_{ij}\right] - w_i A_i - r_i l_i^* \tag{1}$$

$$\text{Subject to: } \sum_j x_{ij} \leq (A_i + \sum_j m^B_{ij} - m^S_{ij}) \qquad \forall\, i$$

$$\sum_j m^S_{ij} \leq A_i \qquad \forall\, i$$

$$x_{ij} \geq 0,\ m^B_{ij} \geq 0,\ m^S_{ij} \geq 0$$

where P_j is the output price received for the jth crop; $f_{ij}(x_{ij} \mid l_i^*, z_{ij}^*)$ is a well behaved production function and satisfies all regularity conditions; E_{ij} is expenditures on non-water/non-land variable inputs, $E_{ij} = P^z_j z_{ij}^*$; r_i is the price of the land input; and $\delta^K_{ij} = f(\tau^K_{ij})$ represents total transaction costs. The first-order conditions for an interior solution to (1) are:

$$P_j f^x_{ij}(x_{ij} \mid l_i^*, z_{ij}^*) - \lambda_i = 0 \qquad \forall\, i, j \tag{2a}$$

$$(P^m + \delta^B_{ij}) - \lambda_i = 0 \qquad \forall\, i, j; \tag{2b}$$

$$P^m - \delta^S_{ij} - \lambda_i - \phi_i = 0 \qquad \forall\, i, j \tag{2c}$$

$$A_i + \sum_j (m^B_{ij} - m^S_{ij} - x_{ij}) = 0 \qquad \forall\, i \tag{2d}$$

$$A_i - \sum_j m^S_{ij} = 0 \qquad \forall\, i \tag{2e}$$

where λ_i is the Lagrange multiplier on the water availability constraint which reflects both the initial allocation constraint under the water-use right and net market purchases, and ϕ_i is the Lagrange multiplier on the water-use right constraint. An interior solution is assured under the assumption that water is a scarce resource which implies that $P^m > 0$. In other words, the water-use right constraint will always be binding at the optimum since no producer will purchase water unless it will be used productively and any producer with an unused portion of their allotment will sell it on the market.

Producers' willingness to buy and sell water implies that the opportunity cost of using water in agricultural production will be defined by the water market price, P^m, net of total transaction costs, $\delta^K_{ij} = f(\tau^K_{ij})$, and the opportunity cost of the water-use right, ϕ_i, in the case of sellers. Producer profits are maximized when the volume of irrigation water applied to each crop (x_{ij}) is chosen such that the value of the marginal product of water for each crop equals the market price of water net of transaction costs. For purchasers of water this is $P_j f^x_{ij}(\cdot) = (P^m + \delta^B_{ij})$ and assumes that the marginal value product of water for each crop equals the market price of water plus transaction costs. For sellers this is $P_j f^x_{ij}(\cdot) = P^m - \delta^S_{ij} - \phi_i$ and assumes that the marginal value product of water for each crop equals the market price of water less transaction costs and the opportunity cost of the water-use right constraint.[9]

The net gains to society from trade (GFT) in water-use rights in a short-term or annual lease market can be defined as the total change in the value of the water-use right, $\Delta V_i(m)$, less total transaction costs for buyers (δ^B_{ij}) and sellers (δ^S_{ij}). Where the change in the value of the water-use right, $\Delta V_i(m)$, includes the change in value for both buyers, $\Delta V^B(m)$, and sellers, $\Delta V^S(m)$, and is measured as the marginal value product of water in its current use:

$$GFT(m) = \sum_i \Delta V_i(m) - \sum_i \sum_j (\delta^B_{ij} m^B_{ij} + \delta^S_{ij} m^S_{ij}) \tag{3}$$

$$\text{where: } \Delta V_i(m) = \sum_i \sum_j P_j f^x_{ij}(\cdot)\Delta x_{ij}$$

When gains-from-trade exist, they can only be realized if buyers and sellers are willing to participate in the market. In the absence of transaction costs, the marginal value product of water, $P_j f^x_{ij}(\cdot)$, has a dual interpretation. From the buyers' perspective, the marginal value product of water represents their maximum willingness to pay. From the sellers' perspective, the marginal value product of water less the opportunity cost of the water-use right constraint, ϕ_i, represents the minimum they must be compensated to consider a potential transaction. Thus, in the absence of transaction costs, the willingness of producers to enter into short-term or annual lease transactions is a function of bid prices, the marginal value product of water in its current use, and the opportunity cost of the water-use right constraint, in the case of sellers. Rents to sellers (buyers) accrue when the market price of water P^m is more (less) than marginal value product of water in its current use.

When transaction costs exist, a producer will sell water if $P_{ij} f^x_{ij}(\cdot) < P^m - \delta^S_{ij} - \phi_i$ and purchase water if $P_{ij} f^x_{ij}(\cdot) > P^m + \delta^B_{ij}$. The net rent to the potential buyer or seller provides a measure of willingness to participate in the market. The net rents that accrue to a buyer, $NR^B (m_{ij})$, or a seller, $NR^S (m_{ij})$, from a water market transaction can be defined as:[10]

$$NR^B (m_{ij}) = [\, V^B (m_{ij}) - m^B_{ij} \, (P^m + \delta^B_{ij})] \tag{4}$$

$$NR^S (m_{ij}) = [\, (P^m - \delta^S_{ij}) \, m^S_{ij} - V^S (m_{ij})]$$

Potential buyers and sellers will only participate in the market if rents are sufficient to cover transaction costs. If the transaction costs are sufficiently large that any one party's share of them outweighs the expected benefits of the transfer, the party will withdraw from the bargain or not consider the transaction (Pearce and Turner, 1990). To the extent that transaction costs can be reduced, market-based approaches to water allocation are likely to be more successful. High transaction costs do not necessarily preclude market activity, particularly when the benefits are high (NRC, 1992). However, since transaction costs drive a wedge between market price and net price paid or received the volume of water bought and sold will be less than in the absence of transaction costs. It suggests that in a relatively water scarce area where water market activity is expected but not observed, the existence of high (or highly asymmetric) transaction costs may be impeding water market development.

7.4 California's Developing Water Market

The gains-from-trade model is employed to assess the potential social gains that could be achieved if water markets were used to reallocate supplies among agricultural producers to meet revised Federal water quality standards for the Sacramento/San Joaquin Delta, the United States' second largest freshwater marsh. Two policy options for meeting these standards are assessed and compared. First is the existing policy of administratively mandated reallocations from current agricultural uses to environmental uses. This approach follows seniority in water-use rights by reducing allocations to those producers with more "junior" water-use rights. The second depends on regional trades in water-use rights which assume that lower-valued water-use rights are purchased to meet the environmental demands. The benefits from regional trade are then calculated as the difference in costs between these two policy options.

Potential social gains from water transfers can only be realized, however, if buyers and sellers are willing to participate in water markets. Participation depends on the existence of positive rents net of transaction costs. Since previous predictions of the potential gains from water markets exceed actual performance, the role of transaction costs in determining the willingness of agricultural producers to enter into water market transactions is examined. This analysis draws on the experience with annual lease markets during the operation of the California State Water Bank in 1991, 1992 and 1994. Market prices for water were determined in a State administered negotiation process between potential buyers and sellers. Administrative transaction costs were explicit in the market price, allowing for a preliminary assessment of the role of transaction costs in the participation decision.

Data on individual water agencies or "districts" enables examination of the willingness of agricultural producers to enter into market transactions by type of water-use right, relaxing the generally imposed assumption in analyzing gains-from-trade that water rights are homogeneous. This permits an assessment of the expected gains-from-trade when seniority in water rights is a constraint. In addition, district specific data more adequately capture the local effects of yield and price variability, as well as agroclimatic constraints, that influence relative water values than do the more typical analyses based on administrative units such as a county or growing region.

The four major agricultural water districts included in the analysis are: Glenn Colusa Irrigation District (GCID), Glenn county; Central California Irrigation District (CCID), Merced county; Westlands Water District (WWD), Fresno county, and; Lost-Hills Water District (LH), Kern county. These districts were selected because of their differences in water-use rights and sources of water, and *a priori* information regarding variation in the value of water in agricultural production. All four of these districts were eligible to participate in the California State Water Bank, although in 1991 policy precluded producers in WWD from participating as sellers and buyers were initially limited to those with critical water needs.

Both GCID and CCID hold appropriative water-use rights for water from the Sacramento and San Joaquin Rivers, respectively.[11] These appropriative rights are senior to those held by the Federal Central Valley Water Project and California State Water Project (SWP). Westlands and Lost-Hills receive their water from contractual entitlements with these Federal and State Projects. Thus, WWD and LH are identified as "contractors" where GCID and CCID are identified as "appropriative rights holders."

In general, producers' average per-acre water allocations are substantially higher in districts that hold appropriative rights than in districts that hold Federal and State water project contract entitlements. Appropriative rights holders also tend to pay substantially less for water, on a per acre-ft basis, than do Federal and State contractors. As a result, the marginal value product of water in appropriative rights districts is expected to be less than in Federal and State contracting districts. These expected differences in the marginal value product of water among potential trading parties suggest that social gains could be achieved if senior appropriative rights holders transferred water to Federal and State contractors who hold more junior rights.

7.5 The Value of Water Rights in the Central Valley

Under the assumption of a constant returns to scale production function, expected net returns to water provide an estimate of the marginal value product of water.[12] Table 7.3 shows estimated marginal value products of water by crop type, water district, and type of property right for the 1991-94 period of analysis. Property rights significantly influence the marginal value product of water, as expected. The district-average marginal value product of water across all crops was over two times larger for the junior water contractors than for senior appropriative water rights holders. The same general pattern is evident when the marginal value products are compared across districts by crop type. These relatively large and systematic differences in the value of water in its current uses among agricultural water districts suggest that substantial potential gains-from-trade in water-use rights may be realized through market-based reallocation. In general, the data suggest that among these four districts, contractors would purchase water while appropriative rights holders would sell. In this situation, the district marginal value product of water for contractors can be interpreted as their maximum willingness to pay for water in a short-term or annual lease market transaction, assuming zero or negligible transaction costs. The marginal value product for appropriative rights holders can be interpreted as the minimum level of compensation they must receive in order to consider a short-term transaction, assuming zero or negligible transaction costs.[13]

Values will vary for other types of water market transactions depending on the terms and length of the contract, personal time preferences, and relative opportunity costs of contracting parties. The value of water in its current use for three other types of water market transactions is calculated: contingent leases, long-term leases, and permanent transfers. These types of transactions are more likely to be employed to facilitate intersectoral transfers or meet permanent reductions in water allocations.

Contingent leases, also commonly referred to as dry-year options, can satisfy the demands of water users who have enough water to meet demand except under dry year conditions.[14] Contingent leases generally specify that a certain quantity of water will be delivered under specific circumstances. For example, a district pays an appropriative water rights holder a fixed sum per acre-ft for the option to purchase water during years when specified dry conditions exist. The terms of contingent leases vary but generally there are limits on how often the option can be exercised during the term of the lease. Contingent leases allow the holder of the water-use right to continue to use the water during normal water years and give the holder of the contingent lease a relatively cost effective means to increase water supply reliability during dry years (NRC, 1992). Long-term leases generally specify that a certain quantity of water is to be delivered each year for the period of the agreement, often ranging from 10 to 25 years.

Table 7.4 shows the value of water for each agricultural water district for different types of market transactions under varying assumptions on transfer length and discount rates. Clearly, the value of water varies not only by type of property right and crop, but will differ significantly by type of transaction. One condition for water markets to develop is that potential market participants have reliable knowledge about water-use values and alternative water supply costs (Michelson and Young, 1993).

7.6 The Gains from a Market Approach to Meet Water Quality Standards

The California State Water Resources Control Board's 1995 "Water Quality Control Plan" that was approved by the United States Environmental Protection Agency established water quality standards for the Sacramento/San Joaquin Delta. The Plan sets limits on the quantity and timing of Delta exports. The Delta is the source from which two-thirds of the State's population and millions of acres of agricultural land receive part or all of their supply (State of California, 1995).

Under the terms of the implementation agreement (the Bay-Delta Accord) agreed to by State, Federal and environmental interests, the CVP and SWP agricultural contractors were assigned responsibility for meeting mandated environmental reallocations. This requires an annual reduction in exports of 493 million m^3 (0.4

million acre-ft) under average or "normal" year climate conditions and 1,356 million m^3 (1.1 million acre-ft) in critically dry years. This solution adhered to seniority in water rights as both CVP and SWP contractors hold rights subordinate to agricultural users with appropriative and riparian water rights.

Policy makers realize that the value of water in its current use for CVP and SWP agricultural contracts is generally higher than that for appropriative rights holders, as the analysis above demonstrates. A long-term solution is being sought that avoids a costly legal battle over property rights yet allocates some responsibility for meeting water quality standards to more senior water rights holders since the social costs of such a policy are expected to be lower. A market-based solution in which CVP and SWP contractors "buy" the rights from more senior water rights holders to meet the mandated reallocations could potentially avoid water rights issues and achieve a lower cost solution.

Table 7.3. The Marginal Value Product of Water in Current Uses by Type of Water Right for Selected Agricultural Districts[a, b]

Crop Type	Appropriative Rights		Contractors	
	CCID	GCID	WWD	LH
	.... $/acre-ft[d]	....		
Row Crops	46	32	100	42
Vegetable[c] and Fruit	46	-	119	-
Permanent	43	-	173	188
All Crops	46	32	106	126

[a] Values are based on a net returns analysis drawing on crop-specific district level data for the 1991-94 period.
[b] Calculated as weighted averages, where weights equal the percentage of total acreage for each crop type.
[c] Includes only annual fruit crops.
[d] 1 acre-ft equals 1,233 m^3.

The analysis in Table 7.5 estimates the potential gains-from-trade that could be achieved with a market-based solution to meet the mandated environmental demands in lieu of administratively mandated reallocations. Table 7.5 compares the cost of administrative reallocations based on seniority in water-use rights with the cost of using water markets to reallocate the required quantity of water. The gain is the difference in the costs of the two policy options or the avoided costs. The cost of the administrative policy is measured as the loss in revenue to CVP and SWP contractors based on the average marginal value per acre-ft of water given up. It assumes that the reductions in Delta exports are shared on an 80/20 percent basis between CVP and SWP agricultural contractors, respectively. In the market-based solution, agricultural contractors purchase water from appropriative rights holders to meet mandated reductions in exports.

Table 7.4. Value of a Water Right by Region and Type of Right Under Alternative Trade Arrangements for Selected Agricultural Districts

Region/ Type of Water Use Right		Annual Lease[a]	Contingent Lease[b]	Long Term Lease[c]	Permanent Transfer[d]
	Disc. Rate (r =)	$(P_j f^x_{ij}(\cdot))$	NPV $(P_j f^x_{ij}(\cdot))$	NPV $(P_j f^x_{ij}(\cdot)$	$P_j f^x_{ij}(\cdot)$ (A_i/acre)
		. . . .	$/acre-ft	. . . .	
North of Delta Appropriative					
GCID		32			
	7%		172	373	
	10%		156	290	1,913
	12%		146	251	
South of Delta Appropriative					
CCID		46			
	7%		248	536	
	10%		224	417	1,357
	12%		210	361	
Contractors					
LHWD		106		1,235	
	7%		571	962	2,083
	10%		516	831	
	12%		484		
WWD		126			
	7%		679	1,468	
	10%		613	1,144	2,729
	12%		575	988	

[a] The annual lease value is the weighted average district-level MVP from Table 7.3.

[b] The contingent lease (or dry year option) value is calculated as the net present value of access to an acre foot of water seven out of 15 years using the weighted average district-level MVP from Table 7.3.

[c] The long-term lease value is calculated as the net present value of water in agriculture for a 25-year period, the length of Federal CVP contracts under current policy, using average district MVP from Table 7.3.

[d] The permanent water-use right value is calculated as the net present value of a per acre water-use right entitlement or contractual allocation under normal water year allocations, using the weighted average district-level MVP from Table 7.3.

This analysis accounts for existing infrastructure constraints that limit exports due to pumping restrictions and seasonal pump capacity (see Table 7.5-footnote (c) and (d)). If these restrictions were relaxed through augmentation of existing pump capacity and/or the construction of a "through Delta" facility to carry additional water south of the Delta, more lower-valued water north of the Delta could be substituted for higher-valued water demanded south of the Delta. Thus, the estimated gain represents a minimum rather than a maximum that could be achieved with a market-based solution.

The analysis assumes that the 493 million m^3 (0.4 million acre-ft) annual requirement to meet the Delta water quality standards would be met with permanent water transfers. Thus, the cost of meeting this requirement is the present value of the per-acre water-use right from Table 7.4. The additional 863 million m^3 (0.7 million acre-ft) requirement during critically dry years is assumed to be met with contingent leases (or dry-year options); this cost is based on the dry-year lease value from Table 7.4.

The cost of meeting mandated environmental demands with a market-based policy approach is approximately one-third that of relying on administrative reallocations, as shown in Table 7.5. A market-based approach would provide significant net social gains unless transaction costs exceeded approximately $0.16/m^3 ($200/acre-ft)[15] However, potential gains-from-trade can only be realized if buyers and sellers are willing to participate in a water market. Since contingent leases have only recently reached the demonstration phase and no long-term leases or permanent transfers have been observed following the changes in Federal and State laws, transaction costs could be limiting market development.[16]

7.7 California State Water Bank: Transaction Costs and Incentives

As a result of the limited water market experience in California, information on transaction costs of any type is extremely scarce. Since it is rents net of transaction costs that determine the willingness of buyers and sellers to participate in markets, lack of transaction cost information impedes analysis of true social gains that could result from trade in water-use rights. However, since some transaction costs were explicit in the State Water Bank prices, the Bank provides an empirical basis to examine water market performance when transaction costs are explicit and non-zero. Because administratively-induced transaction costs were explicit in State Bank prices, rents net of AICs can be derived. If rents net of AICs were positive and expected market activity did not occur, policy-induced transaction costs attributable to the existing institutional requirements for transfers can be suspected of impeding market performance.

The State Water Bank directed water transfers in 1991, 1992, and 1994. In 1991, the price paid to sellers was $0.10/m^3 ($125/ acre-ft). The price to buyers was $0.14/m^3 ($175/acre-ft) for water delivered as far as the SWP Delta pumping plant. This price covered the purchase price of $0.10/m^3 ($125/acre-ft), the costs of contract negotiation, and administration and other terms of the transfer including charges for additional water losses in transit.[17] Prices in 1991 and 1994 were $0.04/m^3 ($50/acre-ft) for sellers and $0.05/m^3 ($67.50/acre-ft) for buyers. The relatively higher prices in 1991 reflect differences in sources of supply (reliance on fallowed cropland as contrasted to groundwater substitution and storage sources in 1992 and 1994) and the relative severity of water shortages. The divergence between the bid prices to buyers and sellers represents the AICs associated with operation of the State Water Bank which buyers were forced to bear.

Given the above prices and specific charges for AICs, potential net rents were calculated following Equation (4) for each of the districts. Net rents per acre-ft of water by property right and crop type are shown in Table 7.6 for the "high" price (1991) and "low" price markets (1992 and 1994). While these rents are net of AICs, they do not account for PICs.[18]

A number of important conclusions can be drawn from Table 7.6. First, potential rents net of AICs indicate that sellers and buyers in all of these agricultural districts

Table 7.5. Potential Gains from Regional Trade in Water Rights to Meet Environmental Demands in the Sacramento/San Joaquin Delta

Method of Meeting Current Delta Requirements	Normal or Average Year Permanent Transfer	Critically Dry Year Contingent Lease
	$ (millions)	
Option 1: Administrative Reallocations		
Following Seniority		
In Water-Use Rights	$396.9[a]	$334.0[b]
Option 2: Regional Water Transfers		
CVP/SWP Ag. Contractors		
Purchase Water-Use Rights	$131.6[c]	$110.7[d]
Potential Economic Gain		
Avoided Cost by Trading in Lieu		
Of Administrative Reallocations	$265.3	$223.5

[a] Assumes that the 0.400 million acre-ft annual requirements will be met with a permanent water-use rights transfer from CVP/SWP contractors shared on an 80/20 basis.
[b] Assumes that the additional acre-ft to meet the 1.1 acre-ft requirement will be accommodated with contingent lease contracts of fifteen-year duration with water transferred seven out of fifteen years.
[c] Assumes that 0.300 million acre-ft is forthcoming from water rights holders north of the Delta, since this is the current constraint on additional transfer capability given infrastructure constraints in average water years (DWR, 160-93, 1994). The remaining 0.100 million acre-ft is assumed to come from water rights holders south of the Delta. In this example, values from GCID for north of the Delta are assumed and CCID values are used for south of the Delta.
[d] Assumes that the additional 0.700 million acre-ft will be accommodated with contingent lease contracts of fifteen-year duration with water transfers seven out of the fifteen years. The 0.600 million acre-ft critically dry year infrastructure pumping constraint in the Delta is assumed to hold. The first .600 million acre-ft is assumed to come from water rights holders north of the Delta; the remaining 0.100 million acre-ft from south of the Delta. Values for north and south of the Delta are represented by GCID and CCID, respectively.

should be willing to participate in annual lease market transactions, given the prevailing values of water, bid prices, and levels of AICs. Second, regardless of the type of water-use right, most districts will have some producers who will be potential sellers and others who will be potential buyers. Producers will be more likely to engage in intra-district transfers than inter-district transfers as long as intra-district expected water market prices, net of transaction costs, received (paid) is more (less) for sellers (buyers) than for inter-district transactions. If intra-district transfers dominate, the volume of inter-district transfers will be reduced. Evidence to date suggests that this is occurring. Current Federal policies that provide CVP contractors right of first refusal to transferable supplies also reduces out-of-project sales. Third, as expected, potential net rents vary by type of water-use right. Finally, while sellers' potential rents net of AICs indicate a potential willingness to participate in both the "high" and "low" price markets, rents net of AICs appear only sufficient to entice producers, in most districts, to buy water in the "low" price market.

Table 7.6. Potential Average Rents Net of Administrative Transaction Costs by Type of Water Right and Crop in Selected Agricultural Districts[a]

Type of Water-use Right/ Crop Type	Sellers		Buyers	
	High Price[b]	Low Price[c]	High Price[b]	Low Price[c]
	.... $/acre-ft[d]			
Appropriate Rights				
CCID				
Row	89	38	*	70
Vegetable/Fruit	22	33	*	*
Permanent	107	30	*	131
All Crops	88	35		80
GCID				
Row	100	18	*	83
Vegetable/Fruit	*	*	*	*
Permanent	*	*	*	*
All Crops	100	18		83
Contractors				
WWD				
Row	33	8	16	80
Vegetable/Fruit	125	41	72	69
Permanent	81	19	47	257
All Crops	64	25	56	123
LH				
Row	100	*	*	*
Vegetable/Fruit	*	*	*	*
Permanent	59	36	*	149
All Crops	73	36		149

[a] For sellers rents are net of AICs but not of the water-use right opportunity cost, ϕ_i.
[b] *High Price* refers to the 1991 State Water Bank buy and sell prices of $175 and $125, respectively.
[c] *Low Price* refers to the 1992 and 1994 State Water Bank buy and sell prices of $67.50 and $50, respectively.
[d] The (∗) represents non-feasible trades as potential rents net of AICs were less than or equal to zero.

The above analysis suggests that substantial potential gains-from-trade could have been realized through market-based reallocations and the positive rents net of AICs should have enticed buyers and sellers to realize these gains. However, only a limited number of actual inter-district annual lease transactions were observed. Table 7.7 compares the estimated potential that could have been bought and sold (the volume of water associated with positive rents net of AICs from Table 7.6) with the actual acre feet transferred (State of California, 1996). The potential transferable supply does not account for the effects of intra-district transfers, producers' likely unwillingness to sell water dedicated to permanent crops, or the economics of groundwater pumping relative to Water Bank purchases.

In general, the results indicate that producers in appropriative rights districts are expected to sell water outside the district and those districts holding State and Federal contracts are the expected buyers. Actual market performance, however, differed significantly. Appropriative rights holders actually made only one sale. In 1992, CCID sold 9.86 million m^3 (8,000 acre-ft). Contractors participated as buyers, but transferred significantly less than would be predicted.[19]

What explains the divergence between potential and actual water market activity? Three critical factors appear to have limited market activity: (1) lower costs of pumping groundwater relative to purchasing State Bank water for some potential buyers, (2) development of significant intra-district spot and annual lease markets limiting inter-district trades, and, (3) high expected PICs, particularly for potential sellers.

The lower relative cost of pumping groundwater has significantly reduced purchases for those districts with groundwater (e.g., WWD and to some extent CCID). In WWD the cost of withdrawing an acre-ft of groundwater was approximately \$0.04/m^3 (\$50/acre-ft) during the period. Substantial intra-district spot market transactions were also evidenced in selected agricultural districts. During 1991, WWD staff estimate that 2,000 to 4,000 intra-district transfers occurred, accounting for an estimated 20 to 25 percent of the district's CVP contract supplies (Archibald and Kuhnle, 1992).[20] This may overstate actual intra-district transactions as landowners with parcels in several locations within the district may have reallocated their total water supplies among their parcels since specific per acre water allocations vary by area within WWD. However, what is clear is that in terms of the number of transactions, more intra-district, than inter-district trading activity took place during the 1991-94 period.

The level and incidence of expected PICs provides a third explanation for the divergence between potential and actual waters market activity. To induce participation, potential rents net of AICs must be sufficient to cover expected PICs. Clearly, the PICs associated with meeting the requirements in California for annual, long-term, or permanent transfers are likely to be substantial as evidenced by the requirements in Table 7.1.

The lack of transfer activity evidence suggests that the residual rent net of AICs was likely insufficient to cover expected PICs associated with current policy. As shown in Table 7.6, district average potential rents net of AICs available to cover PICs ranged from \$0.01 to \$0.03/m^3 (\$18 to \$36/acre-ft) for sellers in the "low" price market and from \$0.05 to \$0.08/m^3 (\$64 to \$100/acre-ft) in the "high" price market. For buyers, district average potential rents net of AICs ranged from \$0.06 to \$0.12/m^3 (\$80 to \$149/acre-ft) in the "low" price market and \$0.05/m^3 (\$56/acre-ft) in "high" price market. The observed ranges of rents net of AICs available to cover expected PICs are relatively low in comparison with actual PICs associated with water transfers in other Western states where less stringent state and Federal transfer requirements exists. Colby (1990) found that actual PICs for a number of water transfers in the West averaged $0.07/m^3 ($91/acre-ft), with considerable variation among states ranging from an average of $0.15/m^3 ($187/acre-ft) in Colorado to $0.04/m^3 ($54/acre-ft) in New Mexico. She found that "PICs are higher where the economic values that may be affected by a proposed transfer are higher – in areas where water is scarcer and water rights sell for a higher price" (NRC, 1992 citing Colby, 1990). Clearly, this is the case in California. While the PICs reported for other states typically reflect the transaction costs associated with permanent transfers, even the short-term interbasin transfer requirements in California are stringent and compare in complexity to interbasin transfer requirements for permanent transfers.

Table 7.7. Comparing Potential and Actual Water Market Activity for Selected Agricultural Users: 1991-94

	Appropriate	Rights	Contractors		
Year	CCID	GCID	WWD	LH	Total
		acre-ft (1,000s) [(a)]			
I. Potential Trades [(b)]					
1991	-228.8	-2398.4	-317.7	-11.3	-956.2
1992	-167.3	-361.3	967.1	35.3	473.8
1994	11.3	-598.0	1194.1	38.1	645.5
Average	-128.3	-452.6	614.5	20.7	54.4
II. Actual Trades					
1991	0	0	88.2	38.8	127.0
1992	0	-8.0	114.1	6.5	112.6
1994	0	0	128.8	n/a	128.8
Average	0	-2.6	110.4	15.1	122.8

[(a)] Negative sign indicates participation as expected seller.
[(b)] Potential acre-ft traded excludes expected sellers producing permanent crops since they are not expected to sell in an annual lease market.

The analysis of the State Water Bank presented suggests that for most potential market participants rents net of AICs were not sufficient to cover expected PICs, particularly among sellers who bear the responsibility for PICs under current policy. The possible exceptions are WWD and LH in the "low" price market years with average rents net of AICs of $0.10/m^3 and $0.12/m^3 ($123 and $149/acre-ft), respectively. These districts actually did purchase water during these years.

7.8 Conclusions and Policy Implications

Consistent with previous research, this analysis of water market potential and performance in California, which accounts for heterogeneous water rights and transaction costs, indicates that substantial potential gains-from-trade in water-use rights among agricultural users exists. However, these potential gains have yet to be realized as buyers and sellers were unwilling to participate in water markets at the prevailing level of rents net of AICs. The analysis of the State Water Bank annual lease market provides some evidence that expected PICs likely stifled inter-district market activity, particularly among sellers in appropriative rights districts who were required to bear the majority of PICs.

The analysis of potential gains from a market-based approach to meet Federal water quality standards in the Sacramento/San Joaquin Delta, in lieu of administrative reallocations, indicates a significant minimum potential social gain of approximately $225 million. It was shown that market-based policies can maintain historical water-use rights and achieve water quality standards at a significantly lower social cost. To achieve social gains through market-based reallocation requires that expected PICs for these largely permanent and long-term transfers are less than $0.16/m^3$ ($200/ acre-ft) on average. While the PICs reported for other western water trades have been less than this level, no transactions of this type have taken place in California.

Under the recent changes in both State and Federal policies that were designed to encourage water trades while accounting for their potential external effects, individual sellers bear the policy-induced transaction costs. The underlying assumption in the policy design was that appropriative rights holders would be the most likely sellers, given their relatively lower marginal value product of water in current uses. It was also assumed that sellers would obtain relatively high net rents providing them sufficient rents to cover these costs. Estimates of the marginal value product of water confirm that potential rents are higher for appropriative rights holders. However, the lower than anticipated level of actual market activity provides some evidence that expected PICs were sufficiently high to preclude market participation at observed prices.

Policy-induced transaction costs are likely affecting the performance of California water markets more generally. Interviews with producers in districts surveyed corroborate this hypothesis. The fact that intra-district water markets, which include AICs but are exempt from most PICs, perform relatively well provides additional evidence that expected PICs constrain market performance. These results suggest that policies influencing transaction costs, particularly PICs, need to be examined to clarify and perhaps to reduce these costs if market performance is to be improved.

While PICs have been identified as likely constraining water market activity in California, limited information exists on the exact nature and level of PICs. Further analysis of the magnitude and incidence of transaction costs, particularly PICs, associated with different types of trades is needed. Previous experience with the costs of adjudicating water rights, particularly the costs of studies to determine consumptive uses, return flows, and groundwater interactions could be utilized to determine the likely cost levels. Other environmental impact studies specific to interbasin transfers could be looked to as indicators of the cost levels of meeting policy requirements.

Providing potential buyers and sellers with the hydrological and environmental impact information might reduce transaction costs and spur market development. Policy makers might also want to reconsider the role of water districts in developing water markets. If the costs of bargaining are too high for individual producers to bear, perhaps districts or other State and Federal agencies should be looked to as the bargaining agent. Districts have greater resources and information to conduct needed impact analyses and are likely to have greater bargaining power with potential urban buyers. Further efforts to integrate surface and groundwater management are also needed if effective and efficient markets are to develop. The State and Federal government could also consider expanding their role in providing information to both buyers and sellers on transactions that do occur as a means of facilitating transfers. Such action has precedence in commodity markets, both domestic and export. It is highly unlikely that the potential for efficiency gains from water markets will be realized without further institutional change.

Endnotes

[1] Title XXXIV of the Central Valley Project Improvement Act in the 1992 Federal Reclamation and Reform Act (PL102-575).
[2] In normal water years, exports will be reduced by about 0.4 million acre-ft and in critically dry years exports will be reduced by approximately 1.1 million acre-ft
[3] See Chapter 8 for a full discussion of the 1995 Water Bank program.
[4] The volume of sales to agriculture in 1991, 1992, and 1994 were 22 percent, 62 percent, and 86 percent, respectively (State of California, 1996).
[5] Of the 1.24 million acre-ft sold to the State Water Bank in 1991, 1992 and 1994, 51 percent was from groundwater; 17 percent from fallowed cropland and the remainder from storage. In 1991, nearly 40 percent of all water supplied to the Bank was from fallowed cropland in two counties. In 1992 and 1994, 85 percent of supplies were from groundwater sources (State of California, 1996).
[6] It is assumed that a water market price emerges as a result of excess demand for water and that individual producers cannot influence the market price.
[7] Transaction costs can be crop specific. For example, where forward contracts between producers and processors exist, the cost of an annual transfer agreement to a seller would increase.
[8] This is an extension of the model developed in Weinberg et al. (1993) which incorporates water buy/sell decisions into producer decisions but ignores transaction costs.
[9] When transaction costs are negligible or zero, producer profits are maximized when the volume of irrigation water applications for each crop (x_{ij}) are chosen such that the value of the marginal produce of each crop equals the market price of water for net purchasers of water ($P_j f^x_{ij}(\cdot) = P^m$) and such that it equals the market price of water less the opportunity cost of the water use right constraint for sellers.
[10] When transaction costs are negligible or zero, the net rents to buyers and sellers are defined as: $NR^B(m_{ij}) = [\, V^B(m_{ij}) - m^B_{ij} P^m]$ and $NR^S(m_{ij}) = [\, V^S(m_{ij}) - m^S_{ij} P^m]$, respectively.
[11] Water users in these districts are referred to as "water rights settlement contractors" as they have agreed to receive a "base supply" of CVP water in lieu of adjudication of appropriative rights. Base supplies constitute 87 percent (or 720,000 acre-ft) for GCID and 100 percent (or 532,400 acre-ft) for CCID and are treated legally as appropriative rights.
[12] Details of the residual methodology can be found in Hearne and Easter, 1995 and a description of methods and data employed in the analysis presented here is contained in Archibald and Renwick (1996).
[13] Producers of permanent crops are unlikely to consider selling a significant share of their water in an annual lease market unless lower cost groundwater is available as a substitute or they adopt water-saving technology such as drip irrigation.
[14] Contingent leases as calculated here are not true options. For an analysis of options see Chapter 8.

[15] Based on a social gain of $223 million for 1.1 million acre-ft of water, shown in Table 7.5.
[16] The State-run Water Supply Options Market in 1995 claimed relatively low transaction costs (Jercich, 1997). In addition to a flat $2,000 charge per contract, buyers paid $2.00 per acre-ft for administrative charges. These costs did not include buyers' or sellers' costs nor all labor and overhead costs for State Department of Water Resources staff involved in the operation of the Bank. Also excluded were PICs.
[17] Additional charges were levied for conveying water to the place of use. Transportation costs were covered under separate contracts.
[18] It may be argued that in the case of the State Water Bank, expected PICs were negligible since the State sanctioned transactions and exempted area-of-origin sources from liability for third-party impacts. However, market participants had to meet other State, Federal, and district-level requirements.
[19] The model predicted that Westlands Water District and Lost Hills Water Districts would participate as sellers in the 1991 market, but both actually purchased water. Interviews with district personnel and producers indicate that most of the water purchased was to irrigate permanent crops in water short areas. In addition, in 1991, Federal law precluded Westlands from participating in the market as a seller.
[20] There is some question as to whether the estimate of 4,000 transfers might also reflect the total number of "filings" by producers that accompany each transfer. Since each transfer requires a form from each participant, the total number of transfers might be 2,000 rather than 4,000.

References

Archibald, Sandra O. and Mary E. Renwick, 1996. "The Potential for Gains-from-Trade in the California Water Industry," Western Regional Science Association 35th Annual Meeting, February 25-29, Napa, California.

Archibald, Sandra O., Thomas Kuhnle, Robin Marsh, Mary E. Renwick, and Barton Thompson, 1992. "An Economic Analysis of Water Availability in California Central Valley Agriculture-Phase II Report," *Center for Economic Policy Research Working Paper Series*, Stanford University, February.

Archibald, Sandra O., and T. Kuhnle., 1992. "An Economic Analysis of Water Availability in California Central Valley Agriculture-Phase III Report," *Center for Economic Policy Research Working Paper Series*, Stanford University, August.

Colby, Bonnie G., 1990. "Transaction Costs and Efficiency in Western Water," *American Journal of Agricultural Economics*, vol. 72(5): 1184-1192.

Hearne, Robert R. and K. William Easter, 1995. "Water Allocation and Water Markets: An Analysis of Gains-from-Trade in Chile," World Bank Technical Paper 315, Washington, D.C., 75 p.

Howitt, Richard E., 1994. "Effects of Water Marketing on the Farm Economy." In Harold O. Carter, et al. *Sharing Scarcity: Gainers and Losers in Water Marketing*, University of California, Agricultural Issues Center.

Jercich, S.A., 1997. "California's 1995 Water Bank Program: Purchasing Water Supply Options," *Journal of Water Resources Planning and Management,* 123(1):1-32.

Michelson, A.M., and R.A. Young, 1993. "Optioning Agricultural Water Rights for Urban Water Supplies During Droughts," *American Journal of Agricultural Economics,* 75(4): 1010-1020.

National Research Council (NRC), 1992. *Water Transfers in the West*, Water Science and Technology Board, National Academy Press, Washington, D.C.

Pearce, David W. and R. Kerry Turner, 1990. *Economics of Natural Resources and the Environment*, Johns Hopkins, Baltimore.

State of California, 1995. *State Water Plan. Bulletin 160-93*, Department of Water Resources (DWR), Sacramento, CA.

State of California, 1996. "Water Bank Transactions, 1991-1994," Water Project Analysis Office, Division of Operation Maintenance, Sacramento, CA., January.

Vaux, Henry J. and Richard E. Howitt, 1984. "Managing Water Scarcity: An Evaluation of Inter-regional Transfers," *Water Resources Research*, 20: 785-92.

Weinberg, Marca, C.L. Cling and J.E. Wilen, 1993. "Water Markets and Water Quality," *American Journal of Agricultural Economics*, 75(1): 278-91.

Zilberman, David and Richard E. Howitt, and D. Sunding, 1993. "Economic Impacts of Water Quality Regulations in the San Francisco Bay and Delta," Dept. Ag. and Res. Econ., University of California-Berkeley, and the Western Consortium for Public Health, Working Paper.

8 Spot Prices, Option Prices, and Water Markets: An Analysis of Emerging Markets in California

Richard E. Howitt

8.1 Introduction

Most of the discussion of water markets revolves around the concept of trading permanent water rights, which implicitly assumes that the water property rights are adequately defined and transaction costs are not a significant barrier to trade. While this situation is relevant to several countries and areas in the U.S., there are many other regions where the transaction costs of trading permanent water rights are prohibitive. Given excessive costs for water rights trades, there is still significant potential for the market reallocation of water, using annual spot markets for times of scarcity and option markets for longer-term contracts.

There are four main reasons why California has, so far, opted to reallocate its water supply through spot and option markets. First, surface water rights in California are dominated by appropriative rights rather than a clear quantitative title. Second, in many watersheds the formal water rights are over-appropriated, leading to the terms "paper water" versus "wet water." Third, given the importance of return flows, environmental impacts and third-party economic effects, these externality effects have to be largely internalized before trades can occur. Fourth, effective internalization of these externalities requires that consumptive use is defined and monitored which involves yet higher transaction costs.

Despite the practice in the electricity, gas, and phone industries of public and private suppliers using a mix of owned rights, option and spot market supplies, the use of spot and options markets for water supplies is regarded as unduly risky. Yet under suitable arrangements, spot and option markets can spread the risk and provide a low cost alternative for reallocating water that may be preferable to permanent transfers of rights. Graphical and analytical methods are used to demonstrate that pure spot markets or permanent rights markets cause all of the risk cost to be born by one party. This polarization of the risk cost leads, under risk aversion, to a larger transaction wedge between trades than a market system that spreads the risks. Anecdotal evidence supports this view, in that most farmers practicing irrigated agriculture who sell their permanent water rights have very small farms or are under severe liquidity constraints. Similarly, water buyers with high-value crops or inelastic urban demands are reluctant to rely on thin and sporadic spot markets for water.

The innovation of water markets is a very recent phenomenon in California and is in a stage of tentative and cautious formation. However, initial attempts to form spot and option markets are encouraging and can provide a basis for comparison with more sophisticated market structures in other resource-based commodities. While water has its specialized institutions and characteristics, there is no reason that water market formation is different from other commodities supplied through networked facilities. Like other public commodities (natural gas and electricity) subject to uncertainty in supply, transportation, and value, water markets can develop a full set of spot, option, and stock price markets within the constraints of physical mobility and transaction costs.

Water is essential for all sectors of California's vibrant economy. It is no exaggeration to state that the eighth largest economy in the world literally runs on water. Given the substantial differences in water endowments among regions of the State and between years, the development of California has necessitated large interbasin transfers. For the first seventy years of water development in California, increases in supply were satisfied by increasingly large storage projects and associated interbasin transfers. (See Figure 16.1 on page 265 for a diagram of the California water system.)

The annual average water use in California is 4,489 million m^3, of which 3,452 million m^3 is obtained from surface delivery systems, some of which move water as far as 1,000 kilometers between basins. The average supply quantity masks a wide range of water supply differences between years and locations in the State due to the physical system and the seniority of the rights under the riparian and appropriative doctrine.

A shift in environmental sensitivity by the electorate coupled with rapidly increasing water development costs brought a halt to the construction of large water projects in 1982, with the defeat of the Peripheral Canal in a State referendum. By 1987, the continued growth in urban water demand coupled with pressures for increased environmental supplies had removed any supply surplus under normal rainfall conditions, a relatively rare occurrence in California. The six-year drought from 1987 to 1992 brought the tight supply situation into bold relief. By 1991 the water supply situation was very serious and it was clear that some form of reallocation was required to minimize the damage to the State's agricultural and urban economies. One of the recommendations of the Governor's Drought Action Team was to establish a State-sponsored drought water bank run by the California Department of Water Resources (DWR).

Once started in 1991, interbasin markets resulted in substantial trade volume in quantity or financial terms. Interbasin trades in 1991 accounted for 101 million m^3 of water with a value of over $111 million (Howitt et al., 1992). The California water market reoccurred in the form of a drought water bank in 1992 and again in 1994 with a widening variety of types of trade and a volume that changed with water availability. In 1995, a potential drought water bank based on sales and purchased options was formed and operated before increased supplies reduced option demand (Jercich, 1997).

Water markets in California are characterized by an emphasis on transfers of water as opposed to water rights. There are several reasons for this. First, most of the ultimate users of water in California do not hold title to the water, but operate under a system of appropriative or correlative rights that are predicated on the "beneficial use" of the water. Beneficial use is broadly defined and includes the transfer and sale of water that is available due to conservation practices or changes in cropping pattern. Second, due to the large fluctuations in water supply and long distances involved in some transfers, it is not economical for purchasers to pay the delivery costs in wet years. Third, externality costs and resistance to water sales diminish with periodic sales. Fourth, the agricultural industry in California is vibrant and growing, thus many regions that are currently open to water sales want to keep their future property rights and options unencumbered.

There are three main types of water rights in California: riparian, appropriative, and correlative. Although the formalities for the sale or lease of an entitlement may vary, in general, most appropriated water can be transferred on a short-term basis provided that the transfer will not adversely affect other rights holders. These transfers may involve changes in the location of use, type of use, or location of delivery. The voluntary transfer of water can occur in a number of different ways. Examples of the types of agreements include the outright sale of an entitlement, fixed-term leases, dry-year options (or contingent contracts), conservation offsets, and water exchanges. The duration of these leases can vary from as short as a single season to as long as a five-year contract.

Riparian water rights holders have a firm title to a proportion of the flows. In some cases, the right has been exchanged for clearly defined quantitative rights from a subsequent development on the river. In either case, the seniority of the riparian or exchange right makes the risks of forfeiture on the basis of beneficial use minimal and thus encourages trading by riparian right holders. Correlative rights that govern all the unadjudicated basins are a significant inhibitor to water trades. The collective nature of groundwater generates significant opposition to groundwater trades which have only occurred in dry years under the condition that groundwater was substituted for surface water, allowing the surface water to be sold.

8.2 A Graphical Analysis of Spot and Rights Markets for Water

Graphical analysis of the economics of water markets is usually shown using the traditional "back-to-back" graphs. Water markets have two characteristics that make the usual approach inapplicable when comparing spot and water rights markets. The two key differences are: first, that water supplies are highly stochastic and second, that the markets are normally introduced after water property (or use) rights have been established. These two aspects of water markets require slight modifications to the usual trade diagram.

Due to the uncertainty that pervades water supply functions, water supplies involve two important characteristics, the expected quantity and the reliability of that quantity. Accordingly, the goal of a water market may be a quantity shift that increases the average quantity without changing the variability of the supply or, alternatively, an increase in reliability by decreasing the downside variation without changing the mean nominal water right. In this section, spot and rights market equilibria are compared under alternative assumptions.

The variability of the surface water supply in California is shown in Figure 8.1. A consolidated index for flows in the four major rivers is plotted annually for sixty years. The quantity is measured in thousands of acre-ft (1 acre-ft = 1,233 m^3).

The flows average 17.1 million acre-ft (21 billion m^3) but very few years are close to the average. The hydrologic system is such that average years are unusual. This places an emphasis on the ability of a supply system or market mechanism to integrate the high flows with the low flows and produce a more reliable yield.

Figure 8.2 is a graphical illustration of trade between two water-using regions "A" and "B". The first departure from the normal trade diagram is that each region is allocated expected water rights X^*_A and X^*_B. The initial assumption is that each region has sufficient water rights for their demands in average and above average years. Accordingly, the demand functions in each region are truncated at X^*_A and

X^*_B. In addition to the average water rights, both regions face a dry-year downside uncertainty in their supply which leaves them with supplies of $X^*_A - \sigma$ and $X^*_B - \sigma$. Clearly region A is a low value water region while region B has high value demands for water. The value σ represents the cumulative probability of a very dry or drought year. While data such as that shown in Figure 8.1 gives rise to a bimodal distribution, dry-year events are usually regarded as discrete events with cumulative probabilities of occurring. For graphical simplicity I have taken the probability of a dry-year water yield as 0.25 and the cumulative probability of a normal or wet year being 0.75.

Figure 8.1. Annual California River Flows

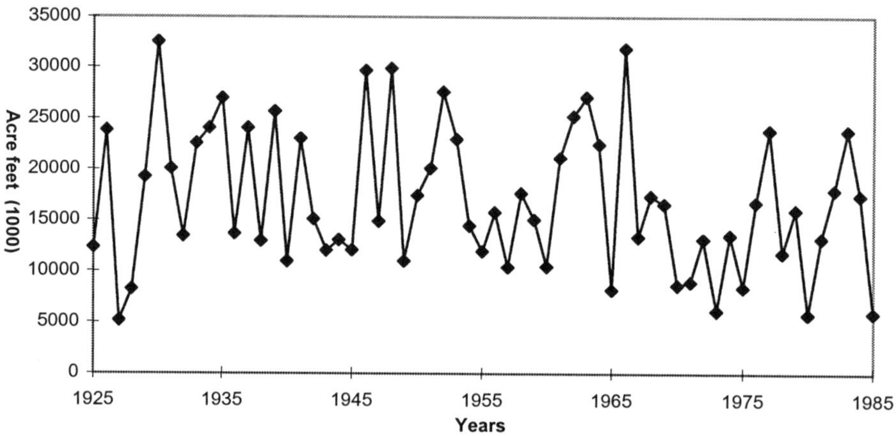

The central graph illustrates the excess demands and supplies from the regions under alternative conditions. The excess demand and supply shown by broken lines occur when there is a dry-year spot market with both regions holding their initial property rights. Under dry conditions both regions face delivery quantities of $X^*_A - \sigma$ and $X^*_B - \sigma$, respectively. A spot market trade from A to B will reach equilibrium at X^*_B and $X^*_A - 2\sigma$ if transaction costs are less than TC.

Using the spot market, region B stabilizes their water supply at X^*_B and region A accepts payment to adjust to greater fluctuations of $X^*_A - 2\sigma$ in dry years. The net benefit is equal to the area between the dry-year excess demand in region B and the dry-year excess supply from region A minus the transaction costs. Even with transaction costs as high as TC, the net benefit is substantial.

Some commentators on water markets are concerned about the problem of "hoarding" by purchasers of water rights. For example, region B may purchase a quantity σ of water rights from region A. This would stabilize their supply at X^*_B in dry years and, in theory, offers the opportunity to sell the water back to region A in a wet-year spot market. After the sale of water rights, the rights in region A are $X^*_A - \sigma$. The wet-year derived demand for region A is shown by the solid line $ED(A)_{wet}$. This derived demand is exaggerated since a bimodal distribution such as Figure 8.1 would produce an upward fluctuation in supply, which could largely offset the wet-year demand by region A. Under these conditions no wet-year trades would occur.

In this example, the wet and normal year supply probabilities are not separated. Under this assumption, the wet-year spot market supply for region B is perfectly elastic at any price above zero since the demand is truncated at X^*_B. Comparison of the two excess demand and supply function shows clearly that the wet-year buy back spot market would need dramatically lower transaction costs than the dry-year market to operate at all. For an equal transaction cost of TC in both markets, the dry-year spot market will completely stabilize the water supply for region B, while a wet-year spot market from B to A will never occur. In short, the differences in excess demand and supplies coupled with spot-market transaction costs cause an asymmetry in the spot markets that may explain the "hoarding" accusations leveled at purchasers of water rights and the reluctance of low water value regions to sell their permanent water rights.

Figure 8.2. Annual Spot Markets

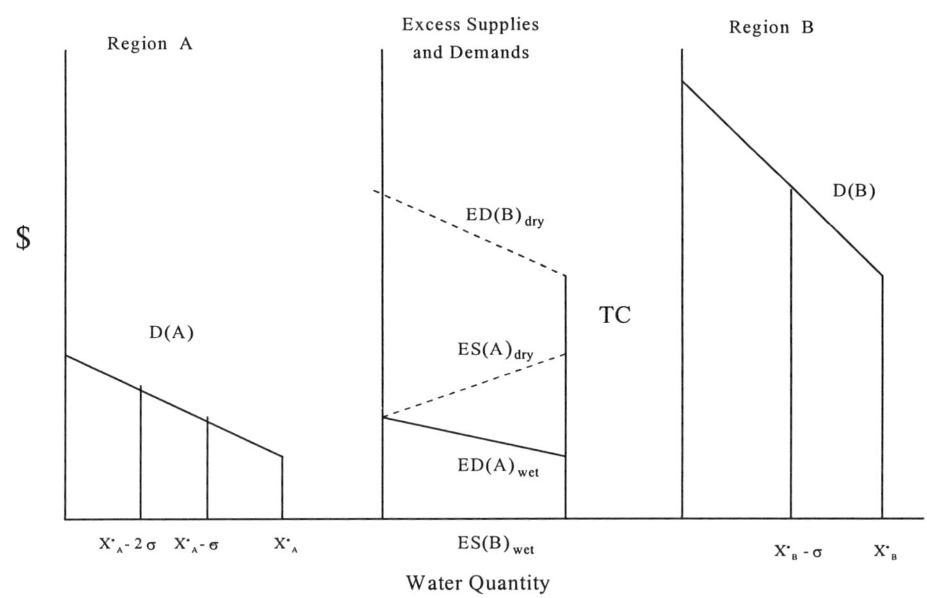

Figure 8.3 shows the same regions with the excess demands and supplies in terms of expected values rather than the conditional values in Figure 8.2. The scenario assumes that σ quantity of water rights could be sold from A to B. Based on the previous analysis, it is further assumed that there is no wet-year lease-back. Under these assumptions the expected excess supply function for A is equal to the expected value of wet and normal year supplies and is shown as E(ES(A)). The expected excess demand for dry-year water is shown as E(ED(B)). It is obtained by weighting the dry-year excess demand in region B by the probability of a dry year. The intersection of the expected excess supply and demand shows that, even with zero transaction costs, the quantity of water traded will not stabilize the supplies for region B and, with transaction costs considerably smaller than TC, no trade will take place.

The examples in Figures 8.2 and 8.3 are motivated by a demand for more reliable water supplies, driven by a shift to more highly valued crops or from agricultural to urban development. Due to transaction costs and the asymmetries of excess supplies and demands, spot markets are shown to stabilize supplies more often and at a lower cost than the sale of permanent water rights. The uncertainty cost of whether the spot markets will exist with water to sell in dry years has been avoided but the role of option markets in reducing this cost will be discussed later.

Figure 8.3. Water Rights Transfer for Supply Stability

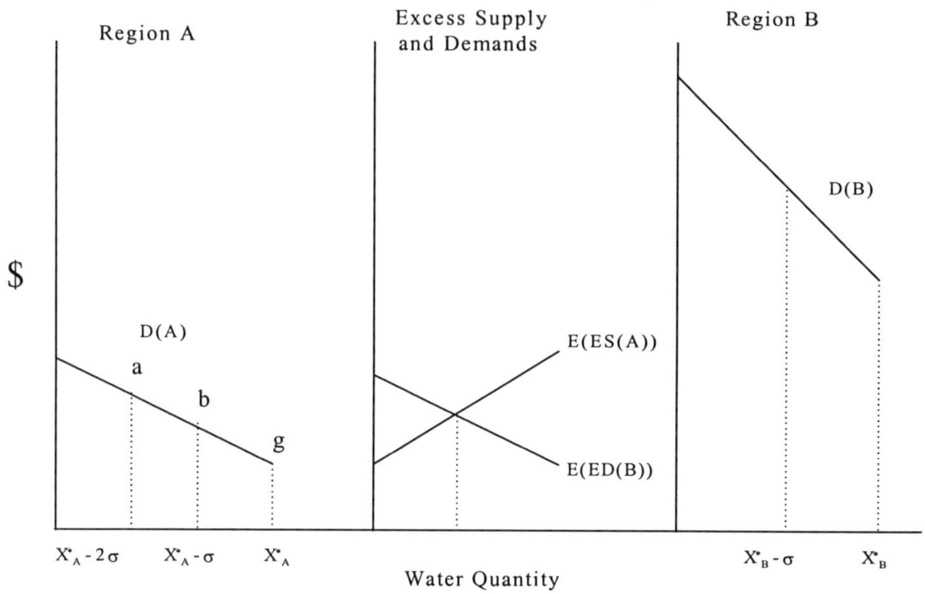

E(ED(B)) = 0.25 ED(B) dry, E(ES(A)) = 0.25 (a-b) + 0.75

Figure 8.4 shows the same regions motivated by a different reason for a water market. The initial water right allocations are the same but the demand in region B has shifted out so that at the original price they will demand X^*_B + TR quantity and under dry conditions the reduction to $X^*_B - \sigma$ will be more costly. The expected excess supply from region A is the same as in Figure 8.3 but the expected excess demand from region B has shifted up substantially due to the higher demand in normal years and the higher value in dry years. It can be readily seen that the expected demand and supply for permanent transfers will reach an equilibrium even with transaction costs that exceed TC in Figure 8.2. Under these assumptions, a dry-year spot market can stabilize the dry-year costs, but does not address the demand shift in normal years. Ignoring other third-party effects, the transfer of permanent water rights is the simplest, and possibly the lowest cost, market structure when the goal is to satisfy permanent demand shifts.

Dry-year spot markets can accommodate permanent shifts in demand at lower third-party costs by increasing the reliable supply in dry years. For example, region B in Figure 8.4 would have complete dry-year reliability if the initial water rights were $X^*_B + \sigma$. Given this base quantity, a dry-year transfer of TR would ensure reliable dry-year supplies for the increased demand D2(B). For water uses where reliability is important, such as in urban areas, this level of initial water rights is common.

Figures 8.2 – 8.4 show that under simplifying assumptions, the efficiency of the types of water markets depends on the source of market pressure. If the demand is for more reliable water in a sector or region due to a change in use, then spot markets probably will be the most efficient. If the source of demand is from a structural shift in the water economy that requires a change in the average supply but not the reliability, then the permanent sale of water rights may be the most efficient type of water market. The effect of annual option markets conditioned on water supply conditions has characteristics that allow an increase in the reliability due to spot markets to be considered as an increase in average supply in the medium term.

8.3 Option Demand Theory and Water Markets

To date, theoretical economic analysis of water markets has largely taken a static equilibrium approach (Hartman and Seastone, 1970, and Vaux and Howitt, 1984). Static analysis is unable to address the central driving characteristics of water markets that are first, uncertainty about supplies and prices, and second, changes in market conditions between and within the water years.

The use of options and spot prices for water has been mentioned and advocated in publications on the economics of water markets (Israel and Lund, 1995; Michelson and Young, 1993) and several types of option prices have been tried. In particular, Saleth et al. (1991) address the problem of market thinness in spot water markets. The authors point out that thin markets may predominate where externalities are important and the formal water rights have over-appropriated the water sources. The paper uses game theory simulations to represent alternative bargaining rules and environments, showing how different rules can be used to offset strategic behavior which will reduce the social benefit of the market. An alternative method of investigating the effect on the market efficiency of thin spot markets using experimental economics is addressed in Chapter 16.

A search of the literature shows only one unpublished analysis, Watters (1995), of the theoretical relationship between the different prices and sources of uncertainty in spot, option sales of water. The formal theory of option pricing originates with the seminal articles by Black and Scholes (1973) and Merton (1973) that showed how to determine the value of options to purchase resources whose future price is stochastic. These articles have the distinction of rapidly spawning a large industry in options and derivatives that is now an integral part of the financial industry. The work in financial economics has generally not translated over to resource economics, probably due to the need to analyze the problems in terms of continuous time stochastic differential equations. The mathematics required may be daunting, or it may be that the continuous time assumption seems inappropriate for natural resource problems.

This brief discussion of the application of option theory to water markets is largely based on a survey article by Smith (1976) and will not attempt to delve into the theorems and Ito calculus used to derive the dynamic stochastic conditions. Hopefully this section will persuade the reader that this theoretical approach provides an appropriate method for analyzing water markets.

Figure 8.4. Water Rights Transfer for Demand Shift

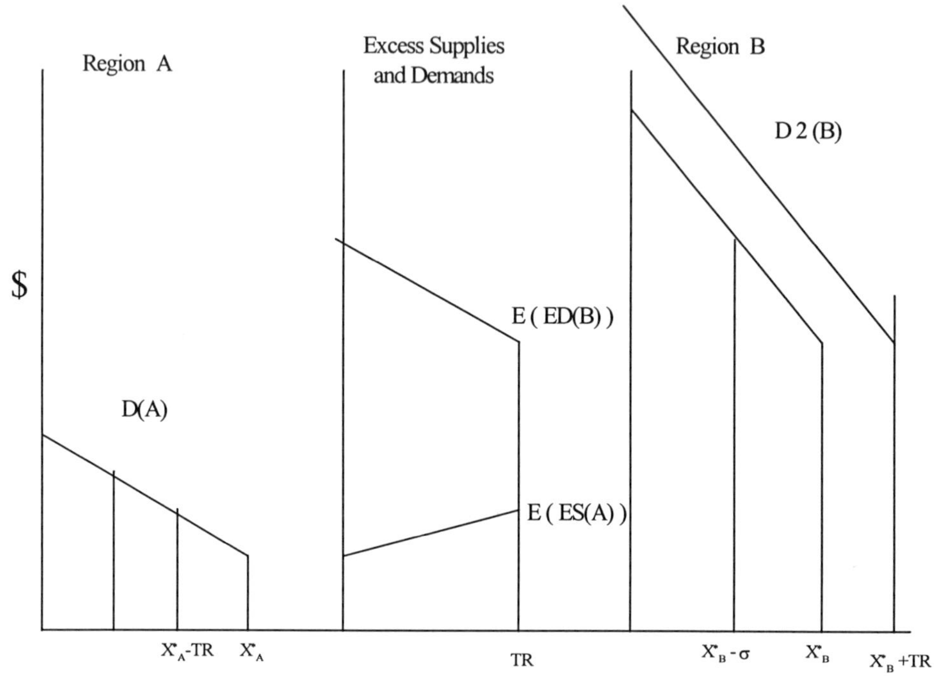

The central features that are common and important to both commodity options markets and water markets are timing and uncertainty. Uncertainty in water markets is due to two principle sources: uncertainty about the quantity of water that will be available at a future time from a specific right, and uncertainty about the value of water in the future given a specified supply quantity. In short-run seasonal terms, the value and quantity are highly correlated, but in the long run, there is an upward trend in the mean value of water due to a combination of increasing demands and supply reallocation for environmental purposes.

The two polar cases of water markets, namely annual spot sales versus sales of permanent water rights, are examples of how the cost of uncertain future water values is shifted between water users. Under a spot market the seller has full information on the realized quantity available before specifying a spot price. Given the full distribution of water rights, the spot market seller always has the fall-back position of using the water in its original lower valued use. The water buyer who has to rely on dry-year spot markets to meet inelastic demands has no fall-back source of supply and thus has to internalize the combined uncertainty of future spot market supplies and the resulting water price. Given the need for longer term supply assurance, the water buyer often wishes to reduce the uncertainty by buying a permanent water right and only having to face the natural variation in supply. The sale of permanent rights shifts the burden of correctly valuing the current worth of future water to the seller. Given the uncertainty of future water values and the long-term increase in real value, buyers of permanent rights who overpay will be compensated later by the upward trend in water values. The seller of permanent water rights who misjudges the capitalized value of their water has no recourse, and is consequently cautious in the current era of rapidly changing demands.

Accordingly, the two polar types of markets, namely spot markets or permanent rights sales, shift the uncertainty from one party to the other and in both cases induce large transaction costs. Risk-averse water owners will hold off committing to a market supply for fear of selling the water rights too cheaply and buyers will be wary of relying on thin and uncertain markets to satisfy firm and inelastic demands for water.

A full option market with the ability to hedge can substitute for a permanent rights market, without the transaction costs and inflexibilities inherent in this type of market. Most of us blithely plug in the coffee pot each morning without worrying that we have no property right to electric power, but only customary use rights. The assurance that power will be available is based on the knowledge that there is a full market for power and our local utility can buy excess power from the grid. Without stretching the analogy too far, the task for water markets in their initial years is to establish sufficient depth in the spot and option markets to compensate for the difficulties and costs of transferring permanent rights.

The central idea of this chapter is that in this emerging stage of water markets both the pure water rights market and annual spot market structures have large asymmetric uncertainties associated with them that will dampen market efficiency and keep the markets thin. These two sources of uncertainty can be termed the supply uncertainty and price uncertainty. The conclusion is that suitable types of option markets can spread the supply and price risk between the current water user and the potential buyer and thus stimulate a fuller and more efficient market. There is usually a public policy role in establishing and enforcing the institutions required for an options market, since these markets have the properties of public goods.

In addition to the role of spreading the risk, option markets may also be more appropriate for two other reasons. In many countries and states the rights to water are defined in terms of customary use rights rather than by a clear legal title. In these cases current water users can be paid to refrain from using their water under specific conditions such as in the California Emergency Drought Water Banks of 1991, 1992, and 1994. Third-party costs of water sales are of concern in many regions where the publicly funded development of water gives the public a legitimate interest in regional public benefits from water use. The third-party costs of water sales can be increased by the uncertainty of annual spot sales, or the finality of the permanent transfer of water rights from some region. Specific examples of these effects in the Western U.S. will be discussed in the next section.

While the uncertain quantities of water that are available from well-defined rights in a given year are a major source of uncertainty, they can be incorporated in terms of the annual scarcity cost that has to be paid for water. That is, the simplified model presented assumes that even in the driest year, water can be purchased from some source at some price. In the extreme, one can think of very highly priced backstop sources of water such as urban waste water reclamation or sea-water desalinization as limiting price increases under extreme shortages. In translating fluctuating supplies into price variability we are moving up and down the demand curve for water, as shown later by the price changes in the three California drought water banks. A second source of uncertainty is how the demand for water shifts over time. A predictable shift over time is called a "drift term" in stochastic dynamic parlance.

For water markets, predictable trends cannot yet be shown thus the difficulties of future price prediction cause both a drift in the mean value but also a component of the price variance.

The formal derivation of the relationship between option prices and future water values requires that the stochastic dynamic relation is expressed in continuous time terms. The continuous time specification is appropriate for water markets since river flow, runoff, or groundwater stocks are variables that can be measured in continuous terms and the current value of water adjusts continuously with them.

If the continuous time specification is inappropriate, a discrete time binomial version of option pricing can be developed using an approach introduced by Cox et al. (1979). This approach has an advantage for seasonal water analysis in that it divides the time horizon into discrete intervals which spread out in a discrete tree diagram and can be solved by recursive approaches. While convenient for empirical approaches, the binomial option price is not amenable to the qualitative analysis that is a central part of this chapter.

The availability of water in California is modeled as an Ito diffusion process in Howitt (1995). Since the value of water in the derived demand is a continuous function of the scarcity of water we can invoke Ito's lemma which shows that a continuous function of an Ito process is itself an Ito process.

The stochastic annual shifts in the water supply shown in Figure 8.1 can be characterized as a random Gaussian process around the mean. Combining this stochastic process with the long-term upward trend in the demand for California water from Vaux and Howitt (1984) enables the stochastic dynamic equation for the value of water to be specified as the following geometric Brownian diffusion process:

$$\frac{dS}{S} = \mu\, dt + \sigma\, dZ \tag{1}$$

In equation (1), S is the value of the water right. Thus the left-hand side gives the rate of return from holding a water right. Equation (1) says that the rate of return from a water right changes over time due to two effects. An average drift upward in value per unit time if μ is positive, plus a random component that is the product of the standard deviation of the process σ and a Wiener process dZ. For those unfamiliar with Wiener processes, Dixit (1993) shows that the Wiener process is the continuous time equivalent of a Markov process. This diffusion process has the properties that the stock value cannot be negative and the variance of the value grows with time.

An option to purchase a unit of water rights at the future time T at price E has the following valuation: if the value of the water right at time T, S(T) is below E then the option is valueless, if not the option is worth S(T)-E in time T. Clearly, the value of the option before time T depends on the probability that the value of the water right will be above E in time T. This probability is a function of the current value of the water right S(t), the exercise price E of the option, the stochastic price dynamics and the time to maturity of the option T-t. The option price can be written in general as:

$$p(t) = f\,(S,E,T,t) \tag{2}$$

Since the only variable that is stochastic is the price of the water right S(t), the change in the price of the option can be derived by applying Ito's lemma to the function f (S,E,T,t) to yield:

$$dp = \frac{\delta f(.)}{\delta S} dS + \frac{\delta f(.)}{\delta t} dt + 0.5\, S^2 \sigma^2 \frac{\delta^2 f(.)}{\delta S^2} dt \tag{3}$$

Equation (3) can be used to calculate a distribution for the rate of return to an option for water in the future. To set an acceptable price for this stochastic asset it is assumed that there is an acceptable trade off between the expected return and risk for the option. Black and Scholes (1973) showed that one could define a riskless holding composed of water rights, some of which were used as a basis to sell options. The rate of return to this riskless portfolio could then be equated to the real rate-of-return r. Using this identity, and another application of Ito's lemma, the fundamental partial differential equation of option pricing for water rights can be derived as:

$$\frac{\delta p}{\delta t} = rp - rS \frac{\delta f(.)}{\delta S} - 0.5\sigma^2 S^2 \frac{\delta^2 f(.)}{\delta S^2} \tag{4}$$

Equation (4) shows that the change in the optimal price for an option to buy water depends on the current real interest rate r, the initial value of the water right S, the variance in the value S over time, and the first and second derivatives of the stochastic option price function, f(S,E,T,t).

Given that the stochastic option price moves in the same direction as the water right value but with a decreasing rate, the first derivative of f(.) will be positive and the second negative. When interpreted in this context, equation (4) says that the rate of change of the option price will be lower for water stocks that start with a higher price, and other factors being equal, a higher variance in the value of the water right will increase the rate of option price change, as will a higher real rate of interest. Significant increases in the storage capacity from the construction of new dams will reduce the option price in two ways, first through the increased water supply, and second through the reduction in the supply variance.

Partial differential equations are notoriously difficult to solve. Black and Scholes (1973) utilize the portfolio assumption to define the option price as the discounted expected option price at time T.

$$f(S,E,T,t) = e^{-r(T-t)} \int_E^{\infty} (x-E)\phi(x)dx \tag{5}$$

where $\phi(.)$ is the log-normally distributed probability density and x is a dummy variable of integration.

Black and Scholes (1973) evaluate the integral to obtain an explicit form of the option value function as the following equation (6):

$$f(S,E,\ T,t) = S(t)N\left\{\frac{\ln(\frac{S(t)}{E}) + (r+\frac{\sigma^2}{2})(T-t)}{\sigma\sqrt{T-t}}\right\} - e^{-r(T-t)} EN\left\{\frac{\ln(\frac{S(t)}{E}) + (r+\frac{\sigma^2}{2})(T-t)}{\sigma\sqrt{T-t}}\right\} \tag{6}$$

where the expression N{.} is a cumulative normal density function.

Equation (6) enables analysts to empirically calculate the optimal price for water options given the following parameters: the current value of water rights S(t) or the rate of return from holding a water right dS(t)/S(t), the instantaneous variance in the value of water rights σ^2, the price of the option E, the real rate of discount r, and the time before the option is exercised (T-t). By using the hedged portfolio construction, the Black Scholes approach avoids the need to specify a measure of market risk aversion.

Equation (6) can be used to show the central theme of the chapter, namely, that option markets can spread the risk inherent in future quantities and values between the buyer and seller. To show the qualitative properties of equation (6), three variables are held constant: the initial value of the water S, the real rate of discount r, and the time period T - t. The joint probabilities of price and quantity risk facing the water right owner and purchaser are combined to form the uncertainty about returns to future water use that is represented by the variance σ^2.

The derivative of (6) with respect to σ^2 is positive and shows that the value of the option increases as the variance increases. The intuition behind this result is that the down side of the option is bounded since the terminal option price cannot be below zero. A larger variance, however, increases the expected value of the upside variation of the value of the water option. It follows that an option purchases the ability to make money on the upside, but leaves the water right owner holding the asset if its value changes substantially on the downside. The derivative of equation (6) with respect to the exercise price of the option E is negative. This effect allows the seller of the option to use the exercise price E to modify the price that would have to be paid for the option given a specified level of uncertainty. Changing the exercise price can be thought of as giving the owner seller of the options the ability to capture more of the upside value of the water right since, with options, they are always stuck with the downside risk.

The derivatives of the other three variables in equation (6) S, T, and r are all positive as would be expected. The time horizon benefit of water market options is obvious from equation (6). In terms of the time horizons of the buyers and sellers, going from the spot water market to the permanent rights market is a jump from the very short to the very long run. In reality, some time horizon considerably less than infinity will probably provide sufficient supply reliability for most potential water buyers. Farmers holding water rights may be prepared to commit to selling during dry years over the next ten years to accumulate funds for some more intensive expansion.

The use of financial economic methods confirms and defines the intuitive conclusion that options are more valuable when the time value of money is high and the uncertainty around the time evolution of the value of water rights is greater. This latter condition characterizes the initial stages of water market formation and suggests that the ability of options to spread the future risks without requiring a permanent and often irreversible transfer of water rights is promising outcome. In particular, the cost of the risk of future water value uncertainty can be explicitly analyzed. Given the ability to decompose these uncertainty related transaction costs, policies to reduce the level of transaction costs can be based on theoretical and empirical estimates instead of the intuitive and anecdotal approach that is more common at the moment.

8.4 Recent Examples of Spot and Option Water Markets in California

There have been small spot sales of water among contractors within local regions of California for many years. The first large set of water sales that resembled a market occurred in 1991 with the formation of the Emergency Drought Water Bank.

The Emergency Drought Water Bank of 1991 was based on annual agreements subject to a fixed payment and sale price. Given the rapidity with which the Bank was formed many agreements were initiated on a verbal basis.

The fixed purchase price for the 1991 bank was $0.10/m^3 ($125/acre-ft), and the selling price from the banks was fixed for all months and regions at $0.14/m^3 ($175/acre-ft). Yet fixing the price at the start of a drought year water market is very likely to lead to excess supplies or demands. In 1991 the State-sponsored market bought substantially more water than the market was prepared to adsorb at the fixed price and consequently purchased $20.5 million of excess supply water that was carried over to the following year. The supply to the 1991 Bank was composed of agricultural land fallowing (50 percent), groundwater exchange (33 percent) and sales of stored surface water (17 percent).

The sale of water based on fallowing agreements has significant advantages. Sale of the water released by fallowing land restricts the amount sold to the beneficial consumptive use on a given area. The area of crop fallowed is inexpensive to monitor and incorporates consideration of the appropriative right, consumptive use, and instream externalities of water on that area of land. Negotiating and enforcing a fallowing agreement is a low transaction cost method of effecting efficient water trades under conditions of imprecise rights and expensive enforcement. In addition to the transaction costs of implementing water sales, fallowing agreements may have a negative impact on the local economy.

Groundwater substitution agreements applied to farmers who had surface water rights and underutilized groundwater rights. The agreement allows farmers to continue to irrigate their fields, using groundwater instead of surface water. A reduction in third-party costs to local agricultural businesses is achieved by this transfer but care is needed that reducing third-party impacts on the local businesses is not at the expense of other externality costs of groundwater stock reduction or land subsidence.

The last option is the release of water stored in reservoirs that would not have otherwise been released. Unless property rights to the water are firm before the trade, the ability to supply water raises the tautological question of whether the water was being put to beneficial use before the trade.

Despite the problems of hurried formation and fixed prices, the 1991 drought water bank was a great success. The bank's water supply was enough to meet demand under conditions of extreme drought, and in doing so, generated a substantial net economic surplus for California's economy. The surplus supply in the bank, that occurred because of fixed prices, was carried over to the following year. The actual quantity of water sold by the Bank was small in comparison to the total use; however the price of the Bank water set a value for all potentially tradable water. Thus the operation of the Water Bank changed the value of most of the water in the State. This ability to increase the value of water without an increase in cost to the farmer is a politically acceptable way of sending the signal to users of the true value of water.

The demand side effect of the 1991 Drought Water Bank can be shown by the dampening of demands for emergency water. In April, 1991, after the last rains that year, water contractors were surveyed by the State Department of Water Resources and stated that their critical "needs" totaled 768.75 million m^3. However, at the end of the water season the amount of water actually purchased from the water bank was only 601.25 million m^3. That is, when faced with the high price of purchased water, consumers revised their needs downwards and actual purchases were 22 percent less than the estimates of "critical needs". Much of this difference was obtained by effective water conservation. In urban areas the average cost of water conservation was estimated at $\$0.22/m^3$. Most urban areas that purchased water at $\$0.14/m^3$ were faced with an additional $\$0.05/m^3$ transportation cost, making the total cost of purchase and transportation of water from the Water Bank comparable to some conservation methods. As a result of the fixed price policy and this reduction in true demand, the 1991 bank carried over 326 million m^3 to 1992.

The Bank generated direct benefits for the State economy by creating a net gain in income and employment by trading water from lower value to higher value uses. Table 8.1, based on Howitt et al. (1992), shows the net effect of the water trades. The overall effect on the net revenue of exporting regions was slightly negative, since the effect of the increased farmer profits did not fully offset the loss of crop production expenditures on the local economy. The average net income loss to exporting regions is calculated as \$12.5 million, which is approximately 5 percent of the net agricultural income. Due to the diversity of regional production, in many regions the reduction is a much smaller proportion of total regional income.

The regional income lost due to crop fallowing in exporting regions is shown on the first line of Table 8.1. The positive effect of the additional income from water sale revenues is shown on the second line of Table 8.1. Clearly farmers would not have sold their water below its value marginal product, but the regional income loss is caused by the difference in expenditure patterns between crop expenditures and increases in net farmer income from water sales.

Table 8.1. Statewide Net Benefits from the 1991 Water Bank

	($ Million)
Exporting Regions	
Income Lost from Crops	76.02
Income Gain from Water Sales	63.27
Export Region Income Loss	-12.75
Importing Regions	
Income Gain in Agriculture	45.40
Urban Consumer Surplus Gain	58.77
Benefits to Importing Regions	104.17
Net Benefits	
Agriculture Benefits	32.65
Urban Benefits	58.77
Value of Surplus Water	13.00
Total Net Benefit	104.42

The value of the water to the importing agricultural regions is calculated using the crop-specific output multipliers for the crops on which the purchased water was applied. As would be expected from the high water price, purchased water in agricultural regions was used for high-valued fruit, nut, and vegetable crops. Thus, while the water sold for agricultural crop production was only 22 percent of the total bought, the income gained from the water sale was 83 percent of the total loss. There are three reasons for this asymmetry in value. First, as mentioned earlier, only half the purchased water was by fallowing contracts which required the reduction or elimination of crop production. Thus, half the water was purchased without reducing the regional crop production in water selling regions. Second, the net return per unit of water was substantially higher for the crops for which water was purchased. Third, the effect of these crops on local income is substantially higher due to the larger amount of local processing and packing involved with the higher valued crops. Due to this large difference in income generation, the net benefits of the water transferred between agricultural regions is substantial.

The urban benefits are sometimes based on very high values of marginal urban water use of \$0.65 to \$0.85/m^3 that are imputed from intensive manufacturing uses. However, these high average values did not reflect the marginal value of urban water in California in 1991. Urban benefits were calculated using elasticities of demand taken from Vaux and Howitt (1984), and a market clearing price at the net cost of water purchased from the bank.

The net benefit to all importing regions is estimated in Table 8.1 at \$104 million. This can be thought of as a lower bound on the benefits since the urban values are low and the excess water bought in 1991 and carried over to 1992 was valued at the price paid for water in the subsequent 1992 water bank.

The sale of water between regions had a positive net effect on employment. Jobs were lost in the water exporting regions but the gain in jobs in the importing regions far outweighed the losses. Table 8.2 shows the estimated impacts on employment by sector. In estimating the effect on urban employment, we face the same problem as with the urban income multipliers. The job multipliers used for the urban sector are low compared to the average industrial use multiplier but reflect a change in water availability to the urban "green" industry of landscaping and horticulture.

Table 8.2 shows that, unlike the income effect, the agricultural industry had a net reduction in jobs due to its sales and purchases with the Bank. However, only 22 percent of the purchased water was resold to agriculture and the State of California as a whole had the benefits of an additional 3,741 jobs from the sale of water through the Bank. In addition there were net gains in agricultural jobs from trade.

These broad measures show that in terms of both income and jobs, the water bank generated substantial net gains to the state and most regions. When considering any resource reallocation, the benefits and losses should be jointly reviewed. In terms of income, the net loss of \$12.75 million has a favorable ratio of 8.2:1 to the net income gain to the state. The equivalent ratio for job gains to losses is lower at 2.3:1, but still shows substantial net gain.

Table 8.2. Statewide Employment Impacts of the 1991 Water Bank

	Number of Jobs
Exporting Agricultural Regions	
Jobs Lost from Crops	3,133
Jobs Gained from Revenues	1,485
Net Effect	- 1,648
Importing Agricultural Regions	
Jobs gained from water imports	1,153
Net Job change in Agriculture	- 495
Importing Urban regions	
Jobs gained from water imports	4,236
Net Statewide Gain in jobs	3,741

The California DWR 1992 and 1994 Drought Water Banks

Since the drought of 1987-91 continued into 1992, the Drought Water Bank was also continued. Given the improved water supplies, the 1992 Bank operated at a lower purchase and sale price and smaller quantities. Water was not purchased by fallowing crops in 1992 and supplies came from surplus reservoir storage (20 percent) and groundwater substitution (80 percent).

The total quantity purchased was 234 million m^3, a reduction of 56 percent, and the price paid for the water was $0.04/$m^3$, a similar reduction of 60 percent. The differences in the price and quantity equilibria between the 1991 and 1992 Water Banks strongly support the contention that both the demand and supply of water in California is price elastic, even under severe drought conditions.

1993 was a much improved water year in California, and a water bank was not needed. However 1994 was once again dry and a third Drought Water Bank was established. Given past bank experience and the similarity with 1992, the 1994 Bank bought 272 million m^3 from reservoir and groundwater exchange contracts. The average purchase price was the same as 1992 at $0.04/$m^3$ ($50.0 acre-ft). The administrative transaction costs of the Drought Water Banks were low, approximately 7 percent (personal communication), and the main reason for the substantial price spread between sellers and buyers was the need to finance the "carriage water" requirement of approximately 30 percent of the delivered quantity required for salinity control in the Sacramento River Delta. In short, the Banks worked well within their restrictions of rigid price levels and regulatory controls on third-party effects.

At the start of the 1995 water season precipitation and river flows were at low levels. To add some security and flexibility to a potential water bank, the California Department of Water Resources (DWR) initiated an option market in December, 1994, (Jercich, 1997). The market took the form of purchasing options to buy water in the event of a drought at the fixed price of $0.003/$m^3$ ($3.50/acre-ft) and requiring deposits on options to purchase water of $.008/$m^3$ ($10/acre-ft). DWR negotiated an exercise price for the supply options that ranged between $0.03/$m^3$ to $0.035/$m^3$ ($36.50 to $41.50/acre-ft). Options for a total of 36 million m^3 (29,000 acre-ft) were purchased by December, 1994, with an exercise deadline of May, 1995. The improved hydrologic conditions after December halted further purchases and made exercising the options unnecessary. The 1994-1995 DWR option water bank was operated with remarkably low transaction costs and was able to return $0.007/$m^3$ ($9.00 acre-ft) of the purchase option deposit to the potential buyers.

The ability to purchase options to be exercised in the event of a dry year is an advance in the direction of water markets with more security in the supply. A criticism of the previous water banks was that they did not allow adequate time for adjustment on both the market supply and demand sides. Introduction of the option market in December should induce a more elastic supply of water to the bank and a price structure that can vary as the extent of water supplies becomes better known between December and April. Currently (1997) the California DWR is holding public hearings on a proposal to establish a six year water purchase program with an annual capacity of 490 million m^3 (400,000 acre-ft). The supply sources for this program would come from excess reservoir capacity and groundwater exchange programs. The groundwater component of this program is receiving substantial criticism in potential water-exporting areas and will probably be removed from the program, reducing the quantity by half. Even with this reduction, the establishment of this option bank is an important advance in California's ability to reduce the impact of droughts. The widespread concern over potential trading of groundwater in California shows that there is widespread concern over the external effects of water trades that can only be allayed by good technical analysis, careful initial trades, and time.

The Palo Verde - Metropolitan Water District Dry-Year Option Agreement

Contracts for water sales that are contingent on the level of water availability have been advocated by economists and water policy analysts. This type of option, which is contingent on scarcity conditions, suits the inherent uncertainty of water supplies (Clark and Abt, 1993; Michelson and Young, 1993; Hamilton et al., 1989).

In 1993 the Palo Verde Irrigation District (PVID), which is situated in California on the Colorado river, entered an agreement with the Metropolitan Water District that paid growers an annual option fee and gave Metropolitan the right to obtain water from PVID growers by fallowing up to twenty five percent of their land during dry years. In a recent two-year test program, irrigators in the Palo Verde Valley fallowed 20,215 acres of land in exchange for a payment of $620 per acre fallowed. The total amount of water conserved was 114 million m^3 of water. The water was stored in a reservoir on the Colorado River (Lake Mead) at a pumping cost of $0.04/$m^3$. MWD has the option to call on this water in any year up to the year 2000. Any conserved water not needed by Metropolitan after 2000 will be forfeited.

Watters (1995) analyzed the Arvin - MWD option trade in the context of financial option market theory. She calculated that the exercise price of the option was $0.11/$m^3$ and that the option premium was $0.013/$m^3$. Using similar data to that in Figure 8.1, Watters (1995) calculated a variance for the return to the water options. Since 2000 is the cut off date applied to both the 1994 and 1995 option agreements, they cover different lengths of time (five and six years, respectively). Using a Black-Scholes model, the theoretical option values are $0.013/$m^3$ for the six-year option and $0.008/$m^3$ for the five-year option. These option values are very close to the actual price paid by MWD.

Arvin Edison-Metropolitan Water District Groundwater Exchange Agreement

The Arvin Edison Water Storage District in the Southern San Joaquin valley is in the process of concluding a joint conjunctive use and water transfer agreement with the Metropolitan Water District of Los Angeles (MWD). The Arvin Edison district uses the groundwater aquifer underlying the district to distribute water to many of its contractors. The district has firm surface water entitlements of 49 million m^3 (40,000 acre-ft) per year from the Federal central valley project, and an additional 384 million m^3 (311,000 acre-ft) of surplus water that is only available 35 percent of the time in wet years. The agreement with MWD enables Arvin to construct the additional infrastructure to capture and store additional flows in the aquifer. The amount defined in the agreement is a minimum of 307 million m^3 and a maximum of 429 million m^3 in storage for MWD. MWD has the option to extract water from the aquifer at rates that do not exceed 92 million m^3 in any year. The total price of the initial 307 million m^3 is $0.106/$m^3$ but after the initial phase the price drops to $0.082/$m^3$. The water is purchased in two stages, with half the price paid to capture and recharge the water and the other half paid when the water is extracted by MWD in dry years. Essentially MWD has ensured themselves of the option to buy water at a low price for dry years by financing the recharge of water in wet years. Arvin gains revenues when the water is extracted, and benefits from increased groundwater and reduced pumping lifts while the water is stored in the aquifer.

These two examples show that there are different approaches to reducing the uncertainty inherent in newly emerging spot markets. One method is to take a formal option where there is uncertainty as to the ability to store the water, as in the Palo Verde case; the other approach is to establish a conditional forward contract such as done with Arvin Edison. There are several other options and forward contracts under negotiation. This seems likely to be a growing trend given the advantage of sharing the future uncertainty by option contracts, and the reduced effect on third parties argued in the next section

8.5 Imprecise Property Rights and Third-Party Effects of Options

In many countries and regions water rights are not held formally by the ultimate user. In California for example, water rights are rarely quantified and defined by title, but are predominantly "usufructuary, a right to the use of water, not a right to own it" (Littleworth and Garner, 1995). Clearly, some type of right is required before trades can commence. In California, the clarification and definition of the conditions under which it is permissible to trade usufructuary rights has evolved as an alternative to the adjudication of tradable permanent water rights (Gray, 1996).

Water rights need to be adjudicated prior to the formation of a market for permanent water rights. An economic objection to this procedure is that the cost of adjudicating may be substantial. Knapp and Vaux (1982) estimate that the costs of adjudicating groundwater in California are in the range of $0.07/m^3. Add to this direct cost the cost of settling the competing claims on the use of water and the transaction costs are likely to become a substantial proportion of the value of the water.

As explained earlier, water options or spot markets can be based on the principle of non-use under specified conditions. The controls required to operate the two spot markets in California and one potential option market were based on existing records of cropping patterns and contracted use rights under appropriative water rights for surface water and correlative rights for groundwater. Instead of measuring the nominal quantity of rights claimed by the seller, the water sales were based on the concept of reductions in consumptive use from a measurable action. Usually this action was the non-irrigation of a specific crop over a given area or the substitution for a surface water source by a valid groundwater source. The key to keeping transaction costs low and avoiding conflict over water rights was to ignore the rights, or the absence of them, and define the quantity sold or optioned in terms of "wet water saved" by the action. Using this indirect method to measure the quantity of water purchased from agricultural areas enables the existing system for measuring crop production to double as an enforcement and measuring mechanism for water trades.

Given that the State is often the ultimate owner of water rights in many countries, and, in addition, has often financed the development of water with public funds, there is a strong and legitimate concern with third-party effects that may be caused by water sales. When there are distinct differences between regions and sectors involved in water trades, appealing to the increase in net national product from water is not politically effective. The third-party externalities induced by water trades can usually be categorized into environmental and economic costs.

In the spot and option markets that have occurred in California over the past six years, environmental externalities from water trades have been contained at a low level by two actions. First, the principle of basing water trades on the net non-use of water, thus avoiding sales of those proportions of applied water that contribute to the common good through deep percolation or return flows. Second, the nature of water sales in California is generally from upstream areas to those further downstream.

Although externalities from reduced stream flow have not been significant, two aspects have had to be regulated. The first was to ensure that the change in timing and points of diversion due to water sales did not lead to temperature changes in the rivers that were harmful to aquatic life. Second, the transportation of additional water across the Sacramento River Delta alters the hydrologic balance between fresh and salt-water barriers. For every unit of water transferred across the Delta, an additional 30 percent had to be released to augment stream flow in the Delta and maintain the normal salt balance. The cost of this "carriage water" was included in the sale price of the water and did not appear to interfere with the operation of the spot market.

Economic third-party effects of water trades have been a notable source of objection to water markets in California. Quantitative evidence of the extent of regional economic impacts is hard to find. Howitt (1994) used county level primary surveys and a simulation model to estimate the aggregate county level income changes attributable to water sales. The results in two counties that supplied 25 percent of the water sales to the Bank ranged from 6.5 percent to 3.2 percent average reduction in income for those county residents not participating directly in the water sales. These low-average third-party costs mask considerable variation within the counties studied. Where sales by fallowing land were concentrated in small areas, several businesses associated with agricultural production suffered a substantial and unexpected reduction in business. This increase in the third-party economic cost was born out by empirical simulations that show a rapidly increasing cost per unit as the proportion of water sold in a local area increases. The key to keeping third-party externalities from water sales at a politically acceptable level is to geographically disperse the sales and provide a means for associated businesses to anticipate when they will occur.

An advantage of option sales over spot markets is that the level of sales and the conditions under which they will occur will be known in advance by most businesses in the area. Accordingly, a supplier to agricultural firms can better plan and anticipate those sales that are, or are not interruptible by water markets.

Permanent transfer of water rights from a region can lead to substantial impacts on the local economy. This type of sale is almost unknown in California, but Howe et al. (1990) show regional losses in farm value of 10 to 21 percent in the Arkansas valley of Colorado. Sales of water under options modulate these impacts in three ways. First, under option contracts, the water remains in farming for the majority of the years, thus keeping the seller on the farm and providing a source of secondary income for associated businesses. Second, since the farmer is still active and resident in the region, the stream of option payments in years that they are not exercised will add to the income in the region. Third, the negotiation of water sales options allows enough time to negotiate third-party compensation where appropriate.

Traditional provisions to prevent or compensate third-party impacts take the form of regulatory restrictions on trades. An alternative method proposed in California to reduce water market transaction costs is to decouple the actual water sale from third-party compensation. The problem is determining how to reduce the uncertainty over individual or district water rights and at the same time ensure that mechanisms are in place to internalize legitimate third-party costs.

California water transfer law has developed in a piece-meal fashion over the past twenty years. A current attempt to combine the existing statutes in the water code together in a consistent body of legislation is known as the Model Water Transfer Act, (Gray, 1996). The Model Act represents efforts of group of academic lawyers and economists to draft legislation that is now under review by lawyers, politicians, and water interest groups. Two themes run through the model act, first is the importance of reducing transaction costs for spot and option markets. Second is the need to internalize different types of third-party costs that water trades may induce.

One approach to internalization in the Model Water Transfer Act is the establishment of a fund for the reimbursement of a clearly defined set of third-party losses. The compensation fund would be financed by levies on interbasin water sales adjusted to balance the level of claims, since many of the initial water endowments were established by the state. The operation of this fund would allow plaintiffs substantial time (two years) to file claims but would at the same time remove the "no injury" rule that enables effective blocking of short term water trades.

The characteristics of water markets in California is that they are usually drought-related and thus require rapid action. Traditional solutions to third-party problems would weaken the property rights by making transfers subject to additional regulation such as the "no injury" rule under which parties who claim injury can petition the Water Resources Control Board for an injunction on the transfer. The key is to define water transfer property rights in a way that decouples the third-party compensation system from the actual water sale. The compensation fund is one way to do it. While this system is open to aggregation inefficiencies and moral hazard costs, it is hypothesized that the reduction in transaction risk will more than compensate for these costs.

The common property nature of California's groundwater resources leaves it as a potential recipient of third-party impacts of surface water sales. Given that groundwater provides 25 percent of the California supply in a normal year and up to 40 percent during a drought, it seems futile to consider transfers of surface water in isolation from groundwater. However, the correlative rights that govern California's groundwater are unquantified and defined only to the extent that they attempt to provide a basis for preventing damage between parties. In this vague status of groundwater rights, California is in the mainstream with many developing countries. Few groundwater basins in California have attempted a quantitative adjudication of their resources. In all cases it has proved drawn out and costly. More importantly, there is very strong opposition to groundwater adjudication on a state level in California. The ability for areas to form local groundwater management regions is defined and occasionally used when there is a dire threat to the aquifer (Blomquist, 1992).

8.6 Summary and Conclusions

The chapter opens with a brief overview of the successful development of dry-year spot markets in California over the past seven years. Spot water markets are established in California and seem to be evolving into dry-year option markets. Markets where permanent water rights are traded are notable by their absence. The polar positions of spot and rights markets for water is emphasized and the effect on risk and third-party transaction costs of these two polar markets is evaluated under alternative reasons for having a water market. The graphical analysis concludes that the desirability of different market forms depends on the source of the market demand. Stabilization of stochastic supplies favors spot and option markets, while permanent shifts in demand may be best accommodated by trading water rights.

Using the theory of option pricing, the ability of the buyer and seller to spread the costs of future uncertainty under a system of option markets for water is demonstrated. The advantage in being able to make a contract for a medium length time horizon under an options market is also noted. The chapter continues by examining empirical measures of performance of several spot markets and some emerging dry-year options agreements. Additional advantages of option markets for water where water availability is defined by use rights rather than property rights are proposed and where third-party cost of water trades have to be internalized.

While other countries will have different institutions, traditions and infrastructures, California has many geographical and economic features that will be found in arid regions that have a potential for gains from water markets. Characteristics that are probably common to all potential water market regions are variations in the water supply on a spatial or temporal dimension or both. In addition, the tensions between growing urban and environmental uses of water supplies and the traditional agricultural uses exist in many growing economies in arid regions. The lesson learned from California's experience with water marketing is that water markets work. Over the past five years, spot and dry-year water markets have become an integral part of the supply system. The evolution from spot towards option markets shows the promise of avoiding the costly and politically difficult alternative of trading permanent water rights.

References

Black F., and M. Scholes, 1973. "The Pricing of Options and Corporate Liabilities," *Journal of Political Economy,* 81(3): 637-659.

Blomquist, W.A., 1992. *Dividing the Waters: Governing Groundwater in Southern California*, ICS Press , San Francisco.

Clark, J.R, and S.R. Abt, 1993. "Temporary Water Transfers for Dry-Year Water Supply," *Water Resources Bulletin,* 29(2): 249-255.

Cox, J., S. Ross, and M. Rubinstein, 1979. "Option Pricing: A Simplified Approach," *Journal of Financial Economics*, September, 6(1): 229-263.

Dixit A., 1993. *The Art of Smooth Pasting,* Harwood Academic Publishers, Chur Switzerland.

Gray, B., 1996. *A Model Water Transfer Act for California,* California Business Roundtable, California Chamber of Commerce, California Farm Bureau Federation, California Manufacturers Association.

Hamilton, J.R., N.K. Whittlesey, and P. Halvorson, 1989. "Interruptible Water Markets in the Pacific Northwest," *American Journal of Agricultural Economics,* 71(1): 63-75.

Hartmen, L.M., and D. Seastone, 1970. *Water Transfers: Economic Efficiency and Alternative Institutions.* Johns Hopkins Press, Baltimore.

Howe, C.W., J.K. Lazo, and K.R. Weber, 1990. "The Economic Impacts of Agriculture to Urban Water Transfer on the Area of Origin: A Case Study of the Arkansas River Valley in Colorado," *American Journal of Agricultural Economics,* 72(5): 1200-1204.

Howitt, R.E, N. Moore, and R.T. Smith, 1992. *A Retrospective on California's 1991 Emergency Drought Water Ban.* California Department of Water Resources, Sacramento.

Howitt, R.E., 1995. "Malleable Property Rights and Smooth Pasting Conditions," *American Journal of Agricultural Economics*, 77(5): 1192-1198.

Israel, M., and J.R. Lund, 1995. "Recent California Water Transfers: Implications for Water Management," *Natural Resources Journal*, 35(1): 1-32.

Jercich, S.A., 1997. "California's 1995 Water Bank Program: Purchasing Water Supply Options," *Journal of Water Resources Planning and Management,* 123(1): 59-65.

Knapp, K.C., and H.J. Vaux, 1982. "Barriers to Effective Groundwater Management: the California Case," *Ground-Water,* Jan/Feb., 20 (1):61-66.

Littleworth, A.L. and E. L. Garner, 1995. "*California Water,*" Solano Press, California.

Merton, R., 1973. "The Theory of Rational Option Pricing," *Bell Journal of Economics and Management Science,* 4(1): 141-183 .

Michelson, A.M., and R.A Young, 1993. "Optioning Agricultural Water Rights for Urban Water Supplies During Drought," *American Journal of Agricultural Economics*, 75(4): 1010-1020 .

Saleth, R.M., J.B. Braden, and J.W. Eheart, 1991. "Bargaining Rules for a Thin Spot Water Market," *Land Economics,* 67(3) : 326-339.

Smith, C.W., 1976. "Option Pricing: A Review," *Journal of Financial Economics*, 3(1): 3-51.

Vaux, H.J., and R.E Howitt, 1984. "Managing Water Scarcity: An Evaluation of Interregional Transfers," *Water Resources Research,* 20 (7): 785-792.

Watters, P., 1995. "Efficient Pricing of Water Transfer Options: Nonstructural Solutions for Reliable Water Supplies," unpublished Ph.D. Dissertation, Department of Economics, University of California, Riverside, California.

9 Institutional and Organizational Arrangements for Water Markets in Chile

Robert R. Hearne

9.1 Introduction

Chile is one of the few countries that has encouraged the use of markets in water resource management. Market allocation in Chile is possible, in part, because a system of transferable water-use rights was reestablished in 1981. These rights are independent of land use and land ownership, thus trading of water rights is fairly unrestricted. The codification of these water-use rights coincided with a series of reforms in the Chilean economy, including privatized land rights and liberalized trade. Although Chile's water rights system is unique, it faces water resource management challenges that are very similar to those in other countries. Increased population and income in Chile's urban areas have created increased urban demand for water. In addition, industrial and residential pollution are overburdening the assimilative capacity of many of Chile's rivers.

Initially, three river valleys were chosen with the assistance of Chile's *Dirección General de Aguas* (Directorate General of Waters) for the analysis of water markets in Chile: the Elqui River, the Limarí River, and the Maipo River. Because water scarcity is an important factor in the incentive to trade water-use rights, river valleys in southern Chile were avoided. The BíoBío Valley in south central Chile was not included in the analysis because other studies were being conducted in that valley. Later, the Azapa Valley was included as a case study in order to investigate additional intersectoral transactions between irrigators and the local municipal supply company (Hearne, 1995). These case studies present general information on water resource use and management in these valleys. They also provide important background information on market transactions that is instrumental in understanding water institutions and organizations.

9.2 Water Management and Transaction Costs

Where water is scarce and legally-defined transferable property rights exist, market trades can be expected when the difference in the value of water between two uses is greater than the costs of transferring the water. In the absence of transaction costs, economic incentives would induce water users to trade water-use rights until the marginal value of these rights was equal across all users. Of course, transaction costs do limit the movement of water and the transfer of water-use rights. Transaction costs for water market transfers include: 1) the cost of the physical infrastructure needed to measure and transport water, including the evaporation and filtration losses during conveyance, 2) the cost of searching and finding willing buyers and sellers and negotiating a contract, and 3) the cost of validating legal ownership of the water-use right, legalizing the contract, enforcing contract provisions, and acquiring the necessary permission from regulatory authorities to transfer water. Because these transaction costs can be large, the number of potential buyers and sellers may be limited — which may result in non-competitive pricing.

In order to reduce the burden of these transaction costs, organizations can be established to construct the water delivery infrastructure, modify and monitor the distribution of river and canal water, expedite the dissemination of market information, maintain public records of water-use rights, and enforce the rights of third parties affected by a transfer of water. These services may be provided by a central government, local governments, community organizations, or private entities. Many governments have made large investments in irrigation infrastructure and water management authorities. Although these irrigation systems were not generally designed to facilitate market transfers, the presence of flexible infrastructure should reduce the transaction costs of market exchanges. Also, user groups, especially water user associations (WUAs) that manage and maintain canals, can play important roles in facilitating and monitoring trades.

In addition, since water use is characterized by a high degree of interdependence, individuals may want to restrict the amount and types of transfers that occur. Changes in the allocation of upstream water and in irrigation practices can impose a negative externality on downstream users. The transfer of water to another area can increase the percentage of canal water lost during conveyance for the remaining users. In Chile, an increase in the number of water-use rights on a canal can reduce the amount of water received per water-use right in the canal during times when river levels are high. This occurs because during periods of high water, unlimited withdrawals from the river are permitted, and the only constraint on water delivery is the carrying capacity of the canal. Also, changes in water use can significantly affect water quality. Thus both government authorities and WUAs may want to regulate water transactions to ensure that they are beneficial to the community of water users and society.

9.3 Water Allocation Institutions in Chile

Chile has a tradition of private development of water resources and private rights to shares of river and canal flows that dates to the colonial era. This tradition has been revived in the National Water Code of 1981[1] which allows private transferable property rights for water use. This water law reversed the 1969 water law, written during a period of land reform, which tied irrigation water to the land and mandated state control over water resources.

Water-Use Rights

The 1981 water law stipulates that water is a national resource for public use but that permanent and transferable rights to utilize water can be granted to individuals in accordance with the law. Water-use rights can be granted by the government upon petition, purchased from an individual owner, or retained based on traditional use. Currently, there is no stipulation that water-use rights must be utilized in order to be retained.

Rights can be defined as permanent or contingent. Permanent rights are granted for use of unallocated sources of supply. In most of Chile's river basins, especially in the north and in the central valley, all sources of supply have already been assigned permanent use rights. Contingent rights are granted for surplus water, that is, water flows that exceed those demanded by permanent rights holders during periods of higher than normal water flows. Reservoir or lake water is not subject to contingent rights since, under most climatic conditions, the regulation of water flow is sufficient to nullify the chance of excess flows of water.

Rights are also designated as consumptive and non-consumptive. Consumptive rights entitle the user to completely consume the water without any obligation to return it to its source. Non-consumptive rights grant the owner the use of the water as long as it is returned to its source at a specified quality and does not interfere with existing consumptive use rights. The law stipulates that rights are to be specified by volume of flow per unit of time. But in reality, rights are defined as a share of stream flow because, in most cases, the high variability of natural river flows prohibits volumetric specification. In order to resolve this inconsistency, rivers are divided into sections, and each canal, intake, and withdrawal point receives a percentage of the water in that section of the river. Volumetric equivalencies of river shares are stipulated for the flow of the river that occurs in 85 out of 100 years. When river flows are insufficient to meet volumetric specifications, water flows are reduced proportionally. However, rights on some rivers have been over-allocated so that water flows are sufficient to meet volumetric specifications in far less than 85 percent of years.

Since consumptive-use rights are granted for the full use of all the water stipulated in the right, downstream users do not have any right to return flows generated from upstream users. This has little effect on the first section of a river since return flows mostly augment downstream sections. Water users in downstream sections of a river divide water that enters through springs, rainfall, and return flows. These rights holders are not protected by law from any change in upstream water use that significantly reduces return flows. There is also no specific restriction on the transfer of upstream water to another basin.

Water-use rights are also required for groundwater exploitation. Individuals can request from the *Dirección General de Aguas* (DGA) a right to groundwater use once they have confirmed the existence of a certain yield at a certain depth. The groundwater-use right is accompanied by a prohibition on other groundwater withdrawals in the protective area specified in the right. Any party with legally entitled rights to water that may be adversely affected by the granting of new groundwater-use rights, can oppose the grant by informing the regional DGA office within 30 days of the publication of the entitlement in the Official Journal. Owners of rights to surface water and WUAs have opposed groundwater exploitation near rivers and canals. If a petition for groundwater is opposed, the Regional Director of the DGA can either grant or disallow the new water-use right.

There is no property tax on water-use rights. But land is taxed according to its productive value, which includes the value of irrigation water. There are seven different categories of agricultural land for tax purposes. These range from high-quality, irrigated, central valley land to non-irrigated land. Thus the concept of separating land and water has not reached the tax code. There is no sales tax on the transfer of water-use rights, but there are fees paid to lawyers, notaries, and the Real Estate Registry.

Water User Associations and Irrigation Administration

Historically, the development of irrigation in Chile has been dominated by the private sector. Over one million hectares (ha) have been developed for irrigation with private investment (Gazmuri Schleyer, 1994). These were mostly small run-of-the-river systems. Starting around 1930, the government began developing major irrigation infrastructure. Many of these investments were never completed and, since 1945, only one major irrigation system (the Paloma Reservoir system in the Limarí Valley) was built by the government. Also, long-term contracts to recover costs from users were denominated in local currency, which has since drastically lost value. In the 1970s, both private and public investment in irrigation was absent due to the uncertainty of agrarian reform and government austerity. With more secure land and water rights and liberalized agricultural policy, private investment in irrigation for high-valued fruit and vegetable crops expanded rapidly in the 1980s.

All privately developed irrigation systems and many of those developed by the state are owned and controlled by independent WUAs. These WUAs are owned and operated by their members, and charge fees based on their capital and operating costs. The WUAs maintain the canal systems, keep records of rights holders, apportion water to individual rights holders according to their recorded shares, and enforce water rights. The 1981 Water Code specifies rules for the formation, governance, and obligations of these WUAs.

There are three different types of WUAs that are recognized in Chile. A water community, *communidad de agua*, consists of water users that share a common source of water. They can be chartered and recognized with formal procedures, but many *communidades de agua* have not bothered to become chartered. Irrigation Associations, *asociaciones de canalistas*, serve irrigators that share a common canal and have a legal status which allows them to enter into contracts and receive financing. *Juntas de vigilancia* (JDVs) are made up of all users and user associations on a common stream or section of a river and are responsible for administering water use in the river. JDVs control the canal intakes that flow from the river and some administer dams for the storage of irrigation water. At the national level, the Confederation of Canal Operators is legally recognized as the representative of most WUAs. According to the Confederation, about one half of all WUAs are legally registered with the DGA.

The Ministry of Public Works, which includes the DGA, has played an important role in water management in Chile. Its *Dirección de Riego* (Directorate of Irrigation) is responsible for planning, supervising construction, and operating public sector irrigation infrastructure. The National Irrigation Commission (CNR) is an inter-ministerial committee chaired by the Minister of Economy with the membership of the Ministries of Finance, Public Works, Agriculture, and Planning. The CNR is the major government entity that determines irrigation policy.

Water Supply and Sanitation

Chile has traditionally had a high level of water and sewerage coverage: 98 percent of urban and 75 percent of rural households have access to piped water, and 80 percent of urban households are connected to central sewerage systems. In the past fifteen years the water and sanitation services have undergone a major transformation. In 1990, the regulatory functions of the former national water supply and sanitation service, SENDOS, was transferred to the newly created Superintendency of Sanitary Services. In addition, SENDOS was decentralized into eleven separate autonomous regional water supply and sanitation companies, along the lines of the metropolitan water supply and sanitation companies of Santiago (EMOS) and Valparaíso (ESVAL). Stock is currently held by the government and CORFO, a publicly owned corporation.

These independent water companies are obligated by law to provide water and sanitation services to the large municipal areas. They are required to deliver full water supply to their concessions 95 percent of the time. Water rates are based on delivery costs, with a fair return on capital, and are reviewed every five years by the Superintendency of Sanitary Services. A premium is charged in the high-use summer months in order to manage demand. These water companies inherited the water-use rights that were traditionally held by the municipalities they serve and more recently by SENDOS. Some of these rights are considered to be priority rights and, consequently, such volumetric withdrawals are not reduced proportionally during times of water scarcity as is the case of other shares. There is a government program, operated through the municipalities, to subsidize potable water supply and sanitation service charges to low-income households.

Environment

The Chilean constitution guarantees citizens the right to live in a pollution-free environment. Water quality standards for both agricultural use and potable water have been adopted, while generalized water quality standards that limit the effluent that can be dumped in a river or stream have not. The amendments to the Water Code proposed in 1993 would have empowered the DGA to set quality standards. The DGA's Environmental Department and the Superintendency of Sanitary Services were coordinating an inventory of point source polluters in the country.[2] Based on the inventory, Superintendency is developing basic standards to ensure the enforcement of a 1916 law concerning industrial effluents.[3]

A major source of water quality problems is the natural sediment flowing from the relatively young and highly eroded Andes mountain range. This mountain range has been deforested for centuries and reforestation efforts are only marginal. Both the mining and industrial sectors are known to produce harmful effluents. Most large mining operations have made efforts to control pollution in order not to disturb the environmental sensibilities of stockholders and the public at large. In southern Chile, a large percentage of water quality problems result from large pulp mills.

In the Santiago area, one particular water course used for irrigation has been, until recently, an open sewer. Outbreaks of cholera and typhoid have led to a controversy over the reuse of municipal wastewater in the irrigation of food crops. Historically, regulations made by public health authorities on the type of cultivation allowed with wastewater irrigation have been enforced inconsistently. However, new vigilance, including efforts to limit the sale of certain horticultural products, as well as the increased demand for produce that is irrigated with safe water, have significantly decreased the risk of food-born diseases.

Formal organizations responsible for environmental management and protection are new to Chile. Two commissions, the Environmental Legislation Committee and the

National Ecological Commission, were formed in the early 1980s but remained relatively inactive. In 1990, the National Environmental Commission was formed. The central role of the latter commission in formulating an environmental agenda for the government was reaffirmed by President Aylwin in 1992. In early 1994, the government passed legislation to require environmental impact studies for new construction and development projects.

Regulation and Conflict Resolution

The DGA is responsible for water resources planning and the development and exploitation of water resources. The agency collects data on water resources, maintains cadastres of water use, and has limited authority to intervene in conflicts between water users. It also grants, upon petition, water-use rights for unclaimed surface and groundwater. These grants are a matter of public record and are published in the Official Journal in both Santiago and the regional capital. During times of drought, the DGA can impound water with compensation to water-use rights holders.

The DGA is also involved in a project to "regularize" water rights. Regularization entails the creation of an inventory of all rights holders, including the rights that have not been formally titled. This process includes the specification of water flows and the water-use rights of irrigators holding traditional water-use rights. Regularization does not involve actual titling, which is done at the local real estate registry at the owners expense.

Although the DGA does have broad authority in water resources management, much of the actual control over river flows is exercised by the JDV. They manage and maintain storage reservoirs, and are responsible for reducing the flow of water to canals during times of water scarcity. These JDV are controlled by the river's water-use rights holders on a particular section of a river. The votes each rights holder possesses depends on the number of shares each owns. Thus in some river basins, irrigators, mining companies, and water supply companies are part of the same JDV. Currently, the court system is the final arbiter of water-use conflicts. But the effectiveness of the courts in conflict management has been limited by judicial restraint and an over-emphasis on formalism (Bauer, 1994). This is a general characteristic of the Chilean legal system, which has not demonstrated the institutional capacity to creatively resolve conflicts.

9.4 The First Section of the Maipo River

The first, or upstream, section of the Maipo River flows from the Andes mountain range through rich agricultural land to the community of Isla del Maipo southwest of Santiago. This is a semi-arid region where average yearly precipitation is only 400 mm. The upper Maipo supplies potable water for 4.5 million inhabitants of the Santiago metropolitan area, powers a few small hydroelectric plants, and provides a source of pleasant scenery and recreation near the city of Santiago.[4] The river also irrigates approximately 100,000 ha of urban, suburban, and rural land. The major irrigated crops include permanent fruit crops, such as kiwi, pears, peaches, and grapes, as well as horticultural products, traditional grains, and pasture.

The first section of the Maipo is divided into 8,133 shares, each representing at least 8 liters/sec, 85 percent of the time, with flows much larger during the critical months of December and January. Most of this water is used in agriculture, including 760 shares which are conveyed across the Mapocho River (a tributary of the Maipo) and used to irrigate lands north of Santiago. The *Junta de Vigilancia de Río Maipo: Primera Sección* is responsible for supervising the distribution of water, but it remains mostly inactive. There are five large WUAs which, together with EMOS, control 85 percent of Maipo's water in the first section (see Table 9.1).

Table 9.1. Major Water Users in the First Section of the Maipo River

Water User/ Water User Association	Number of Shares
Sociedad del Canal del Maipo	2,428
Asociación del Canal del Maipo	1,660
Canales Unidos de Buin	1,010
Canal Huidobro	650
Canal Pirque	630
EMOS[a]	565
Others	1,190
Total	8,133

[a] EMOS also owns shares that pertain to other canals.

EMOS serves much of Santiago's population with its 1,369 shares. Most of these shares are withdrawn from the river at either EMOS' own intake gate or through the adjacent Canal San Carlos. Without the use of storage, EMOS's 1,369 shares deliver 5.2 m^3/sec for the required 95 percent security. The company also owns both the El Yeso Reservoir with 255 million m^3 and Laguna Negra, a natural lake. Because of this storage capacity, EMOS's shares can deliver about 10 m^3/sec at 95 percent security.

EMOS made a number of unsuccessful attempts between 1990 and 1993 to purchase Maipo River water. The company claims that it has never refused an asking price, but was able to purchase only 33 shares of Maipo water, mostly from urban developers, with prices averaging $10,000 per share.[5] EMOS is willing to accept the costs of changing fixed flow dividers but, because of the expense, it is waiting to purchase a much larger quantity of rights before it physically transfers the water. In the past, EMOS was able to extract water-use rights from urban land developers in exchange for connection to EMOS' water supply. A recent court ruling prohibits this practice. Thus, EMOS appears to be maintaining a passive role in the water market until it completes studies on groundwater development, on improved management of the El Yeso Reservoir, and on lowering the intake at Laguna Negra. EMOS has estimated that it will need the equivalent of 3,000 shares of Maipo River water by the year 2020.[6] In the shorter term, EMOS is scheduled to open a new water treatment plant in 1998, but it does not have the crude water supply to meet this plant's capacity.

The company's most obvious source of water-use rights are urban canal users serviced by the *Sociedad del Canal del Maipo* (SCM). This organization delivers 1,667 shares of Maipo River water to mostly urban "irrigators" within the city of Santiago and 761 shares to farms on the north side of the Mapocho River. These urban "irrigators" are owners of small plots in urban residential areas (According to the DGA, the 1988 *cadastre* of the Maipo River has pages upon pages of these users, many of whom do not have legally inscribed rights). Also, because the SCM is able to finance itself with a small (15,000 kw) hydroelectric plant, it does not charge its users for water delivery.[7] Thus, there is absolutely no demand management for a very large percentage of the upper Maipo River. It is also possible that some of these many water-use rights owners are unaware of their ownership. The canal system has adjustable water control gates and can easily move water from one area to another and provides water "on demand" without charges. The SCM has purchased 1.16 shares of consumptive water-use rights in order to protect the flow of water through its hydroelectric plant.[8] EMOS itself is a member of the SCM, and it has placed its treatment plants below the SCM's hydropower plant.

There is a strong incentive for EMOS to seek the cooperation of SCM in its attempts to purchase Maipo River water-use rights. EMOS can also negotiate with other large water-use rights owners such as the mining company *Disputada de Las Condes,* which owns 153 shares that it apparently does not use and the Potable Water Company *Lo Castillo,* which owns 135 shares that it holds in reserve and currently does not employ.[9] Many of these shares were acquired in a September, 1981, auction of 313 shares in which *Disputada de Las Condes* acquired 107 shares, *Lo Castillo*, 80 shares, and EMOS, 26 shares. Given that EMOS can exploit a large aquifer downstream of Santiago, the current water situation in Santiago is not grave. However, in the presence of many low-valued urban "irrigators" and non-consumptive use, EMOS should be able to use the market more effectively than it has in the past to obtain water supplies for its future needs.

One reason that EMOS might be slow in using market mechanisms to secure supplies of crude water is its heritage as a state enterprise with certain prerogatives. For instance, there is the hope that in the case of chronic water shortage, the government will interfere to assure EMOS's water supply. Another reason why sellers have not responded to EMOS's attempts to purchase water in the Maipo is the public knowledge of high prices recently paid to sellers of water-use rights in the Mapocho River. There, the water company *Lo Castillo*, which services Santiago's wealthy northeastern suburbs, has driven up the price of water rights. However, in 1993, *Lo Castillo* decelerated its buying program and began to investigate groundwater sources.[10]

Another reason for the lack of trading is the cost of physically changing canal flows through fixed flow dividers that exist in many of the canal systems other than SCM. This transaction cost makes marginal transfers of water expensive. In addition, farmers are comfortable with the current water/land input ratio which is a result of the last phases of the land reform. Furthermore, within the agrarian culture in this area, land and water continue to be seen as inseparable.

Temporary transactions between neighboring farmers have been known to occur in a few canal systems. According to WUA managers, these arrangements mostly involve an annual use of a share of the river in exchange for payment of the yearly WUA service fee.[11] This type of arrangement occurs when a farmer withdraws from production, but does not wish to sell the water-use right because the value of the land would decline. The WUA may facilitate this type of transaction in order to ensure timely payment of the annual service fee.

9.5 The Elqui Valley

The water in the Elqui Valley, except for water irrigating 1,500 ha, is controlled by the *Junta de Vigilancia de Río Elqui* (JDVRE). This WUA delivers 25,000 shares of water to 118 canals along the Elqui River and its tributaries, the Claro, Cochiguaz, and Turbio Rivers. One fifth of these shares remain inactive since the owners do not pay their fees to the JDVRE. The inactive shares are essentially divided up among the remaining rights holders. Each share is supposed to deliver 1 liter/sec of water, although Aninat et al. (1993) reported that at 80 percent probability, a share of the Elqui River delivered 0.45 liter/sec in May and 0.57 liter/sec in January. The JDVRE also owns and controls the La Laguna Reservoir which stores 40 million m^3. The *Dirección de Riego* has plans to construct a new dam at Puclaro with a reservoir capacity of 175 million m^3 to provide a more secure water supply in dry years.

The Water Supply and Sanitation Company of Coquimbo (ESSCO), the water utility in Region IV, currently owns 768 shares of the Elqui River. Of these, 280 shares pertain to the traditional water supply of the twin cities of La Serena and Coquimbo. Another 270 shares were "expropriated" in 1986, when SENDOS, the government water supply service, demanded the cessation of the water rights that pertained to land that was by then "urbanized." In addition, ESSCO purchased a total of 225 shares during 1992-93. One of these purchases was of water rights from areas that were recently urbanized. Another purchase was from an individual who lost land in bankruptcy but was able to protect the water rights. The third purchase was from two grape producers who sold excess water, amounting to 60 percent and 70 percent of their shares, in order to invest in drip irrigation technology. ESSCO is confident that it can rely on the water market for its future needs and plans to purchase 1,300 additional shares by 2020.

As discussed in the next chapter, ESSCO purchased water-use rights of the Elqui River in order to satisfy its crude water needs instead of investing in the proposed Puclaro Dam project. It turned down an offer of 50 million m^3 of water storage capacity sufficient for its required 95 percent security requirement. ESSCO is also investigating groundwater as a source of water to help meet the seasonal demand for water from the summer tourist population. ESSCO's water-use rights have a priority and the water from their original water-use rights is not reduced proportionately in dry years. There is some confusion as to how many shares should receive this priority but this confusion is somewhat inconsequential, because so far ESSCO has always received priority on all of its rights. Yet if all of ESSCO's new shares receive a priority, this would imply that all other water users will receive even less water in dry years as ESSCO continues to buy more shares.

Two of ESSCO's transfers of water entail moving water from the point of purchase in the river to ESSCO's intake downstream. To account for losses during the transfer, the JDVRE reduces these shares by one-half percent for each kilometer of the transfer. Thus the 225 "net shares" that ESSCO has purchased in the past few years were originally 292 "gross shares." This calculation was not designed to account for return flows to the river, although average flows at points near the mouth of the river are higher than those at the convergence of the tributaries 80 kilometers upstream. This reduction in shares, for loss of water in transit, is a transactions cost that only ESSCO and one tourism developer have paid. As of 1993, ESSCO had not invested in modifying infrastructure. It has solicited permission from the regional DGA to transfer the purchased rights. ESSCO has paid a lawyer 2 percent of purchase price to help in these transfers. The lawyer has also attempted to facilitate other purchases, but a few potential water-use rights transfers have been canceled because of insecure rights and unsuccessful title searches.[12]

ESSCO has played a leading role in the Elqui River water market, but there are other transactions involving the transfer of water to large grape growers who want to increase their security of water supply. There are also some small transfers of water to developers of small residential and tourist housing along the valley and on the coast north of La Serena.

In one transaction, the transfer of water-use rights out of a canal caused a conveyance externality that limited the ability of the canal to serve the remaining users. One irrigator sold his water-use rights after a neighboring farmer had sold shares and water flows through the canal were no longer adequate for irrigation. Also, irrigators using Canal Herradura believe that the upcoming transfer of 7 percent of its water to ESSCO will significantly alter the ability of their canal to deliver water.[13] According to the WUA serving Canal Bellavista, the largest canal in the valley, which delivers 3,800 shares of water through a system of fixed flow dividers to 5,200 ha, there have been no transfers of water-use rights in this canal.

The importance of DGAs efforts to organize WUAs and regularize water-use rights is demonstrated by the confusion over water-use rights ownership in the area of the *Estero Derecho* of the Claro River. Water in this tributary is not controlled by the JDVRE and, in this ideal area for grape production, some buyers and sellers of water rights suggested that their traditional rights had not been respected. Some records were reportedly destroyed in a fire. This resulted in distrust of the local WUA which has made it difficult to buy and sell water-use rights in the area.

9.6 The Limarí Valley

The major part of the irrigation infrastructure in the Limarí Valley was provided by government-supported investment in irrigation infrastructure.[14] Both the Cogotí and Recoleta Reservoirs were built in the 1930s by the national government. They are both currently owned and operated by WUAs. The Paloma Reservoir was built in the 1960s by the national government and is currently owned and operated by the *Dirección de Riego* in close cooperation with the WUAs that receive Paloma water. The *Dirección de Riego* also owns and operates the large trunk canal system connecting Paloma to the canals of the individual WUAs. The Punitaqui Canal, a small secondary canal system constructed along with Paloma, also belongs to the *Dirección de Riego*. Currently the *Dirección de Riego* provides water to irrigators free of charge. Except for the Punitaqui system, Paloma was built to improve the security of irrigation for areas already irrigated, including those that receive water from the Recoleta and Cogotí Reservoirs. For this reason, rights to water stored in Paloma have yet to be established for individuals and privatized, although this water is distributed to shareholders of the WUAs.

These WUAs are well organized and, together with the *Dirección de Riego* officials who operate the Paloma Dam and canals, cooperate in operating the interconnected system (see Table 9.2). Despite the fact that the government has yet to charge irrigators for the Paloma Reservoir and its canal system, the WUAs feel that they can better operate the system. Because of this, efforts to turn over operation of the Paloma Dam and canal system to the users were initiated by the WUAs in 1993 and, in principle, supported by the *Dirección de Riego*. The irrigators and canal managers involved in these efforts feel that by establishing an autonomous enterprise to manage the Paloma system, they can better develop the hydroelectric and tourist capacities of the dam and reservoir.[15] None of the three large dams in the Limarí Valley has a hydroelectric generation station.

Table 9.2. Major WUAs in the Limarí Valley

Water User Association	Area Irrigated(ha)	Source of Water
JDV Río Grande y Limarí	13,107	*Río Grande* above Paloma Reservoir and Paloma Reservoir
*AC Palqui-Maurat-Semita**	2,200	*Río Grande* above Paloma Reservoir
JDV Río Cogotí	2,113	*Río Cogotí* above Cogotí Reservoir
AC Recoleta Reservoir	15,000	*Río Hurtado* Recoleta Reservoir and Paloma Reservoir
AC Cogotí Reservoir	12,000	Cogotí Reservoir and Paloma Reservoir
AC Camarico	5,500	Paloma Reservoir
JDV Río Guatalame	953	Cogotí Reservoir
AC Punitaqui	1,000	Paloma Reservoir
JDV Río Hurtado	3,325	*Río Hurtado* above Recoleta Reservoir

* AC is *Asociación del Canal*

Fees paid by irrigators to their WUAs vary considerably. Some charge a fixed fee for each share of water owned, some charge for each cubic meter of water delivered, and some charge for both. Also, some WUAs require farmers to supply labor for maintaining primary and secondary canals, while others do not. Although distribution losses are a function of the distance between the water source and the farmgate, there are no extra fees for more remote water users. One WUA, the *Asociación del Canal Palqui-Maurat-Semita*, has purchased water-use rights in order to increase the amount of water it can distribute to its shareholders.

The reservoirs in this valley are interconnected, with Cogotí and Recoleta upstream of Paloma. It is very easy to transfer water from upstream to downstream sections of the river. But, because of storage limitations above Paloma, the transfer of water from below the Paloma Reservoir to canals upstream has been prohibited. This is important since the highest valued use of water is in the production of table grapes in the hot, sunny uplands. One WUA is investing in a project to increase the security of water supply above Paloma, but this WUA will continue to prohibit transfers of rights from one area to another. It is also possible to keep in reserve shares of water in a WUA that charges low service fees and transfer these shares to other canals to be used during dry years.

In this valley, two important factors contribute to low transaction costs and facilitate market transfers of water. These are the volumetric denomination of water-use rights and well-organized WUAs. Volumetric denomination involves both the preseason knowledge of availability of water, which is possible with 1,000 million m^3 of storage capacity, and the use of gates with flow meters throughout the system. The Cogotí and Recoleta Reservoirs are operated to provide water throughout a three-year drought. The Paloma Reservoir has a greater storage capacity and after the last three-year drought was still one-third full.

During April of each year, the WUAs, the DGA, and the *Dirección de Riego* meet to determine the volumetric equivalent of each share based on actual storage and predicted rain. Depending on the WUA, during an average year, shares delivered range from 3,000 to 7,000 m^3. This system allows for the quick transfer of water on a short-term basis (a spot market). Also, within certain limitations, water-use rights can be easily transferred permanently from one point to another. This wide range of transactions is possible because of the WUA's ability to effectively operate the storage system and the availability of gates to regulate water deliveries.

In the past three years there has been a rapid accumulation of land and water rights by large fruit exporters, especially in the area of El Palqui, upstream of and adjacent to the Paloma Reservoir. This accumulation is in contrast with the gradual consolidation of land and water rights that has taken place in much of Chile in the years subsequent to land reform. Grape growers report that in the last few years, many small- and medium-sized farmers have forfeited land and water rights to fruit exporting companies in lieu of debts owed to the companies.

In the Limarí area there are four or five major fruit exporters. Most of these, including the largest (Unifrutti), are multinational. Small-scale producers enter into financing and marketing contracts with these firms. The contracts are for a fixed multi-year period. The company provides financing for land preparation and high-technology drip irrigation, technical assistance, and agricultural inputs. The contract specifies that the debt be paid in grapes, priced according to Philadelphia market prices. Producers are not free to sell to rival exporters until their debt is paid.

The production of table grapes was very profitable from 1984-91. But low prices, poor harvests, and the seemingly continual expansion of production has reduced profitability. Some small-scale producers are willing to accept the blame for their loss. Many bought vehicles with the loans. Others say that large producers receive favorable marketing treatment. Low grape prices in 1992 and 1993 created financial problems for many small producers and their resulting forfeiture of land and water-use rights has allowed the export firms to consolidate both water and land holdings.

9.7 The Azapa Valley

The Azapa Valley, located in the lower section of the San Jose River Basin in Chile's far northern region, supports 1,327 ha of irrigated farmland and supplies water to the city of Arica. The valley, bordering the extremely arid Atacama Desert, is very dry and has negligible rainfall in the lower reaches. Arica is a commercial and tourist center with a small fishmeal processing industry and clothing factories. Two sources of surface water for this valley include an interbasin transfer of water from the Andean highlands through the Azapa Canal, which currently averages 796 liters/sec and small springs which have an average flow of 305 liters/sec. Both of these flows are diverted to the Azapa Canal for irrigation. Also, the Azapa Valley contains an unconfined aquifer with 180 wells extracting an estimated existing total storage of 302 million m^3 in 1993.

Parallel, and in close proximity to the Azapa Valley, is the Lluta River, one of the few rivers in northern Chile to have a constant flow. Unfortunately, water in this river has a high degree of natural contamination (salt, boron, sulphur, and sulfates) and is only fit for irrigating pasture, maize, and onions. Other small sources of brackish water on the coast north of Arica are used by Arica's textile industry.

In the Andean highlands near Bolivia, another relatively moist basin feeds Lake Chungurá, the highest lake in the world, and the Lauca River which drains into Bolivia. This is a unique and fragile ecosystem which supports the Lauca National Park and is home to pastoral Aymara Indians. Since 1962, water from this basin was diverted at Lake Cotacotani, through the Lauca Canal to a hydroelectric station at Chapaquina and to irrigators in the Azapa Valley. A more recent project to transfer water from Lake Chungurá was physically completed and later enjoined from operating after a legal suit was won by an environmental interest group.[16] A separate project to rehabilitate wells in the national park is under development in order to increase the water supply for the city of Arica. This project was not opposed by the *Corporación Nacional Forestal* (National Forest Company) which manages the park.

Irrigated crops in the Azapa Valley include olives and horticultural products. The latter are grown largely in greenhouses with drip irrigation. These crops are sold at a premium price because they can reach the market in Santiago before other competitors. Olives are exported to the world market directly from Arica.

Irrigators who receive water from the Azapa Canal belong to the *Comunidad de Aguas del Canal Azapa* (COMCA). This WUA has recently received its legal charter. Until 1989, the Canal Azapa was managed by the *Dirección de Riego*, but water-use rights for this water have been distributed and COMCA is in the process of purchasing the Canal Azapa from the government. The Lauca Canal is still managed by the *Dirección de Riego*. The WUA which controls the Lluta River, the *Junta de Vigilancia de Río Lluta*, was in the process of receiving its legal charter in 1994. While it was in the process of obtaining its charter, this WUA did not permit transfers of water-use rights.

Until 1993 the city of Arica faced a severe water shortage. In some areas of the city, residents received water for only a few hours of the day. The government then declared that potable water supply was to be given top priority and different mechanisms were instituted to ensure the short- and medium-term supplies of crude water for the Water Supply and Sanitation Company of Tarapaca (ESSAT), the regional water utility: the central government released capital for ESSAT to finance an expansion of its water supply; the DGA "discovered" unused wells for ESSAT to exploit via rental; and the *Dirección de Riego* allowed ESSAT to use its wells, without payment, until further supplies were secured. ESSAT also put pressure on farmers to rent wells by invoking the government's appeal to farmers to release water for Arica and the possibility that the state could impound water during periods of severe water shortages. The rentals were lucrative for farmers because ESSAT digs or rehabilitates the well, installs and operates the pump, delivers water free-of-charge for a few hours per week to the farmers, and pays a rent for the use of the well. ESSAT pumps continuously, while farmers would use water only a few hours per week. Thus ESSAT has significantly increased pumping in the Azapa Valley, despite indications that the Azapa aquifer is being drained at an alarming rate and that pumping costs are rising.

The depletion of the Azapa aquifer is a matter of great controversy. A 1994 Japan International Cooperation Agency study concluded that at current withdrawal and recharge rates, the remaining life of the groundwater storage is less than 20 years. Farmers in the valley have begun to complain about the increased costs of pumping and reduced flows in the valley's springs. The DGA has the authority to impose restrictions and limit water extractions from aquifers that are determined to be in danger of depletion as was done in the Copiapo Valley. But, given the government's stated priority on potable water supply, the DGA is not likely to act to limit ESSAT's pumping.

The success of ESSAT in increasing the supply of water available to the city of Arica can be attributed to the government's political desire to bring water to the city. This government action demonstrates the flexibility of the 1981 Water Code and the reluctance of the government to use its powers of emergency expropriation.

Finally, although the local water market has been used to enable ESSAT to increase its short-term supply of water, there is a noted absence of permanent market transfers of high quality water from the Azapa Canal to ESSAT.

9.8 Observations and Conclusions

Institutional arrangements in Chile have played a crucial role in the development of markets for water-use rights in local areas. Both the security of water-use rights and a flexible infrastructure have been important in these markets. Chilean irrigators are generally content with the codification of their traditional water-use rights. Since water-use rights are a tangible asset, which do not currently face a property tax, irrigators benefit from their ownership even when the market for these rights is inactive.

WUAs play an important role in facilitating the market reallocation of water, especially in the Limarí Valley, where trading is active, and in the Elqui Valley, where intersectoral trading is important. Many of these WUAs have been able to adapt to the needs of their members for services that facilitate or impede transactions. One possible explanation for the lack of intersectoral trading in the Maipo River is the failure of EMOS, the local municipal water supply company, to gain the cooperation of the *Sociedad del Canal del Maipo* in its plans to purchase water-use rights.

There is little danger that the current allocation of water-use rights will lead to the exploitation of urban water consumers. Indeed in the valleys of Elqui, Limarí, and Azapa, there is a general deference to urban water supply. Potential disputes are resolved in favor of municipal water supply without much conflict. One important reason for this is that there is little rural-urban competition in these areas. The urban areas of La Serena/Coquimbo, Ovalle, and Arica serve as service centers for the local agricultural areas. Most large irrigators and leaders of WUAs have houses and businesses in these communities, and do not want them to be short of water. Thus, in the Elqui Valley, ESSCO receives a priority on water delivery during periods of low water flow. In the Limarí Valley, ESSCO obtained groundwater rights, adjacent to the Limarí River near Ovalle, without significant opposition from the JDV that manages this river. In the Azapa Valley, the aquifer is being rapidly depleted in order to supply water to the city of Arica. Despite the fact that water users have been able to avoid serious conflicts in these valleys, the absence of institutions for intersectoral discussion and conflict resolution is apparent.

In the Maipo, Elqui, and Limarí Valleys, there was no indication that market allocation of water has contributed to environmental degradation, reduced water quality, or caused a decline in the use of return flows. In the Azapa Valley, ESSAT's increased access to groundwater rights has led to an accelerated decline in aquifer storage. Despite this, there is little indication, from these case studies, that the 1981 Water Code needs to be modified in order to preserve the environment.

However, accompanying Chile's economic growth is an increasing demand for environmental amenities, including freshwater-based tourism and recreation. Since many of these environmental amenities are public goods that are not well defined in a system of private property rights, the DGA or some new organization arrangement such as river basin management corporations may be needed to take action to protect these amenities.

The current 1981 Water Code does include some mechanisms to control certain negative externalities. For example, although the water code does not recognize the rights of users of return flows, the DGA has the limited authority to deny permission to change river intakes. The DGA could use this authority to limit any significant change in the location of water use that would adversely affect downstream users. Also, when aquifers are being drained by over-pumping, the DGA has the authority to restrict the withdrawal of groundwater. It has done this in the Copiapo Valley, but not in the Azapa Valley due to the government's stated mandate of securing water for Arica.

The need to develop a flexible institutional capacity for river management is demonstrated in the Limarí Valley where WUAs are acting independently to try and form a river management company that will allow them to take control of the Paloma Reservoir. These WUAs have experience in managing the Cogotí and Recoleta Reservoirs and have demonstrated that they have the ability to manage large canal systems. But they have yet to accept the challenge of developing the hydroelectric generation capacity of their dams. This suggests that current WUAs, designed specifically to address the needs of irrigators, may lack the incentives needed to ensure sustainable intersectoral development, especially in the presence of interdependencies in consumptive and non-consumptive use. Thus the government may need to provide WUAs with incentives to form river basin management corporations that can help manage river basins in a more integrated manner.

Acknowledgments

The research was supported by the World Bank's Agricultural Policies Division. The author extends thanks for the collaboration of colleagues at the World Bank, University of Minnesota, and the *Pontificia Universidad Católica de Chile*, especially K. William Easter, Gershon Feder, Ariel Dinar, and Guillermo Donoso. A special thanks to all those in the Chilean government and private sector who provided information and insight, especially Fernando Peralta, Samuel del Canto, Gustavo Manriquez, Humberto Peña, Luís Jorquera, Pablo Anguita, Juan Ignacio Domiguez, and Eduardo Venezian. Also gratitude is extended for the dedication and camaraderie of data collectors Pablo Venezian, Carlos Bonilla, Jose Cancino, Carlos Anis, and Jeronimo Concha.

Endnotes

[1] D.F.L. Number 1122, published in the Official Journal, Santiago, October 29, 1981

[2] Interview with Carlos Salazar, DGA, June, 1993.

[3] Memo from Terrence Lee, CEPAL, March, 1995.

[4] Other sources of potable water for the Santiago area include the Mapocho River and groundwater.

[5] Meeting with the management of EMOS, June, 1993. All values shown in U.S. dollars unless otherwise indicated.

[6] Meeting with officials in the EMOS Planning Department, November, 1993 and February, 1994.

[7] Meeting with the secretary of SCM, November, 1993.

[8] Record of *Conservador de Bienes Raices*, Puente Alto, Chile

[9] Meeting with the manager of *Lo Castillo* November, 1993.

[10] Meeting with the manager of *Lo Castillo,* June and November, 1993.

[11] Interview with the manager of *Canales Unidos de Buin*, November, 1993

[12] Meeting with the manager of ESSCO's planning department, December, 1993.

[13] Interview with the president of the Association for Canal Herradura, December, 1993.

[14] ESSCO supplies water to the city of Ovalle, and pumps water from right next to the Limarí River, using groundwater rights granted by the regional DGA. The grant of groundwater rights was initially unopposed by the *Junta de Vigilancia de Río Limarí*, which believed that the DGA was only granting permission for an experimental well. Given this source of water, there is little incentive for ESSCO to acquire new water-use rights for Ovalle.

[15] Currently, a private individual owns the non-consumptive water-use rights that could be used for power generation at the Paloma Dam.

[16] Meetings with DGA, *Dirección de Riejo*, and *Corporación Nacional Forestal* officials, February, 1994.

References

Aninat, Mendez y Merino Ltda, 1993. "*Estudio de Opciones de Financiamento de Obras de Riego: Informe Final," Dirección de Riego*, Santiago, Chile.

Bauer C., 1994. "*Derechos de Propiedad y El Mercado en una Institucionalidad Neoliberal: Efectos e Implicancias del Codigo Chileno de Aguas de 1981," Revista de Derecho de Minas y Aguas,* Escuela de Derecho, Pontificia Universidad Católica de Chile, Santiago, Chile.

Gazmuri Schleyer, R., 1994. "Chile's Market Oriented Policy: Institutional Aspects and Achievements," in *Water Policy and Water Markets,* World Bank Technical Paper # 249, L. LeMoigne, K.W. Easter, W. Ochs, and S. Giltner (eds.), the World Bank, Washington, D.C.

Hearne, R.R., 1995. "The Market Allocation of Natural Resources: Transactions of Water-Use Rights in Chile," Ph.D. Dissertation, University of Minnesota.

10 Economic and Financial Returns from Chile's Water Markets[1]

Robert R. Hearne and K. William Easter

10.1 Introduction

This chapter builds on the previous chapter and presents an analysis of water markets in Chile with an emphasis on areas of relatively frequent trades. It provides an estimate of the gains-from-trade and focuses on the exchange of water-use rights. This is important because the exchange of water-use rights requires more elaborate institutional arrangements than does a simple exchange of water. The latter, sometimes referred to as a "spot market" or informal market, occurs when the owner of a legal or prescriptive right to a certain volume or flow of water sells a portion of that water, sometimes outside of legal sanction, to a neighbor in a simple transaction. These exchanges are for a finite period of time – sometimes for only a few hours of irrigation. Although the unit of sales may not be metered volumetrically, both buyer and seller usually have good information on the volume involved. A more permanent transaction involves the exchange of the water-use right itself. This generally requires legal sanctions to maintain the security of the right after the transfer. These transfers are generally permanent, but can be for a finite, but extended period of time – at least one irrigation season. The burden of uncertain supply falls on the purchaser of the right.

The market exchange of water and water-use rights is a relatively rare phenomenon, and thus the economic analyses of actual exchanges are scarce. A few studies have simulated the effects of market and administrative transfers in the western U.S. (Vaux and Howitt, 1984; Dinar and Letey, 1991; Whittlesey and Houston, 1984; Hamilton et al., 1989; Weinberg et al., 1993; and Rosen and Sexton, 1993). Other studies have relied on data from actual trades in order to assess the efficiency of water markets. Maass and Anderson (1978) evaluated the water market in Alicante, Spain, and found that the market system produced greater net increases in regional income than the rotation systems. Chang and Griffin (1992) estimated the gains-from-trade for transfers from agriculture to municipal water supply. Finally, Colby (1990) estimated the transaction costs required to obtain approval of water transfers and the litigation costs of third-party challenges and suggested that they are not overly burdensome.

10.2 Calculation of Gains-from-Trade and Economic Rents

In order to assess the performance of water markets and the importance of transaction costs in Chile, four river valleys, the upper Maipo in Chile's central valley, the Elqui and Limarí Valleys in north-central Chile, and the Azapa Valley in the far north of Chile were selected (see Hearne and Easter, 1995). These valleys were identified based on prior information concerning active trading of water-use rights in these areas. Attempts were made to identify transactions of water-use rights in these areas for the years 1986-93. These years were chosen because of more active water trading, because they include both wet and dry years, and because farmers were expected to be able to more accurately recall information on recent transactions. Water transactions in conjunction with land transactions were generally not considered because they usually do not involve a change in water-use patterns. Finally, since few transaction were found in the Azapa Valley and the upper Maipo, gains-from-trade were only estimated for the other two valleys.

A survey instrument was developed to solicit information from farmers who have participated in water market transactions. This information included: 1) individual characteristics of buyers and sellers, 2) water transactions, including price, quantity, and transaction costs, and 3) farmers' crop yields. In order to maintain the confidence of the respondent in the survey procedure questions on incomes and tax rates were avoided.[2] Individuals making transactions were identified from records of water user associations (WUAs), from local real estate registries, and from other transacting parties. Agronomists from the *Universidad Católica de Chile* were recruited as enumerators. Enumerators attempted to contact all individuals who had made transactions in the areas selected for analysis. Managers of urban water supply companies and industrial users of water were also interviewed (Hearne, 1995).

Budget Analysis

Farm budget analysis is used to calculate the net returns and shows the maximum that a farmer could pay for water and land and still break even. The net returns are thus an upper bound to a farmer's willingness to pay for irrigation water. Farm budgets are used for both an economic and financial analysis of water markets. Net returns are divided between land and water according to their relative market values in crop production (solicited in a farmer survey). The financial analysis is performed to determine individual incentives to participate in water markets. In contrast, the economic analysis provides the net return to society from market transfers. In an economic analysis, transfer payments are excluded and prices are adjusted for economic values.

Farmers' per hectare yields are based on survey data. In the case where farmers did not answer this question, average yields from other surveyed farmers were used. For major crops, such as table grapes, missing observations on yields were infrequent. Output price and cost of production data were taken from representative crop budgets used by the *Dirección de Riego* and the Department of Economics of the *Universidad Católica* of Chile in the financial and economic appraisal of irrigation projects in the Elqui and Limarí valleys, dated 1991 and 1992.

Because a time series of input prices for crop production was not available, relative input and output prices were assumed to remain constant. Farmers' costs for labor to clean common canals and the water fees paid to WUAs are included as production costs, because they are the cost of water delivery (WUAs are non-profit organizations, and they determine their own fees in accordance with their capital and operating costs). The labor cost for on-farm canal cleaning is included in the labor cost of the crop budgets. Water costs are allocated among crops in proportion to their relative water use as determined in the farmer survey.

The cost of debt servicing for initial crop development is included for all permanent crops. These costs are also taken from the crop budgets. Net revenues are discounted by 7 percent in order to account for management services. In order to account for the risk absorbed by a farmer, a 15 percent commercial loan rate was used for both interseasonal finance and crop development.

In Chile, agricultural taxes are fairly complex. A simple option for farmers is to pay a land tax of 10 percent of the assessed value. However, because land assessments are generally not up-to-date, the assessed values reported in the survey are probably well below the market values; therefore, reported market values were discounted by 50 percent before calculating the 10 percent land tax. The land taxes were deducted from net revenues in the financial analysis but they were not deducted in the economic analysis because taxes are transfer payments and are not a cost to society.

Value of Potable Water Supply in the Elqui Valley

The value of water-use rights in the Elqui River for potable water supply can be determined by estimating the cost savings for suppliers who can forego investing in more costly alternatives. Since the demand for water in this service area will continue to rise the least-cost alternative for meeting this demand, in the absence of effective water markets, is the appropriate methodology for assessing the value of water. Groundwater in this area is very deep and of poor quality and would not be a cost-effective alternative, except as emergency supplies during summer months.

The primary alternative for providing additional water would be the construction of the Puclaro Dam in the Elqui Valley. In fact, ESSCO, the municipal water supply company which services the valley's urbanized coastal area, opted out of a proposed joint venture with the Ministry of Public Work's *Dirección de Riego* to construct the Puclaro Dam on the Río Elqui. At a cost of $23.5 million (all values are in U.S. dollars unless designated Chilean dollars, CH), ESSCO was offered sufficient capacity in the proposed reservoir to meet its needs for secure water supplies well into the next century.[3] ESSCO's refusal of this offer coincided with its current program to purchase water-use rights on the Elqui River.

Information for the calculation of the value of delaying the construction of this water project is taken from a 1991 feasibility study of the Puclaro project (CIAPEP, 1991) and from conversations with ESSCO's planning department. According to the CIAPEP study, the demand for water in the urban coastal area is growing at a rate of 2.1 percent per annum. At this rate ESSCO's current supply of water will be insufficient to meet its needs by the year 2011 (which also is a reasonable date to have expected, in 1992, the full completion of the Puclaro Dam project). Following a procedure outlined by Moncur and Pollack (1988), the value of a water-use right is equal to the costs saved in delaying the need for introducing a more expensive water supply. The value, Φ_t, is the 1992 value of the 570,000 m^3 of water needed in 2010 to cover the increased demand for water, as shown in Equation 1.

$$\Phi_t = \frac{C_1}{e^{r(T-t)}} \qquad (1)$$

where: C_1 = CH$1,006 million (annual payment for a 35 year loan of $23.5 million with an exchange rate of CH$350 = U.S. $1.00); t = 1992, T = 2010 and r = 12 percent.

Thus, Φ_t = CH $116 million which is the 1992 value of the water needed to fulfill ESSCO's unmet water demand in 2010. One interesting feature of this calculation is that the volume of water needed in future years is not a factor in the determination of its value. This is because the Puclaro reservoir would be large enough to meet any reasonable increase in demand.

Once the value of the quantity of water needed to delay the investment in the more expensive technology is determined, it is necessary to convert this value into a per-share value. It is assumed that 2.5 shares of the Elqui River are needed to ensure each liter/sec, because in 95 out of 100 months, the lower Elqui River delivers at least 0.4 liter/sec per water-use right.[4] Thus, Φ_t is divided by 45.3 shares to reach a value of CH $2,561,220 for each of the first 45.3 shares purchased in 1992. The value of all net shares of Elqui water purchased by ESSCO in 1992-93 are listed in Table 10.1.

Table 10.1. Value of Water Rights for ESSCO's Purchases in the Elqui Valley

	Year of Purchase[a]	Number of Net Shares	Value of a Share (Chilean dollars)
1992	2010	45.3	2,561,220
1992	2011	21.9	2,204,690
1993	2011	25.0	2,475,505
1993	2012	47.7	2,157,305
1993	2013	48.5	1,881,798
1993	2014	36.6	1,643,589

[a] ESSCO purchased 67.2 shares in 1992 and 157.8 shares in 1993.
[b] Projected demand based on current (1993) growth in water use.

Many of the water-use rights transactions in the Elqui Valley involved the transfer of water from agriculture to potable water uses outside of ESSCO's service area. The most likely alternative source of crude water supplies for all of these residential and tourist developments is the same source as ESSCO's. Therefore, the avoided cost of water for ESSCO is also the value of a water-use right to others needing potable water supplies downstream of the proposed Puclaro Dam.

Economic Analysis of Gains-from-Trade

For the analysis of gains-from-trade, financial prices taken from representative crop budgets were adjusted to present economic values based on information from the 1991 CIAPEP study of the Puclaro Dam. Because Chile has a relatively open economy with low tariffs, free exchange rates, and open markets, the difference between economic and financial prices is small. For exportable goods, such as table grapes, a percentage of the exporter's commission is added to the farmgate price in order to account for this transfer payment. For imported inputs, financial prices are adjusted for both import tariffs (11 percent) and the foreign exchange premium (10 percent). An adjustment was not made to differentiate between the economic and financial costs of constructing the Puclaro Reservoir, because the adjustment factor is very close to one (0.991) and the percentage of imported goods is relatively small.

Gross gains-from-trade were calculated by subtracting the value of water to the seller before a sale from that of the buyer after a purchase, as in Equation 2.

$$GFT_t^{b,s}(w) = \{V_t^b[w(W_{t-1}^b + w)] - V_t^s[w(W_{t-1}^s - w)]\} - TC^{b,s}(w) \quad (2)$$

where:

$GFT_t^{b,s}$ = the gains-from-trade to society of a transfer in time t, of a water-use right w, from seller s, to buyer b;

$Tc^{b,s}(w)$ = the total transaction costs of a transfer of w from s to b;

$V_t^b[w(W_{t-1}^b + w)]$ = the buyer's value of w, as a percentage of the average value of total post-trade water-use rights $(W_{t-1}^b + w)$; and

$V_t^s[w(W_{t-1}^s - w)]$ = the seller's value of w, as a percentage of the average value of total post-trade water-use rights $(W_{t-1}^s - w)$.

Efforts were made to identify and interview both buyer and seller of each transaction. When only the buyer was interviewed, the value of water from sellers on the same canal or area for the same year was used to substitute for the other party. When only the seller was interviewed, the value of water to similar buyers was used.

In cases where the seller of a water-use right was known not to have used the water prior to the transaction, the right was valued at the weighted average value of water to neighboring buyers and sellers of water. This is because unused water is generally distributed to other water users along the same canal or river. This would tend to place a high value on the transferred water-use right, because the weighted average emphasizes the value of water to the very profitable larger-scale farmers who are buying water and consequently make the estimate of gains-from-trade conservative. Because of the interconnections in the Limarí Valley, average values of water for above and below the Paloma Reservoir were used in the analysis. In the Elqui Valley, water was divided among users of the Río Claro and Río Cochiguaz, users of the Río Elqui, and users of the Herradura Canal.

All values were adjusted to June, 1993, values using Chile's consumer price index. Transaction costs were obtained in the survey and include: fees for attorneys, notaries, and engineers, payments for modifications of canals or gates, the costs of soliciting information, and the value of the time involved in the process. A summary of total transaction costs for the Elqui and Limarí Valleys is given in Table 10.2. Net gains-from-trade were calculated by subtracting the transaction costs for both buyers and sellers from gross gains-from-trade.

Financial Analysis of Economic Rents

Financial prices were employed in the crop budgets to analyze individual economic rents for buyers and sellers. For this case, any water-use right not used by the seller prior to the transaction was valued at zero. The seller's net rent is the sale price, adjusted to 1993 values using Chile's consumer price index, less both the value of the water to the seller and the seller's transaction costs, as shown in Equation 3.

$$NR_t^s(w) = P_t^{W(b,s)}(w) - [V^s(w(W_{t-1}^s - w)) + TC^s(w)] \quad (3)$$

where: $P_t^{W(b,s)}$ = the transfer price of w negotiated between seller s and buyer b;

$NR_t^s(w)$ = the net rent to seller s of a sale of w to the buyer b, and

$TC^s(w)$ = seller s's transaction costs.

To the buyer, the net rent is the difference between the value of water to the buyer and the sum of the buyer's purchase price and transaction costs, as shown in Equation 4.

$$NR_t^b(w) = V_t^b(w) - [P_t^{W(b,s)}w + TC^b(w)] \tag{4}$$

The sum of rents to buyers and sellers does not equal the calculated gains-from-trade because economic gains differ from financial rents. Also, for some transactions, both buyer and seller were not interviewed and financial rents were only calculated and reported for interviewed parties.

Table 10.2. Transaction Costs of Water Sales

	Elqui Valley		Limarí Valley	
	Buyers	Sellers	Buyers	Sellers
		U.S. $		
Total Transaction Costs as a Percentage of Water Price	0.21	0.02	0.05	0.02
Costs of Attorneys, Notaries, and Obtaining Legal Inscription of Rights as Percentage of Total Transaction Costs	0.59	0.79	0.16	0.34
Costs of Engineering and Modifying Canal Infrastructure as a Percentage of Total Transaction Costs[a]	0.20	0.03	0.64	0.62
Opportunity Cost of Time Invested as a Percentage of Total Transaction Costs	0.20	0.18	0.18	0.00
Costs of Gathering Information on Buyers and Sellers as a Percentage of Total Transaction Costs	0.00	0.00	0.02	0.04

[a] Estimated expenditures for engineering services, modification of gates and canals, and indemnities for transferring water through canals. Includes expenditures through 1993 and does not include planned expenditures for modifications needed in the future.

10.3 Results

In the Elqui Valley, transactions were infrequent but there have been significant intersectoral transfers as well as a slow transfer of water-use rights within agriculture. In the Limarí Valley, with its well-developed system of irrigation infrastructure and well-organized WUAs, transactions were fairly frequent.

The Elqui Valley

The Elqui Valley in Chile's Region IV supplies water for 18,700 ha of farmland as well as potable water for a medium-sized city (250,000 inhabitants) with a very large summer population. A small tourist industry in several small communities in the scenic valley complements the large coastal tourism boom. Major crops include table grapes, pisco grapes, other fruit crops, potatoes, and pasture. One small mine in the upper reaches of the basin utilizes water from a tributary of the Elqui River. Rainfall is scarce in this region with average yearly precipitation of less than 120 mm. Wet and dry years follow a cyclical pattern, with two and three year droughts common. As discussed in Chapter 9, the Elqui River is divided into 25,000 shares where each share is supposed to deliver 1 liter/sec in a good year, but, on average, actually delivers only about half this amount.

In the Elqui Valley, a population of 47 permanent transactions for the period of 1986-1993 were identified with the assistance of WUAs in the valley. For 41 of these transactions, representing a total of 706 shares of the river, either the buyer or seller or both were interviewed. In total, 14 individuals selling 491 shares and 18 buyers purchasing 467 shares were interviewed. Some individuals were involved in more than one transaction.

Ninety percent of the shares sold in this valley had not been used in recent years by their sellers. Some of these sellers had surplus water, others had rights to water along canals that did not convey water efficiently, others had quit farming, and still other sellers owned rights to water but did not own land. In general, water that is not used by its owner is used by neighboring farmers in the same canal or section of the river, especially in years with low rainfall.

Purchase prices of shares of the Elqui River are quite variable, reflecting differences in transaction costs and the particular conditions of buyers and sellers. ESSCO's average purchase price in 1992-93 was near $1,100/ share. During the same period, small plot nonfarm developers in the lower Elqui Valley were paying $2,500/ share for water from a conveniently located canal. Exchange prices between individual irrigators ranged from $250 to $1,000/share. Often these disparities in prices reflect the individual circumstances of the seller. Transaction costs for most transfers were relatively low because this is a narrow valley with many short canals flowing directly from the Elqui River and its tributaries (see Table 10.2).

The economic analysis of gains-from-trade in the Elqui Valley shows that the average gross gains-from-trade for the 706 shares were $846/share (see Table 10.3). With average transaction costs of $56/share, net gains-from-trade are $790/share. ESSCO has purchased 292 shares, which amounts to 28 percent of its current water-use rights. The estimated net economic gains from these transfers ($658/share) are relatively modest. As discussed above, these estimates may be low because much of the water sold to ESSCO was not used by their previous owners and the procedure of using the value of water from neighboring farmers may be biased by more productive farmers.

There are some small transfers of water to developers of small residential and tourist housing along the valley and on the coast north of La Serena. The gains-from-trade from these transfers are higher than those for transfers to ESSCO. This reflects the fact that the value of water to the sellers of these shares is relatively low. The large gains-from-trade exchanges among farmers reflects the high value of water to a few profitable farmers buying water-use rights.

10.3. Economic Analysis of Gains-from-Trade in the Elqui Valley

Elqui Valley	Number of Shares	U.S. $ per share[a]
Total Gains-from-Trade	706	846
Net Gains-from-Trade	706	790
Trades with ESSCO Total Gains-from-Trade	292	675
Trades with ESSCO Net Gains-from-Trade	292	658
Other Intersectoral Trades Total Gains-from-Trade	63	1,160
Other Intersectoral Trades Net Gains-from-Trade	63	1,139
Intra-Agricultural Trades Total Gains-from-Trade	351	934
Intra-Agricultural Trades Net Gains-from-Trade	351	839

[a] The June 1993 exchange rate of CH $ 403 = U.S. $ 1.00 was used.

Table 10.4. Financial Analysis of Economic Rents to Buyers and Sellers in the Elqui Valley

Elqui Valley	Number of Buyer or Sellers	Number of Shares	Mean Net Rent per Share U.S.$[a]
All Sellers	14	491	1,156
All Buyers	18	452	3,047
Individuals Selling to ESSCO	4	292	1,071
ESSCO	1	225[b]	3,104
Other Sellers of Intersectoral Transfers	3	32	1,041
Other Buyers of Intersectoral Transfers	6	40	4,890
Other Sellers	7	167	1,327
Other Buyers	11	194	2,603

[a] The June 1993 exchange rate of CH $ 403 = U.S. $ 1.00 was used.
[b] Some of ESSCO's shares are reduced upon purchase by the *Junta De Vigilancia de Río Elqui* in order to account for conveyance losses.

The financial analysis of individual rents in the Elqui Valley shows that the net rent to 14 individuals, selling 491 shares of Elqui water, averaged $1,156/share (see Table 10.4). In contrast, the net rent to 18 buyers, purchasing 452 shares, averaged $3,047/share. In general, individuals and firms buying water-use rights received larger net rents than water sellers. Intersectoral trades produced higher net rents than trades between farmers. An important reason for these high economic rents is the fact that many of the water-use rights sold in this valley were valued at zero because they were not employed by their owners at the time of sale.

The Limarí Valley

South of the Elqui Valley in Region IV is the productive Limarí Valley, which contains the Limarí River and its tributaries the Cogotí, Grande, Guatalame, Hurtado, and Rapel Rivers. This valley has 50,000 ha of irrigated farmland producing table grapes, pisco grapes, horticultural products, basic grains, and pasture. It is a dry area with mean annual precipitation of 140 mm. There is one small city, Ovalle, with a population of 80,000 that draws water from the Limarí River. A central feature of this valley is the presence of a large interconnected system of three interseasonal storage reservoirs: Paloma (750 million m^3), Cogotí (150 million m^3), and Recoleta (100 million m^3). This storage, along with gated canals and well-organized WUAs, allows for volumetric specification of water-use rights.

In the Limarí Valley, a population of all water transactions was not developed. Instead, data collection was concentrated on areas with frequent transactions and individuals with many transactions. Attempts were made to balance interviews in these areas with other areas of less market activity. In the areas of the Río Hurtado and Río Grande which irrigate a total of 5,400 ha above the Recoleta and Paloma Dams respectively, only a few transactions were identified. All other transactions were identified in areas below the reservoirs. As discussed in Chapter 9, each WUA assigns a different amount of water to each share.[5] Because of this, all water-use rights were converted to an average volumetric denomination, based on survey results.

The low transaction costs and frequent trades in the valley can be attributed to both modern infrastructure and well-developed WUAs. Because of reservoir storage and gated canals, water is delivered to farmers on demand and water transfers are easy. Thus, the frequency of transfer is high and the market for water-use rights is active. Individuals easily separate water from land and farmers make marginal water-use decisions.

In total, 37 farmers selling water-use rights accounting for 2.0 million m^3/year were interviewed along with 19 buyers purchasing 7.2 million m^3/year. The economic analysis of gains-from-trade in the Limarí Valley reveals that the average gross gains-from-trade was $2.47 for each m^3/year transferred (see Table 10.5). After subtracting transaction costs, the average net gains-from-trade declines to $2.40 for each m^3/year. Trades that involved the transfer of water to large grape producers created the highest net returns.

The financial analysis of individual rents shows that the average net rents to 37 individuals selling water rights amounting to 1.7 million m^3/year were essentially zero (Table 10.6). This reflects the need of many farmers to sell their water-use rights to help make debt payments, even though the water had about the same value to them in crop production as the price they received. Net rents to 19 individuals purchasing water-use rights that deliver 7.5 million m^3 averaged \$1.78 m^3/year.

Table 10.5. Economic Analysis of Gains-From-Trade in the Limarí Valley

Limarí Valley	Volume m^3/year	U.S. \$/ m^3/year[a]
Total Gains-from-Trade	9,218,000	2.47
Net Gains-from-Trade	9,218,000	2.40
Total Gains-from-Trade (Purchases by Large Table Grape Producers)	5,834,000	2.85
Net Gains-from-Trade (Purchases by Large Table Grape Producers)	5,834,000	2.84
Total Gains-from-Trade (Other Purchases)	3,384,000	1.81
Net Gains-from-Trade (Other Purchases)	3,384,000	1.65

[a] The June, 1993 exchange rate of CH \$ 403 = U.S.\$1.00 was used.

Rents for buyers are generally higher than those for sellers. For example, rents to large table grape producers, purchasing 78 percent of the water-use rights, are high relative to all sellers which reflects the high value of water to these profitable farms. Much of this difference in individual rents from market transactions is due to the short-term financial stress of many sellers and the market power enjoyed by buyers. Although there are many potential buyers in the valley, frequently, for an individual canal, one or two large-scale farmers are able to determine the price on the canal. Yet transaction prices are generally well-known in this valley and little advantage is derived from asymmetric information.

Prices range from \$3,000 for a right with an average volume of 4,500 m^3/year in the table grape producing area above the Paloma Reservoir to \$500 for a share with the same volume below Paloma. As discussed in Chapter 9, this difference in price reflects both the premium placed on water in the hot, dry sunny uplands, and the prohibition on transferring water-use rights from below the Paloma Reservoir to canals above the dam. Without this prohibition it would be relatively inexpensive to pump water directly from the Paloma Reservoir to the adjacent grape producing areas irrigated by upstream canals. Also, the value of reservoir storage is illustrated by the fact that a water-use right in the Limarí Valley is worth more than a water-use right in the Elqui Valley that in some years delivers five times as much water.

Many of the permanent transfers of water-use rights in this area involve large acquisitions of both land and water by a few large table grape exporters. The land and water transactions are generally reported separately because of the mobility of water rights in this valley. Some of these transactions entail a shift away from traditional crops to higher-valued fruit crops.

Table 10.6. Financial Analysis of Economic Rents to Buyers and Sellers in the Limarí Valley

Limarí Valley	Number of Buyers or Sellers	Volume m^3/year	U.S. $/ m^3/year [a]
All Sellers	37	1,708,000	0.00
All Buyers	19	7,510,000	1.78
Sellers to Large Table Grape Producers	20	857,000	0.08
Large Table Grape Producers	3	5,834,000	2.05
Other Sellers	17	851,000	-0.08
Other Buyers	16	1,677,000	0.84

[a] The June 1993 exchange rate of CH $ 403 = U.S. $ 1.00 was used.

10.4 Observations and Conclusions

Chile's heritage of privately developed irrigation, and its traditional allocation of river water by shares, has created a favorable climate for the establishment of transferable water-use rights. These rights have secured water supply to irrigators and thus provided an enabling environment for investment in permanent fruit crops and improved irrigation technology.

The overall growth in the value of Chile's agricultural output during the last decade can be attributed to various reforms in both input and output markets. The effects of the 1981 Water Code on the agricultural sector cannot be fully separated from the effects of liberalized trade and secure land rights. But, the agricultural sector that is highly dependent on irrigation has expanded without new investments in irrigation infrastructure.

This research has demonstrated that the market transfer of water-use rights does produce substantial economic gains-from-trade in both the Elqui and Limarí Valleys in north-central Chile. These economic gains occur in intersectoral trades and in trades between farmers, and they produce rents for both buyers and sellers. Yet buyers, especially large table grape producers in the Limarí Valley and individuals buying water-use rights for potable water supply in the Elqui Valley, receive higher rents then sellers. In the Elqui Valley, total and net gains-from-trade per share were within the range of recent transfer prices of $1,000/ share. In the Limarí Valley, gains-from-trade from shares of water-use rights were 3.4 times the transaction prices of $3,000 for a share of water delivered from the Cogotí Reservoir (one share delivers 4,250 m^3 in an average year).

One of the most interesting results of this analysis is the relatively modest economic gains from intersectoral trade in the Elqui Valley. Although, the value of water in municipal water supply is high, the value of water to profitable farmers is also high. When water is transferred away from these profitable farmers, the economic gains from this reallocation are small. Even if water is not used by its owner, it is generally used by other farmers. If these farmers are profitable, then the economic gains of the reallocation are small, even though the financial gain to the seller is large.

In two of the four areas studied, market transactions were too infrequent to permit the estimation of gains-from-trade. Despite the fact that these valleys were chosen for analysis because of expectations that they had relatively active local water markets, only the Limarí Valley had active trading. In this valley, transaction costs are low and trade between farmers is facilitated by the presence of reservoir storage, gated canals with flow meters, and well-organized WUAs. In the Elqui Valley, the presence of many short canals flowing from the river also reduce the costs of physically changing the flow of water.

In large canal systems with fixed flow dividers, the cost of changing stream flows might be prohibitive, especially for trades among farmers, as suggested by the scarcity of trading in the Canal Bellavista system in the Elqui Valley and the large canal systems in the upper Maipo Valley. Sales of marginal amounts of surplus water are far from typical. Many investment and crop selection decisions have been based on the joint land/water allocation that resulted from the final stages of the land reform.[6]

The proposed Puclaro Dam project, on the Elqui River upstream of La Serena, is an example of how the presence of a market alternative to water allocation may reduce political pressure to invest in large water storage projects. This project has been proposed in different forms since 1956. As recent as 1989, potable water was considered to be the most important benefit of this dam. But when ESSCO did not agree to collaborate with the Directorate of Irrigation in paying for part of the construction costs, the political importance of the dam declined.

This research demonstrated the economic and financial gains from using markets to reallocate water in the Elqui Valley instead of investing in a large water storage project. It also showed how public investments in water storage and delivery systems in the Limarí Valley have the external benefit of reducing the transactions costs involved in market trading. In the presence of a system of transferable water-use rights, further analysis of large water storage projects should consider the value of storage and control gates in facilitating the reallocation of water through markets.

Acknowledgments

Much of this information was obtained during a six-month mission to Chile during 1993 and 1994 with funding provided by the World Bank's Agricultural Policies Division. See the end of Chapter 9 for the full acknowledgments.

Endnotes

[1] This chapter draws heavily on Hearne and Easter, 1997.

[2] In general, respondents had no incentive to bias responses. Nevertheless, confidentiality of individual survey responses was promised and respected. Because fees paid to notary republics are based on transaction prices, there might be an incentive for buyers to report low transaction prices at the registration of the title. However, transaction prices were often public knowledge and many buyers wanted transaction prices to be known in order to attract more sellers.

[3] Conversation with the *Dirección de Riego*, Santiago, June 1992. The *Dirección de Riego* also offered reservoir water at a lower price but without the security of delivery that ESSCO is required to have. Municipal water supply companies are, by law, required to have a 95 percent security in water delivery.

[4] The estimate of 2.5 shares/ liter/sec is taken from a review of the *Junta De Vigilancia de Río Elqui's* records of water flow in the river and is consistent with ESSCO's stated

plan to purchase 1,300 additional shares of the Elqui River by 2020.

[5] Because of the volumetric denomination, it is dangerous to directly compare the volume of water rights in the Elqui River to those of the Limarí Valley. In the former, 15,750 m^3 are delivered in an average year, but without regard to the time of delivery. In the latter, water is delivered to farmers on demand at specified times.

[6] During the final stages of land reform, land expropriated prior to 1974 was divided into family farms equivalent in value to 8 irrigated ha and water rights traditionally assigned to the land were distributed to the land owners.

References

Chang, C., and R.C. Griffin, 1992. "Water Marketing as a Reallocative Institution in Texas." *Water Resources Research*, 28(March):879-890.

CIAPEP (*Curso Interamericano en Preparación Y Evaluación de Proyectos*), 1991. *Embalse Puclaro. Instituto de Economía. Pontificia Universidad Católica de Chile*, Santiago.

Colby, B.G., 1990. "Transactions Costs and Efficiency in Western Water Allocation." *American Journal of Agricultural Economics,* 72(5):1184-1192.

Dinar, A., and J. Letey, 1991. "Agricultural Water Marketing, Allocative Efficiency, and Drainage Reduction." *Journal of Environmental Economics and Management,* 20:210-223.

Hamilton, J.R., N. Whittlesey, and P. Halverson, 1989. "Interruptible Water Markets in the Pacific Northwest." *American Journal of Agricultural Economics*, 71:63-75.

Hearne, R.R., 1995. "The Market Allocation of Natural Resources: Transactions of Water-Use Rights in Chile." Ph.D. Dissertation. University of Minnesota.

Hearne, R.R., and K. William Easter, 1997. "The Economic and Financial Gains from Water Markets in Chile," *Agricultural Economics*, 15:187-199.

Maass, A., and R.L. Anderson, 1978. *...and the Desert Shall Rejoice: Conflict, Growth and Justice in Arid Environments.* MIT Press. Cambridge, MA.

Moncur, J.E.T. and R.L. Pollack, 1988. "Scarcity Rents for Water: A Valuation and Pricing Model." *Land Economics,* 64:62-72.

Rosen, M., and R. Sexton, 1993. "Irrigation Districts and Water Markets: An Application of Cooperative Decision-Making Theory." *Land Economics,* 69:39-53.

Vaux, H.J., and R.E Howitt, 1984. "Managing Water Scarcity: An Evaluation of Interregional Transfers." *Water Resources Research,* 20:785-792.

Weinberg, M., C. Kling, and J. Wilen, 1993. "Water Markets and Water Quality." *American Journal of Agricultural Economics,* 75(2):278-91.

Whittlesey, N. and J. Houston, 1984. "Water Markets for Stream Flow Augmentation." in *A Critical Assessment of Forecasting in Western Water Management*, J. Cassidy and D. Lettenmair (eds.), American Water Resources Association, p. 139-146.

11 Opportunities and Constraints to Improved Water Markets in Mexico[1]

Robert R. Hearne

11.1 Introduction

Throughout the world, the growing demands for potable water, irrigated agriculture, and environmental services is putting increasing pressure on finite supplies of freshwater resources (World Bank, 1993). Because of increasing demand and finite supplies, new methodologies and new systems for more efficient and equitable distribution of water between competing users need to be developed. Traditionally, centralized, state managed, command-and-control systems have been employed to ensure equitable distribution of water and to provide subsidized water delivery services to farms and cities. But poor state management, increasing fiscal pressures on central governments, increasing concern about the environmental effects of large catchment and irrigation systems, and the continual growth of urban populations have led certain governments and agencies to rethink the role of government in water resources management. A new paradigm of decentralized management, user control of water delivery services, transferable water-use rights, and water markets has emerged. Often this paradigm remains as an idea propelled by economists and policy advisors, with little analysis of how this paradigm has functioned in the field (Easter and Hearne, 1995; Rosegrant and Gazmuri Schleyer, 1994; Rosegrant and Binswanger, 1994).

In 1992, the Government of Mexico initiated a new federal water law which decentralized water resources management and allowed the market transfer of water-use concessions between individual irrigators. This new water law coincided with a series of policy reforms which started in the late 1980s and included the privatization of communal land holdings, *ejidos*, the transfer of the operation of canal systems to water user associations (WUAs), the revision of the role of the National Water Commission (CNA), and liberalized trade. These reforms were expected to improve water resources management through greater user participation in irrigation management, as well as to increase irrigators' incentives to improve water-use efficiency.

At the time of its proposal, the 1992 Federal Water Law was considered to be the first step in the establishment of formal water markets (Easter and Hearne, 1995; Rosegrant and Gazmuri Schleyer, 1994). Mexico's system of transferable water-use concessions was compared to the system of water-use rights in Chile where limited transactions have been shown to produce economic gains and to reduce the need for the construction of a new dam and reservoir (Hearne and Easter, 1995 and 1997). However, Mexico's 1992 Water Law maintains a central role for the CNA in the management of water resources and the regulation of transfers of water-use concessions. Mexico's long tradition of national control over water resources may have impeded the development of the institutions, such as WUAs, that can facilitate the transfer of water.

This chapter addresses the opportunities and constraints to improved water resource use and allocation through the market incentives that result from transferable water-use permits. The second section reviews water allocation institutions in Mexico. This is followed by three case studies of water allocation that illustrate how allocation decisions were made in two regions of northern Mexico. The chapter concludes with a summary and policy recommendations.

11.2 Water Institutions in Mexico

National control of both land and water has been a key feature of the Mexican landscape since before independence. Indeed the Papal Bull of 1493 awarded both the land and water of most of the hemisphere to the Catholic King and Queen of Spain. The land reforms which followed the 1910 Mexican Revolution were the apex of this state control. This land reform redistributed land to the peasants in the form of state-owned *ejidos*. Correspondingly, all water resources were nationalized. The national character of the current CNA can be explained by the importance that the framers of the 1917 Constitution placed on strengthening a national program of land reform with irrigation development (Roemer, 1994; Cummings and Nercessiantz, 1994).

The 1917 Constitution, which is still in effect, stresses that the nation is the original proprietor of land and water and that rights or concessions to use natural resources may be granted to private individuals. Furthermore, all private property is subject to appropriation by the nation for reasons of conservation and improved distribution of wealth. The Mexican nation's original control of natural resources is in many senses similar to "eminent domain." However, the nation's direct ownership of subsoil resources and water is emphasized directly in the 1917 Constitution (Roemer, 1994).

Food production requirements, assistance to the new system of *ejidos*, and the national desire to settle the extensive unpopulated areas of northern Mexico led to a national role in irrigation development. Before the 1910 revolution the total amount of irrigated area in Mexico was close to 700,000 ha. Currently, there are 6 million ha of irrigated land in Mexico. More than half of this area, 3.2 million ha, is in 80 nationally developed irrigation districts, which vary in size from 3,000 to 270,000 ha.[2] In virtually all of these systems, large parts of the newly irrigated land was reserved for *ejidos*.

These irrigation systems have enjoyed the paternalistic support that the government has granted the *ejido* system. Not only were capital costs of irrigation construction never recovered from farmers, operation and maintenance expenses (O&M) were heavily subsidized (Cummings and Nercessiantz, 1994). Irrigation development complemented a more general government intervention in the agricultural sector. The government and government-supported parastatals guaranteed producer prices, subsidized fertilizer, agrochemicals, and seed, supported farm credit and crop insurance, and controlled imports.

Most of the large irrigation systems were constructed during the period from the 1930s to the 1970s by the CNA's predecessors, the National Irrigation Commission and the Ministry of Water Resources (SRH). These institutions were mainly dedicated to the construction of irrigation systems. In the 1970s, the SRH reaffirmed central control of water with an extensive program of national water planning that culminated in the 1975 National Water Plan. Later, after the SRH merged with the Ministry of Agriculture to form the Ministry of Agriculture and Water Resources (SARH), the CNA was established in order to reinforce water policies and strategies (SEMARNAP, 1995).

The CNA was later included as a semi-autonomous part of the Ministry of the Environment, Natural Resources, and Fish (SEMARNAP, 1995).

In all of its forms, the CNA has maintained a dominant, centralized role in Mexican water resources management. Although it has accepted overall responsibility for the planning and management of water resources, the CNA has retained an orientation toward construction, irrigation development, and support for land reform (Cummings and Nercessiantz, 1994). Since 1980, the provision of potable water and sewage services have been the responsibility of state and municipal governments. However, CNA is responsible for delivering bulk water supplies to local potable water services.

The 1972 National Water Law continued to stress the federal government's control over water resources. Water was allocated to individual users through concessions[3] which were to last for 50 years. These concessions were awarded under a system of priorities, with domestic use receiving the top priority followed by agriculture and then all other users. Water-use concessions were not transferable and were defined by both type of use and the land on which the water was used.

The construction, operation, and maintenance of irrigation infrastructure has always been a federal responsibility and, in contrast to the large, federally-sponsored reclamation and irrigation projects in the U.S., irrigation projects in Mexico are considered to be single-purpose projects with no financial contribution from other sectors and users. Consequently, financing for both the construction and maintenance of irrigation systems was severely limited. Traditionally, both irrigators and the government contributed to operation and maintenance (O&M) expenditures, yet these expenditures were not adequate. In the early 1960s, farmers' contribution to O&M expenditures was over 60 percent of the total outlays. But over time, both the government and the irrigators reduced their contributions. By the end of the 1980s, irrigators paid only 18 percent of O&M expenditures (Roemer, 1994; Cummings and Nercessiantz, 1994; Gorriz et al., 1995). Because of the limited expenditures on O&M, canal systems in the irrigation districts fell into disrepair, with water delivery unresponsive to farmers' needs. Overall conveyance efficiency was reduced to 30 percent, about half as much as irrigation systems in California and Arizona (Gorriz et al, 1995).

The 1992 Mexican Federal Water Law was designed to impose market-based incentives on water resources management and to decentralize much of the management of Mexico's irrigation districts. This new law coincided with a new 1992 Agrarian Law and with the 1993 North American Free Trade Agreement (NAFTA). Together, the effect of these reforms is a sharp contrast in the incentives facing farmers. Most importantly, the 1992 Agrarian Law allowed greater freedom for farmers in *ejidos* or *ejiditarios* to sell, rent, sharecrop, or mortgage their allocated parcels of land. This law in effect declared an end to the redistribution of land, while allowing *ejiditarios* the benefits of private property. Furthermore, the Ministry of Agriculture began to gradually eliminate subsidies to agricultural inputs and price supports for all agricultural commodities except for maize and beans.

Under the 1992 Mexican Federal Water Law, water is national property. Private transferable water-use concessions are granted to individuals, WUAs, and incorporated firms for a period of up 50 years. Although concessions are renewable, the CNA has to approve these renewals. Water-use concessions are volumetric and based on consumptive use. In the irrigation districts, the CNA allocates concessions to the WUAs organized at the level of irrigation units, or *módulos*, which in turn allocate concessions among the users according to their own procedures. The initial allocation of water-use concessions is based on the historic use of the water. The federal government is authorized to create a restriction on water use in order to ration water during droughts, prevent the overexploitation of an aquifer, preserve water quality, restore an ecosystem, and protect sources of potable water.

The title of concession is granted by the CNA. This title contains the legal foundations for its expedition, the name and home address of the concessionaire, the location for the extraction of water, the volume, the initial projected use, the place for discharge of return flows, with the necessary specifications for volume and quality, the duration of the concession, and the obligations and rights of both the CNA and the users.

In times of scarcity, the water required to meet volumetric water concessions is not available. There is no system of prioritized volumetric rights such as those found in the western U. S. nor a tradition of proportional reduction such as the one in Chile. However, the bylaws of each irrigation district should specify a rationing system to distribute water when volumetric requirements cannot be met. When these bylaws are respected and enforced, growers have some security that a certain percentage of their water concession will be available.

The 1992 Water Law obligates the CNA to create a Public Registry of Water Rights (REPDA) of all concession holders. All the titles of concession should be registered with the REPDA. This allows the CNA to control the assigned volumes, as well as to record the information needed to grant future concessions. The REPDA also certifies public and juridical acts of registration, extension, suspension, termination and transmission of water rights, as well as of the permits for sewage waters utilization. This registration of water-use concessions allows any individual access to information on the allocation of water. Ideally, it guarantees and gives legal validity to registered concessions, facilitates water resource planning and programming, and is a readily available instrument for water users who may want to defend their rights during conflicts.

The 1992 Water Law also creates a system of water fees to be paid by the owner of the concession according to the use of the water. This water fee is assessed for both water extracted and for quantity and quality of discharges. These water fees are designed to support the activities of the CNA. However, the fee for irrigation water has been zero.[4] If water fees are not paid in three consecutive years, then the CNA can declare that the water is not used and can rescind the concession. Municipal and state government authorities responsible for potable water and sewage services are required to pay fees for water delivery and wastewater discharges.

Although the registration of water-use rights should increase the security of water delivery, individuals who register their rights are subject to the water-use fee. Given the difficulty in monitoring extraction of water, especially from groundwater sources, the registration of concessions has been slow. According to the CNA's 1994 annual report, the total identified number of water users having some "right" on water volumes was 206,500 but the number of registered users was only 26,375. The volume of registered water as a percentage of total identified water use is shown in Table 11.1.

The 1992 Water Law stipulates that the responsibility for the management of irrigation districts should be accepted by the users. As of the end of 1994, 76 percent of the land area irrigated in Mexico's irrigation districts had been fully or partially transferred to user control. Primary catchment and main delivery canals remain under the control of CNA. The law also stipulates the conditions under which the transfer of management of irrigation systems to the users will occur. Further, CNA procedures for the transfer of water have been adopted. Initially, the large irrigation districts are divided up into *módulos*. The determination of the final size and shape of these *módulos* is made by CNA officials with consideration of irrigation infrastructure, community, and economies of scale (Gorriz et al., 1995). The irrigators who wish to form a WUA need to establish a not-for-profit organization, or *asociación civil.* Although each WUA is governed by a General Assembly and Executive Council, the structure and organization of these groups is greatly complicated by the presence of *ejiditarios* and small private farmers in the same system. Because of this complication, many different representation systems have been established in different *módulos*.[5]

Table 11.1. Registration of Water-Use Concessions

Use of Water	Registered Volumes as Percent of Total Volumes
Irrigation	55 %
Urban	65
Industrial	30
Hydropower	90
Sewage	50

The Executive Council of the *módulo,* the local CNA engineer controlling the Irrigation District's storage system and headworks, and the Irrigation District's Hydraulic Committee all play important roles in the determination of a farmer's water delivery schedule and crop selection. The Hydraulic Committee is, according to law, comprised of representatives of all the WUAs in the district. However, local CNA engineers and agricultural extension officers may also be considered part of the hydraulic committee. In each district, the availability of irrigation water is estimated by the CNA at the beginning of the cropping season. This information is made available to farmers who are then required to submit a cropping plan to the WUA. Based on farmers' cropping plans, the WUA submits an irrigation plan to the district's Hydraulic Committee and to the CNA which can minimize conveyance losses by delivering water in bulk under a coordinated schedule.

Most irrigation districts use a rotation system to deliver water according to a prearranged schedule. This is because these systems were initially designed in the 1940s and 1950s to irrigate grains. However, many WUAs have introduced an arranged demand system that allows farmers to place daily requests for irrigation water. This "modernization" process requires not only investments in canal infrastructure and gates, but also changes in farmers' irrigation techniques.

As long as the terms of the concession are not changed, transfers of water only require the notification of the Public Registry of Water Rights. In the event that a transfer of a concession affects a third party, authorization is required from the CNA. Transfers of a concession outside the district requires the approval of the general assembly of the WUA as well as authorization from the CNA. The benefits of a transfer of water outside of the district are reserved for the district, not for the water user.

11.3 Case Studies

The San Juan River - Monterrey Area

In a 1952 agreement, water from the San Juan River, which originates in Nuevo Leon and flows north through the Marte R. Gomez Dam in Tamaulipas, was allocated to irrigate 76,951 ha in the Lower San Juan Irrigation District Number 26 in the Río Bravo area of Tamaulipas, near the U.S. border. This is a coastal, semitropical area with maximum temperatures reaching 35°C , average annual rainfall of 370 mm, and annual evaporation of 2,031 mm. The district adjoins a string of small, industrial border cities, centering on Reynosa with a 1990 population of 265,000.

In this district, 13,231 ha belong to the "social" sector of 1,407 *ejiditarios* and 63,720 ha belong to 3,535 private owners. The average *ejido* plot is 9.40 ha while the private farms have an average size of 18.02 ha. Traditionally, the major crops in this district are: 1) maize, with 50 percent of the cultivated area, 2) sorghum, with 20 percent of cultivated land, and 3) cotton, with 30 percent of cropped area. This district had a fairly secure water supply with water from the Marte R. Gomez Dam which has a storage capacity of 932 million m^3, some additional water from the Río Bravo/Río Grande, and a canal network of over 1,100 km.

During 1993, the management of the irrigation district was transferred to thirteen WUAs. These *módulos* range from 1,974 ha to 9,269 ha. Water fees to recover the costs of operation and maintenance of the canal system range from $31.45/ha/season to $17.90/ha/season with expected delivery of 3,500 m^3/ha. Estimated cost recovery during the 1994-95 irrigation season was 87 percent but dropped to less than 50 percent during the 1995-96 because of water shortages.

The district's security of water supply was drastically reduced when the 1952 agreement on the use of the San Juan River was abridged with the construction of the El Cuchillo Dam in Nuevo Leon that was never filled. This new project resulted from an agreement which was signed on May, 1990, "in order to satisfy urban and industrial water demands of the city of Monterrey, State of Nuevo Leon, and to preserve those for multiple uses at Irrigation District 26 in the State of Tamaulipas." The dam, which has a storage capacity of 1,300 million m^3, will also supply irrigation water to the small Irrigation District 31 "Las Lajas." The city of Monterrey is an important industrial center with a growing population of over 3,000,000. Fifty percent of the population are recent immigrants. The agreement, which was signed by representatives of the Federal Government as well as the Governors of Nuevo Leon and Tamaulipas, stipulates that 60 percent of the water that is to be diverted from El Cuchillo is to be replaced by treated wastewater. This wastewater will flow from Monterrey, via the Pesqueria River, to Marte R. Gomez Dam and to the irrigators of District 26. The CNA in Nuevo Leon will operate the reservoir to suit the needs of the city of Monterrey. CNA will pass water to irrigators in Tamaulipas, but only to suit its own needs.[6]

The completion of El Cuchillo Dam in October, 1994, corresponded with the initiation of a prolonged drought which has postponed any discharge from the reservoir. Furthermore, the wastewater treatment plants, which were to be the source of water to

Irrigation District 26, had not yet been completed in April, 1996. Current wastewater flows from Monterrey are being diverted from the Pesqueria River to other irrigated land. Thus, irrigators in the lower San Juan have operated with a severely curtailed water supply. Given the powers that the CNA had been granted in the 1992 Federal Water Law, this reduction of water flows to the irrigators is not extraordinary for a developing country, but it does reduce the security that is provided by a legal water concession.[7]

In *Módulo III-3* of this Irrigation District, water rationing during droughts follows a traditional system. In wet years, water is divided according to the land area irrigated. In dry years, water is divided equally among users by equivalent volume, independent of the land area which they irrigate. Therefore, under drought conditions, a farmer with 10 ha would receive the same volume of water as the owner of 100 ha. But because of the limits of conveying and metering water in the canal system, water is allocated by irrigation turn, during which all the canals in an area are flooded. So in a dry year, the owner of 100 ha would receive one irrigation turn for the whole season and the owner of 10 ha would receive the equivalent volume of water but in two irrigation turns. During an irrigation turn, canals have constant flows of water and irrigation laborers work around the clock.

Since water is not available to the farmer on demand, but according to a schedule, crop selection is not made independently by individual farmers. The Executive Committee of the *módulo* and the Hydraulic Committee of the Irrigation District, which includes farmers, government extension agents, and CNA officials, set crop selection priorities and schedule water deliveries. As a consequence, with the exception of a small stand of oranges, only sorghum was planted during 1996 in this *módulo.*

Because the Irrigation District is located in Mexico's industrialized northern border region, farmers have alternative employment opportunities. It is estimated that within the Irrigation District more than 35 percent of land and water has been leased by *ejiditarios* to small property owners. The annual cost of leasing both land and water varies between $115 and $160/ha. This can be compared to a market value for land which ranges from $1,000 to $3,000/ha. Although the growing industrial and residential areas along the border traditionally receive their water supply from the lower San Juan district's canals, no transactions from irrigators to other sectors have been recorded. Because of the traditional priority of municipal use over agricultural use, the residential and industrial users have security in their supply of water and, under present conditions, do not need to consider transfers of water concessions from agricultural users.[8]

The situation in the Irrigation District 26 corresponds well with that of District 25, located on the lower Río Bravo, farther east and downstream from 26. This district receives water from a series of large international dams (Amistad and Falcone) with 45 percent of the releases going to Mexico. Because of the drought, the water traditionally used for irrigation is being diverted for urban and industrial uses in Reynosa, Río Bravo, and Matamoros, which have a large concentration of processing and assembly plants. During the 1995-96 season, irrigators in District 25 were promised no water at all for a surface area of 203,250 ha. After planting more than 140,000 ha of sorghum, these farmers received additional water and were able to irrigate 80,000 ha during the winter season.

Immediately adjacent to farmers in District 25 and District 26 are farmers in Texas that share the same climate and the same international water sources. On the U.S. side of the river there are also water shortages. But a well-documented process of proportional reductions of water rights and water markets has provided security of water delivery to Texas farmers (Chang and Griffin, 1992). Also in Texas, pressurized irrigation is common with less dependence on large canal systems than Mexico. South of the border farmers grow traditional grains and, during droughts, mostly sorghum. In the Texas Rio Grande Valley, high-valued crops such as fruits, vegetables, and cotton are irrigated. Currently, agricultural markets are not fully integrated, but under the NAFTA agreement Mexican farmers should enjoy the same market access as U.S. farmers.

The Lagunera Region

The Irrigation District 17, the Lagunera, is located in an interior basin in Mexico's north-central region within the states of Durango and Coahuila. This is a semiarid region with average annual precipitation of 277 mm, average evaporation of 1,929 mm, and maximum temperatures of 44°C from June through September. The district surrounds the industrial cities of Torreon in Coahuila, and Gomez Palacio and Ciudad Lerdo in Durango with a total population of over 1,200,000.

The district has a total of 223,674 ha but only 112,696 of these are irrigable. This land is divided among 35,084 *ejiditarios* and 2,734 private owners. The average *ejido* farm has 2.34 ha of irrigated land and another 2 ha of dryland. Similarly, the average privately-owned farm has 11.13 ha of irrigated land and some other dryland.

This district is divided into seventeen modules that are irrigated by the Nazas River and three more modules irrigated by the Aguanaval River. Major crops are cotton, alfalfa, beans, sorghum, walnuts, and maize. Recent trends show an increase in the cultivation of higher valued crops such as melons, grapes, alfalfa, and watermelons.[9] Water to irrigate this district comes from the Lazaro Cardenas Dam with a capacity of 2,779 million m^3 and from the Francisco Zarco Dam with a storage capacity of 368 million m^3. Annual surface flows from these sources averaged 1,348 million m^3 for the four years prior to the 1995-96 drought. There are over 2,500 tubewells in the irrigation district and additional tubewells to supply water for residential and industrial uses. Total withdrawals of groundwater is estimated at 600 million m^3 per year.

Of the seventeen *módulos*, management responsibility has been partially transferred to WUAs in nine of these. Water fees for O&M expenses range from $15/ha/season to $27.14/ha/season, with expected volumes of 5,000 m^3/ha. Water is distributed on a rotation basis with four irrigation turns scheduled during the 1995-96 drought year.

Alfalfa remains an important crop because of the large dairy industry in the region which has an annual output worth an estimated $95 million. However, the CNA does not allow the use of surface water to irrigate alfalfa, even when reservoirs are full. Given the high returns to alfalfa, CNA feels that farmers can afford the cost of pumping groundwater for this purpose. Groundwater is used to irrigate alfalfa as well as to provide for residential and industrial water supply. Groundwater levels have fallen at an annual rate of between 0.5 to 1 meters, which implies that the ban on surface irrigation of alfalfa will eventually threaten the water supply for the Torreon urban area.

During the 1995-1996 season, a drought period, water availability limited irrigation to only 26,000 ha. This water was divided equally between the two states, with 93 percent assigned to *ejiditarios* and 7 percent to private owners. To minimize conveyance losses, only a few *módulos* close to the main canals were to be irrigated.

Those who own land outside this area could lease their water-use concessions to farmers in *módulos* that were to be irrigated or they could "request" the use of land in irrigated areas and move their water-use concessions to this land.[10] Land and water are leased together, but the water is considered to be the valuable resource. During this season, the cost of leasing water increased from a range of $4.00 to $5.10/thousand m^3 ($57-71/ha) in an "average" year to $10.20/thousand m^3 ($143/ha). This can be compared to land values ranging from $1,857 to 2,143/ha. During this year, more than 140 million m^3 of water was leased in this fashion, usually from *ejiditarios* to private owners. This corresponds to 35 percent of the *ejiditario* land that was allocated irrigation water during this season.

Furthermore, in an emergency measure, the CNA, in consultation with state governors, other political leaders, and WUAs, has mandated that only cotton will be irrigated with canal water during the 1995-1996 season. This is said to be a political decision made to maintain the employment of agricultural labor, since cotton is a labor-intensive activity.[11] In mandating the crop selection for an entire surface irrigation system, the CNA is applying a liberal interpretation of its substantial powers to restrict water use in times of scarcity. Although this is not a unilateral decision, and certainly not taken lightly, it does reinforce centralized control over what in other circumstances would be a decision made by individual farmers. Thus, the CNA is reverting to its historic role of overseer of irrigation systems and sponsor of irrigation based *ejidos*. It also defies the long-term consequences of destabilizing the production of alternative crops, such as melons, that may require stable marketing contracts, and perennial crops, such as grapes and walnuts that may be severely damaged by lack of irrigation water.

The Mexicali Valley

The lower Colorado River enters Mexico with an annual flow of 1,850 million m^3. It then irrigates the Río Colorado Irrigation District 14, which is located in the Mexicali Valley in the states of Sonora and Baja California. This is an arid area with average annual precipitation of 59 mm, average annual evaporation of 2,251 mm, and maximum temperatures reaching 45°C during July through September. The district borders the city of Mexicali, with a large concentration of industry and a 1990 population of 438,000.

The Irrigation District includes 136,600 ha of surface irrigation and 71,365 ha which are irrigated by 1,100 million m^3/year pumped from 725 deep tubewells (236 private wells irrigate 18,271 ha, while 489 "federal" wells irrigate 53,100 ha). The irrigators include 7,067 *ejiditarios*, with an average farm size of 17.9 ha, and 7,627 private owners, with an average farm size of 14.2 ha.[12] The Mexicali Valley's main crop has been cotton, except from 1992-94 when they shifted to wheat because of low cotton prices. Although the fiber's quality is the best in Mexico, water requirements reach 13,700 m^3/ha, reducing farmers' profits. Light-textured soils, high salinity levels, and poor irrigation technology are considered reasons for this high level of water use.

Salinity has been a critical issue in the lower Colorado River Valley and in the Mexicali Valley. Concern for salinity has been complicated by a dispute between the U.S. and Mexico regarding salinity monitoring. A 1973 accord regulated the salinity levels of the transboundary water flows but this water was to be monitored at the Imperial Diversion Dam located 40 miles north of the border and upstream from Yuma

County's discharges. Thus the water that actually reaches Mexico has saline levels that sometimes makes it unsuitable for most agriculture.

The management of Irrigation District 14 has been transferred to 22 *módulos*, ranging in size from 13,260 ha to 4,758 ha during the period from 1993 to 1995. However, in 1996, water fees were not completely "negotiated" with the irrigators. There are three types of fees: 1) water service fees for surface and groundwater, which average \$3.64/thousand m^3, 2) an annual "rehabilitation" fee, which averages \$ 7.50/ha, and 3) a "soil" use fee for private wells, which averages \$2.03/ha/year. Estimated cost recovery for O&M during the 1994-1995 season was 61.1 percent. So far, these fees do not cover the cost of the maintenance and repair of the system's deep tubewells. The canal system includes: 470 kilometers of main canals, of which 74 percent is concrete-lined; 2,432 kilometers of secondary canals, of which 77 percent is concrete-lined; and 1,662 kilometers of open ditch drainage.

In addition to irrigation water, the district delivers 100 million m^3 of water annually to the borderland cities and urban centers in northwestern Sonora and northern Baja California, including Mexicali and Tijuana. This water does not form part of the 1,850 million m^3 allocated to the district from the Colorado River but is pumped from 67 deep wells located to the east of Irrigation District 14 at the Mesa de San Luis. Annual water yield from these wells is not to exceed 197 million m^3 annually. But by the year 2000, annual water requirements in these urban areas are expected to reach 270 million m^3. The projected future growth of Tijuana and the border area cities provides Mexico with a powerful motivation to initiate water transfers from agriculture to urban uses. The scarcity of alternative water sources and the confined land area leaves the CNA with little opportunity for further catchment and delivery system development.

Due to its unique mix of surface and groundwater, as well as private and federal tubewells, a market for water-use concessions is relatively well developed in the Mexicali Valley. Indeed, water-use rights have been sold and leased without the accompanying piece of land, an activity that is quite rare in Mexico. In 1994-95, prices for annual leases have ranged from \$100/ha to \$214/ha or a price per 10,000 m^3 of \$70 to \$150. Recorded leases during this season show 13,749 ha leased to "enterprises" and 46,702 ha leased to farmers. This amounts to an estimated 857.5 million m^3 of water transferred during the annual period.

11.4 Conclusions and Observations

In many respects the 1992 Mexican National Water Law provides an excellent opportunity for market forces to provide incentives for an improved and more flexible allocation of water. By giving user groups the opportunity to manage their own canal systems, the law allows for greater accountability of irrigation services to the farmers. In addition, by giving individual users the freedom to buy, sell, or lease water-use concessions, the law provides a mechanism for water to move to more productive uses, while maintaining water security. The law also provides a balance between market forces and state control. The continued role of the federal government in water resources management is ensured through the CNA and through the federal control of water.

The 1992 Water Law does provide the CNA with a many discretionary powers, in regulating transfers of concessions, restricting the use of water resources, renewing concessions, and in determining the parameters for protecting water quality. Indeed the

CNA's powers to restrict water use as written in Title Five of the Water Law are quite formidable. Of course, the role of the CNA can easily be justified, especially in times of emergency water shortages. But, the benefits of secure supplies of water to concession holders can be diminished if these discretionary powers are not utilized with moderation and consistency.

Three case studies presented an early picture of the implementation of the 1992 Water Law in three northern valleys, with special attention to water markets. Because two of these case studies featured water allocation problems during the third year of a three-year drought, it is very easy to criticize water managers. However, it is appropriate to assess water management during times of water scarcity. In both the Lagunera and the Río San Juan Irrigation Districts, engineering solutions to water scarcity problems were provided at the expense of allowing individual actors with secure water-use rights to negotiate solutions among themselves.

In the Lagunera, the CNA chose to minimize conveyance losses in secondary canals and limit irrigation to only a few *módulos* near the main canals. Furthermore, cotton was selected as the only crop to be irrigated by the entire surface irrigation system. The avoided productivity losses from conveyance losses may compensate for the cost of farmers moving from their own unirrigated land to the land that CNA chooses to irrigate. But the hidden cost of lost opportunities to develop alternative crops may be quite high and not fully considered during the centralized decision making process.

The diversion of water from the Río San Juan Irrigation District to the city of Monterrey was initiated before the 1992 Water Law. This diversion is an important contribution to the growth of Monterrey and the economy of northern Mexico. But the reluctance of the negotiators to fully compensate irrigators with either the same quantity of water as that diverted or improvements in their irrigation infrastructure demonstrates that water-use concessions can be quite tentative. The future implementation of a system of wastewater reuse is a positive feature of this diversion, especially since wastewater reuse would be a difficult solution for independent actors to negotiate among themselves. However, the precedent that water can be diverted away from irrigation systems without full compensation threatens the security of water supply to all irrigators in Mexico.

One of the sources of insecure water supplies is the volumetric definition of water-use concessions. Unless a system of proportional reductions or well-defined priorities is in place, volumetric specifications are almost meaningless in times of water scarcity. Such a system may be in place when irrigation districts have firm bylaws on the allocation of water during droughts. But these systems need to be respected as they are in the U.S. and in Chile where systems of timed priorities and proportional reductions provide the owners of water-use rights the security that their share of water will be available. In Mexico, water scarcity allows the CNA to impose emergency command and control regulation that effectively eliminates the security of a water-use concession.

In order for water markets to effectively reallocate marginal amounts of water, systems that properly measure and divide flows need to be available. In irrigation systems that rely on a strict rotation system, transfers of water present a particular challenge. In Mexican systems where irrigation turns are not strictly measured by periods of time, the simple trades of hours of irrigation that are practiced in South Asian *warabundi* systems cannot be implemented. Furthermore, in irrigation systems where conveyance losses require water managers to severely restrict the land area to be irrigated in times of drought, the free movement of water is severely restricted.

Therefore, in order for Mexico to take the full advantages of water markets, additional investment both in the capacity of users and canal managers and in the canal infrastructure may be needed.

Because the benefits from transfers of concessions from an irrigation district to an outside user are reserved for the district, the potential for intersectoral transfer of water are limited. Such intersectoral transfers may be limited to the type of transfer that was made between the city of Los Angeles and the Imperial Irrigation District of California. In this trade, the city received water-use rights but compensated the irrigation district by agreeing to pay for improvements in the water delivery system. Although this type of transfer limits the incentives received by individual farmers, it does provide an opportunity for a mutually beneficial transfer (Easter and Hearne, 1995; Rosen and Sexton, 1993). In the case studies presented, transfers of irrigation water to municipal and industrial users, with compensation in the form of improved irrigation technology, could prove to be beneficial to all parties.

Acknowledgments

Much of the information for this paper was gathered during a visit to Mexico in March and April, 1996, with funding provided to the International Institute for Environment and Development by the Swedish International Development Agency. José Trava was a valued collaborator who coauthored an earlier version of this paper. Useful information was provided by Sam Johnson, Lilain Saade, Raul Rivas, Enrique Palacios, Andres Roemer, Ernesto Romero, Rogelio Gonzalez, Fernando Gonzalez Villareal, Gernino Martinez, Gustavo Hinajosa, David Garcia, Alejandro Cruz, Carlos Hernandez, and Luz Elena Martinez. Mark Rosegrant, Josh Bishop, K. William Easter, and Nick Johnstone provided helpful comments to previous drafts.

Endnotes

[1] This chapter is based on Hearne and Trava (1997) paper.

[2] A further 1.8 million ha is in smaller irrigation units of 50 to 200 ha. These have traditionally have been collectively or privately managed.

[3] Although "concession" is the proper word in both English and Spanish, most people generally call these *derechos* or rights.

[4] With the transfer of canal management to WUAs, irrigators have been required to pay a fee for management, operation, and maintenance. Part of this fee is allocated to the CNA for the operation of headworks, main canals, and drainage networks.

[5] Meeting with Sam Johnson, International Irrigation Management Institute, March, 1996.

[6] CNA has made a single transfer of 200 million m^3 of water to the Marte R. Gomez Dam. The CNA staff in Monterrey state that water was transferred in order to prevent evaporation in El Cuchillo. It is more likely that the water was transferred to placate the claims of the irrigators of the Lower San Juan Irrigation District.

[7] In Chile, one of the most important benefits of a system of transferable water-use rights is the security that farm managers have in water supply (Hearne and Easter, 1995).

[8] As part of the 1990 "Special Agreement" which allowed the transfer of San Juan River water to Monterrey, the State of Tamaulipas was to locate alternative sources of water for the industrial and commercial cities of Camargo, Diaz Ordaz, Reynosa, and Río Bravo.

[9] Much of the information for this section comes from interviews with investigators from the National Agriculture and Forestry Research Institute and IIMI .

[10] The agronomist from the National Agriculture and Forestry Research Institute (INIFAP) with whom I spoke in April, 1996, suggested that the value of land without irrigation water was sufficiently low so as to allow the CNA to temporarily allocate parcels of land in the irrigated *módulos* to farmers who wanted to transfer their water-use concessions.

[11] The Lagunera, and especially the cities of Torreon and Gomez Palacio, is an area where the closure of old industrial plants has caused high unemployment.

[12] The size of the *ejido* farms is atypical of a Mexican irrigation district in which most farms are no more than 10 ha. Part of this is attributed to President Lazaro Cardenas who wanted to create prosperous "social" farmers when the government expropriated lands from the Colorado River Land Company. Also, in 1942, the President General, Manual Avila Camacho, decided that the *ejido* parcels in the Mexicali Valley should be no less than 20 ha in size (larger than private owners' parcels!).

References

Chang, C. and R. Griffin, 1992. "Water Marketing as a Reallocative Institution in Texas." *Water Resources Research*. 28:879-890.

Cummings, R. and V. Nercessiantz, 1994. "The Use of Pricing to Enhance Water Use Efficiency in Irrigation: Case Studies from Mexico and the United States," in *Water Policy and Water Markets: Selected Papers and Proceedings from the World Bank's Ninth Annual Irrigation and Drainage Seminar, Annapolis Maryland, December 8 - 10, 1992*. World Bank Technical Paper 249. World Bank, Washington, D.C.

Easter, K. William and R.R. Hearne, 1995. "Water Markets and Decentralized Water Resources Management: International Problems and Opportunities." *Water Resources Bulletin*. 31:9-20.

Gorriz, C., A. Subramanian, and J. Simas, 1995. *Irrigation Management Transfer in Mexico: Process and Progress*. The World Bank Technical Paper Number 292, World Bank, Washington, D.C.

Hearne, R.R. and K. William Easter, 1995. *Water Allocation and Water Markets: An Analysis of Gains-From-Trade in Chile*. The World Bank Technical Paper Number 315, World Bank, Washington, D.C.

Hearne, R.R. and K. William Easter, 1997. "The Economic and Financial Gains From Water Markets in Chile." *Agricultural Economics*, 15:187-199.

Hearne, R.R. and J.L Trava, 1997. *Water Markets in Mexico: Opportunities and Constraints*. Environmental Economics Programme Discussion Paper, 97-01, IIED, London.

Roemer, A., 1994. *An Interdisciplinary Approach to Water, Federal Law, and Institutions: The Mexican Case.* Ph.D. Thesis. Graduate School of Public Policy, University of California at Berkeley.

Rosegrant, M. and H. Binswanger, 1994. "Markets in Tradable Water Rights: Potential for Efficiency Gains in Developing Country Water Resource Allocation." *World Development.* 22(11):1613-1625.

Rosegrant, M. and R. Gazmuri Schleyer, 1994. *Reforming Water Allocation Policy Through Markets in Tradable Water Rights: Lessons from Chile, Mexico, and California.* EPTD Discussion Paper No. 6. International Food Policy Research Institute.

Rosen, M. and R. Sexton, 1993. "Irrigation Districts and Water Markets: An Application of Cooperative Decision-Making Theory." *Land Economics.* 69:39-53.

SEMARNAP, 1995. *National Water Planning in Mexico: Experiences, Results, and Perspectives.* SEMARNAP. Mexico.

World Bank, 1993. *Water Resources Management.* A World Bank Policy Paper, World Bank, Washington, D.C.

12 Water Markets in India: Economic and Institutional Aspects

R. Maria Saleth

12.1 Introduction

Water scarcity is an acute problem for a monsoon-dependent country such as India, since a large part of its farm output, income, and employment emanates from a relatively small irrigated area. How farmers cope with such water scarcity at the micro-level is as important, if not more, as how policy-makers deal with it at the macro-level. Local-level water scarcity does motivate farmers not only to improve on-farm water-use efficiency but also to develop new approaches for inter-farm water sharing. An eminent case is the spontaneous emergence and growth of water markets in many parts of India. Although buying and selling of water are nothing new in India,[1] the recent water markets are significant in that they occur in an entirely different economic, institutional, and technological environment. While modern water trading practices in India have been traced to the 1920s, a systematic documentation of them began only in the late 1960s (e.g., Patel and Patel, 1969; Shah, 1985 and 1993; Shah and Raju, 1988; Copestake, 1986; Kolavalli, et al., 1989 and 1990; Saleth, 1991 and 1994; Shankar, 1992; and Janakarajan, 1994).

Even though Indian water markets are localized, fragmented, and uneven across regions, they are growing in magnitude and gaining in significance. This chapter attempts to: (a) provide an idea of the magnitude and value of Indian water trading, especially at the national level, (b) outline the technical and institutional environment within which Indian water markets are operating, (c) describe their major economic and institutional features, (d) evaluate their efficiency, equity, and sustainability implications, and (e) suggest the legal and institutional changes needed to make them an efficient institutional option for groundwater management.

12.2 Water Trade: Magnitude and Value

In the Indian context, water markets typically emerge due to technical indivisibility and the need to cope with power shortages. Farmers often install lift irrigation systems (LIS), i.e., wells, pumps, and water conveyance networks, with a water delivery capacity exceeding their own needs. Since LIS involve large investments (ranging from $860 to $11,430,[2] depending upon the well type and depth, pump capacity, and conveyance system), the excess pumping capacity and surplus water represent an economic loss not only to LIS owners but also to society in general. This excess pumping capacity can be used to provide water for adjacent farms which have insufficient water or higher extraction costs, thereby benefiting not only the buyers but also society at large. It is under these conditions that typical markets have emerged in India.

Water markets have been established in groundwater irrigated areas, in canal irrigated areas with unreliable surface supplies, and in areas with lift-based surface irrigation. The wells are either dugwells or tubewells, the energy source is either electricity or diesel, and the water conveyance system consists of open channels, hoses, or underground pipelines. The following are some of the characteristics of Indian water markets. First, most water sales do not involve any reduction in irrigation by sellers. Although Shah (1993) cites cases in Gujarat where farmers sometimes abandon cultivation to market water, they are the exception rather than the rule. As a result, the private opportunity cost of water becomes either undefined, or when defined (i.e., in terms of the impact of current sale-related water withdrawals on the future pumping costs of both the sellers and their neighbors), it remains outside the economic calculation of sellers under the current systems of groundwater ownership and power pricing. Second, although water sales are primarily for irrigation, sales for non-irrigation uses are not uncommon. Some water sales for brick kilns were noted in Uttar Pradesh (Shankar, 1992) and the same is true for urban domestic use in Tamil Nadu (Palanisami, 1994; Janakarajan, 1994). Third, while the government does play a catalytic role through electrification and concessional credit support, it is the private farmers who develop and control groundwater irrigation and water markets in India. Finally, there is an issue with considerable legal implications concerning whether the sellers are selling the excess pumping capacity (i.e., rental market for LIS), the water, (i.e., water market), or both.

While there is no systematic national-level estimate of the magnitude and value of water trading, a few micro studies provide some idea, especially at the village or regional levels. For instance, a survey of tubewells conducted in Nadia and Malda districts of West Bengal found 60 percent of the well owners selling water (Agro-Economic Research Centre, 1988). Of the two West Bengal villages surveyed by Kolavalli and Atheeq (1990), the village in Nadia district had 85 percent of well owners selling water, while the village in Puralia district reported no water selling. In a 7-village study of Tamil Nadu's Vaigai basin, water was sold by only 25 percent of the well owners studied in wet areas (i.e., canal or tank commands) and by 15 percent in dry areas (Janakarajan, 1994).

In terms of the area irrigated with purchased water, Shah (1993) estimates it at 80 percent in northern Gujarat, while a 16-village sample study in Allahabad district, Uttar Pradesh, found it to be about 60 percent (Shankar, 1992). In contrast, in the Vaigai basin, the area irrigated with purchased water seems to be no more than 30 percent of the total irrigated area (Janakarajan, 1994). There are also studies which report no water trading in their sample areas (Shah, 1993).

At the national level, the area irrigated through water markets is estimated to be nearly 50 percent of the total gross irrigated area with private LIS (Shah, 1993). Unfortunately, this estimate does not allow disaggregation by region, well type, or energy use. While there is no current disaggregated data, the data from the national sample survey conducted by the National Sample Survey Organization (NSSO) in the mid-1970s is somewhat disaggregated. To better appreciate the implications of this data for water marketing, two points are critical. First, even though the data is outdated, it is the only source of information that provides a basis for a national-level estimate of both water trading and its economic magnitude. Second, although the data is for pumpset rentals, a closer examination reveals that pumpset rentals inherently involve water sales in the case of all immobile pumpsets permanently fitted to dugwells and tubewells or connected with electric power lines. Pumpset rentals can occur independent of water selling only in the case of mobile pumpsets (e.g., diesel pumps fitted on a lift-based surface system or filter points) which can be moved with negligible costs to pump water from sources other than those of their owners.[3] Since Table 12.1 distinguishes pumpset rentals by water and power sources, it is possible to isolate pumpset rentals that must involve water sales from those occurring without water sales.

Table 12.1. Pumpset Rentals and Their Distribution by Water Source and Energy Use: 1976-77

States	Total Pumpsets Reported ('00)	Pumpset Rentals		Pattern of Rentals by Water Source (%)			Pattern of Rentals by Energy Use (%)		
		Number	%	Wells	Tube-wells	Others[1]	Elect-ricity	Diesel	Both[2]
Andhra Pradesh	5,971	407	6.8	62.9	19.4	17.7	42.2	57.7	60.0
Assam	2	2	100.0	0.0	0.0	00.0	0.0	100.0	00.0
Bihar	3,547	878	24.8	4.1	70.6	25.3	30.5	58.6	10.9
Gujarat	4,933	116	2.4	94.0	6.0	0.0	84.5	15.5	20.0
Haryana	2,914	138	4.7	2.9	91.3	5.8	51.5	48.6	0.0
Jammu & Kashmir	6	1	16.7	100.0	0.0	0.0	100.0	0.0	0.0
Karnataka	2,669	165	6.2	68.5	0.0	31.5	40.6	57.6	1.8
Kerala	1,038	386	37.2	38.1	0.0	61.9	14.3	84.2	1.5
Madhya Pradesh	3,423	833	24.3	33.5	0.4	66.2	7.1	92.9	0.0
Maharashtra	7,424	1,050	14.1	68.9	0.0	31.1	31.2	67.3	1.4
Orissa	171	69	40.4	5.8	0.0	94.2	5.8	94.2	0.0
Punjab	5,544	170	3.1	4.1	95.9	0.0	25.9	71.8	2.3
Rajasthan	4,228	328	7.8	54.0	5.5	40.6	21.3	77.1	1.5
Tamil Nadu	11,096	267	2.4	87.6	5.6	6.7	78.3	18.7	3.0
Uttar Pradesh	8,722	224	2.6	8.0	89.3	2.7	5.4	94.6	0.0
West Bengal	1,959	974	49.7	0.8	45.3	53.9	4.1	95.1	0.8
Total	63,647	6,008	9.5	35.2	27.8	37.0	24.9	72.7	2.4

Source: NSSO (1984 and 1985).
[1]It includes lift irrigation from surface water bodies like rivers, ponds, etc.
[2]The category 'both' includes wells having both electric and diesel pumps--the latter are used mostly as stand-by.

Of the total estimated pumpsets in Table 12.1, almost 10 percent involved pumpset rentals. Since about 63 percent of these rentals occurred with dugwells and tubewells, the majority of them must happen as part of water selling. This is particularly true in the case of both the Indo-Gangetic plains and hard-rock regions in western and southern India. Since pumpset rentals in the 'others' category (i.e., surface water-based lifts or filter points) are over 50 percent of the total rentals in Assam, Kerala, Madhya Pradesh, Orissa, and West Bengal, the likelihood of the rentals occurring as pure rentals is greatest in these states.

Based on the power source of rented pumpsets at the national level, about 25 percent of the ones relying on electric power clearly involve water selling because of the inherent immobility of electric pumps. Even in the case of rented pumpsets with diesel power, a substantial proportion also involve water selling since their mobility is physically limited or economically costly. The overall implication is that pumpset rentals involve water selling more than they do pure pumpset renting except possibly in the states of Assam, Kerala, Madhya Pradesh, Orissa, and West Bengal.

With increasing water scarcity, growing productivity of groundwater irrigation, and expanding LIS since the mid-1970s, the scale of water trading must have increased significantly and its regional composition and energy base must have changed substantially. Even if we assume conservatively that pumpset renting/water selling has doubled over the past two decades, roughly 3 million pumpsets (i.e., 20 percent of the currently observed 14.2 million pumpsets) are likely to be involved in water trading. If each pump irrigates two ha of buyers' land, the current additional irrigated area due to water markets would be around 6 million ha, representing about 15 percent of the total area under groundwater irrigation. If we consider a net addition to output of $230/ha/year,[4] the total value of output due to water sales would be $1.38 billion/year – representing 30 percent of the total plan expenditures on groundwater irrigation since 1950 (GOI, 1992b). Since the total value of water trade even for Gujarat – a state known for widespread water markets – is at about $140 to $170 million/year (see Shah and Raju, 1988), the estimated figure above can be taken, at least, as a lower bound for the total value of water trade at the national level.

12.3 Water Market Environment

Indian water markets are operating within an environment defined by an interplay of technical, resource-related, and institutional factors. Of the 86 million ha of gross irrigated area, about 42 million ha are irrigated from groundwater sources. Since irrigation and cropping intensities are higher for groundwater irrigation, its productivity impact is greater than that of other irrigation sources. Groundwater irrigation dominates in Gujarat (70 percent), Madhya Pradesh (50 percent), Maharashtra (56 percent), Punjab (57 percent), Rajasthan (61 percent), Tamil Nadu (45 percent), and Uttar Pradesh (66 percent).

Groundwater irrigation is made possible by 9.8 million electric and 4.4 million diesel pumpsets fitted on some 5 million tubewells located mainly in the Indo-Gangetic and deltaic regions, 10 million dugwells spread essentially in the hard-rock region covering western and southern India, and 0.17 million surface water-based lifts mostly in the deltaic regions. While there is no current data on the ownership, capacity, and energy use pattern of LIS, an analysis of the data from a national level sample survey conducted during 1976-77 (NSSO, 1984 and 1985) suggests that over 90 percent of the total LIS are privately owned. About one-fifth of them are jointly owned by two or more farmers because of either sub-division succeeding land inheritance or investment sharing due to cost/risk factors. While electric and diesel pumps are, more or less, equal in number at the national level, diesel pumps are relatively more significant in Uttar Pradesh (69 percent), Orissa (86 percent), West Bengal (94 percent), and Madhya Pradesh (56 percent). Diesel pumps have greater mobility and thus are also used as a stand-by mechanism during electrical power shortages in many parts of India. The capacity of about 71 percent of the pumpsets exceeds five horsepower (hp).

The farm power charge and supply policies have an important influence on water markets. The unit rate system where power charges are based on metered consumption is disappearing fast in India as most states have switched from unit-rate to a flat-rate charged annually on the basis of hp. The lump-sum charge leads to a marginal pumping cost of almost zero and causes average pumping costs to decline with water sales, encouraging water trading. Unfortunately, most of the positive effects on sales are attenuated by both deteriorating power quality (e.g., load shedding, availability at odd hours, etc.) and restricted power supply that limits farm power availability to only 4-6 hours per day. Moreover, the lump-sum charges encourage overpumping of vulnerable aquifers.

In practice, the legal and regulatory regimes governing groundwater in India are not much different from those in other countries. While groundwater is popularly considered a 'common pool' or 'open-access' resource, existing Indian laws seem to link groundwater rights with land rights. For instance, based on the 'dominant heritage' principle implied in the Transfer of Property Act IV of 1882, the Easement Act of 1882 allowed private usufructuary rights in groundwater by viewing it as an easement inseparably connected to the dominant heritage, i.e., land. When this legal linkage between land and groundwater is operationalized, it amounts to a correlative rights system where rights to groundwater are in proportion to land ownership. This form of groundwater rights is either implied or explicitly stated in many subsequent laws and policies. In fact, the 1976 National Commission on Agriculture actually recommended such a system for India [Government of India (GOI), 1976].

Although the National Water Policy of 1987 recognizes the need to limit individual and collective water withdrawals, it fails to specify the required institutional mechanisms. The Model Groundwater Bill of 1992, which postulates a kind of groundwater permit system, also fails to set withdrawal limits (GOI, 1992a). However, since the Model Bill requires the mandatory installation of water meters, it does leave room for an eventual introduction of a water quota-based permit system. The Bill did induce some legal/administrative initiatives in states like Karnataka, Maharashtra, and Tamil Nadu. Prior to the 1992 Bill, Gujarat had a law based on an amendment to the Bombay Irrigation Act of 1976 that limited tubewell depth to a maximum of 45 metres (m) but only in the Mehsana region. The ineffectiveness of this law can be gauged based on the fact that the actual depth observed for Mehsana averaged 60-100 m (Table 12.2).

There are no explicit policy statements on water markets in India, although there have been a few actions aimed at discouraging them. For instance, in the initial years, the state electricity boards thought about discouraging water markets since they involved an illegal use of electricity (Shah, 1993). Since the National Commission on Agriculture has postulated a correlative system of groundwater rights, it visualized only rental markets for LIS but not water markets. This is based on the idea that "if a farmer constructs a private tubewell which yields more water than the share of his holding, then it should be possible for the farmers (i.e., those without their own wells and pumpsets) having contiguous holding to avail of their share of water on payment of share cost" (GOI, 1976). Following this reasoning, both the National Sample Survey Organization and the National Council of Applied Economic Research treated water sales as the rental of wells and pumpsets in their surveys conducted in the mid-1970s (Saleth, 1996). The Model Groundwater Bill seems to be against water markets when it states that "small and marginal farmers will not have to obtain a permit if the well is proposed to be sunk for *exclusively personal purposes excluding commercial use*" and "the commercial use of water could be a basis for refusing a permit to *any user*" [emphasis added] (GOI, 1992a p. 3-4).

Despite these legal and policy actions, the control over groundwater at the field level continues to be governed by a *de facto* system of rights[5] where larger farmers with higher pumping capacity and deeper tubewells have a greater control over the resource than others. In the face of weak legal provisions, other regulatory approaches based on well-spacing and depth restrictions as well as power supply manipulations have been tried but with little success. The well-spacing and depth norms are enforced only in areas where annual groundwater withdrawal exceeds 85 percent of the annual recharge.

While the well-spacing norm prohibits new wells within a radius of 200 m in most parts of India, the norm can be as high as 680 m in areas with deep tubewells and serious depletion problems (Shah, 1993). Similarly, there are also depth restrictions. For instance, in Gujarat, the depth of tubewells cannot exceed 100 m. Since these spacing and depth restrictions take effect only when a farmer applies for a concessional loan or well permit and electric connection, they mostly restrict resource-poor farmers. While a restricted power supply policy provides some regulatory respite, it is of little value in the face of larger pumping capacity, multiple wells, and diesel pumps. Thus, current policies reinforce rather than regulate the *de facto* system within which current water markets are operating.

12.4 Economic Features of Water Transactions

Among the most important economic features of Indian water markets are the method of water payment, the linkages between the market environment and water rates, and the effects of water scarcity and power charges on water rates.

Methods of Water Payment

Although water rates on an hourly basis are common in most water markets, payments are also based on area irrigated and the number of irrigations. While the hourly rates vary from about \$0.10 to \$1.30 for electric pumps and \$0.20 to \$0.60 for diesel pumps, the area-based rates vary from \$6.40 to \$105/ha (Table 12.2). The per irrigation rates vary from \$1.50 to \$1.75/ha in Bihar and Uttar Pradesh and from \$0.50 to \$0.75/ha in Haryana, Punjab, and West Bengal (Saleth, 1991). In some cases, water rates also vary by volume. For instance, in Banaskantha District, Gujarat, water is sold either at a 'full-flow' rate of \$3.40/hr (i.e., \$0.10/hp/hr) or at a 'half-flow' rate of \$1.70/hr (\$0.05/hp/hr) (Shah and Ballabh, 1993). Apparently, these rates are used in the context of pipeline systems where 'full-flow' involves the sale of the whole flow to a single buyer whereas 'half-flow' involves sales to two buyers, each with an outlet open at the same time.

In the Vaigai basin, Tamil Nadu, the water rates vary not only by water discharge but also by water quality (e.g., almost \$0.30/hr for good quality water but only \$0.14/hr for saline water) (Janakarajan, 1994).

Although payments for water are mostly in cash, either immediately or at the end of the crop season, payment-in-kind involving a 'water rent' in the form of a given share of the crop output also occurs. This 'water rent' varies from one-third of the output in Andhra Pradesh and Tamil Nadu to 50 to 66 percent in Gujarat (Asopa and Dholakia, 1983; Shah, 1993; Janakarajan, 1994). Though rare, water is also paid – either in part or in full – in terms of labor (Shah, 1993; Janakarajan, 1994). Another interesting mode of payment for water is observed in Madurai district, Tamil Nadu. Here, water sales occur in conjunction with large-scale intra-basin water transfers effected through lengthy underground pipelines financed by private initiatives and supported, in part, with bank loans. The buyers have to deposit with the sellers a refundable loan of about $700 which the sellers use primarily to service other loans or to buy additional rainfed land. Under this arrangement, water is supplied *in lieu* of interest payment. Since the monthly interest rate in the area varies from 24 to 120 percent, water payments vary from about $170 to $850/ha/year.

These different pricing methods are used in different situations to meet specific objectives. For instance, the hourly payment is used for diesel pumps since diesel consumption is linked to hours of operation. In contrast, though hourly rates are common for electric pumps, especially under unit rate-based power charges, area rates are used for specific crops or seasons in order to minimize monitoring costs, especially under flat rates for power. For instance, in Nadia district, West Bengal, area-based rates are used for paddy (rice), whereas hourly rates are used for other crops. Similarly, high monitoring cost and low demand for night irrigation also make area-based rates more appealing in the case of water sales for night irrigation (Kolavalli and Atheeq, 1990). While hourly or area-based rates are used in the case of regular water sales, per irrigation rates are charged for sporadic sales of supplemental or conjunctive irrigation. Area-based rates, especially with crop-sharing arrangements, are more appealing in areas with severe water scarcity and greater production risk. The terms of payments are dependent on the power source and irrigation frequency. Immediate payment is needed for diesel pumps with sporadic water sales whereas monthly or seasonal credit is normally allowed for electric pumps with regular water sales.

Market Environment and Water Rates

As illustrated by Table 12.2, water rates are influenced by water depth, cropping pattern, energy source, and power charges. Several interesting patterns are visible irrespective of whether one uses the simple hourly rates or those normalized for pumpset capacity. The hourly rates in the three Gujarat cases are far higher (ranging from $0.40 to $1.30) than the rates in other cases (ranging from $0.10 to $0.60). The higher water rates in Gujarat probably reflect the effects of the unit rate-based power charges as well as the water depth, pumping capacity, and cropping pattern. Even among the non-Gujarat cases, the level and spread of water rates observed in the hard-rock regions (Andhra Pradesh and Tamil Nadu) are higher ($0.10 to $0.60) than those ($0.11 to $0.14) in the Indo-Gangetic regions (Punjab, Uttar Pradesh, and West Bengal). The higher level and spread of hourly rates in the hard-rock regions probably reflect the joint effects of higher water scarcity and wider hydro-geological variations.

While higher rates are often attributed to power tariffs and monopoly behavior, water scarcity and productivity-related factors also have a strong influence. For instance, the water rate in the north Kheda region is low because the water market is operating in the canal region which has a better groundwater supply. In contrast, the rates in Mehsana and Sabarkantha – two regions known for their rapidly receding water table – are very high. Similarly, the differential pattern of water rates in the hard-rock and Indo-Gangetic regions also shows the influence of relative water scarcity.

Table 12.2. Water Market Environment and Water Rates, India (Continued on next page).

Source No.	Year	Location (District and State)	Water Table Depth (m)	Method and Level of Power Charges[2] (Rs)	Electric Pumps: Average Pumping Capacity (hp)	Electric Pumps: Water Rates (Rs/ hour)	Electric Pumps: Water Rates (Rs/hp/ hour)	Electric Pumps: Crops	Electric Pumps: Water Payment (Rs/ha)[3]	Diesel Pumps: Water Rates (Rs/ hour)	Diesel Pumps: Ratio of Price to Costs[4]
[1]	Around 1984	Mehsana and Sabarkantha, Gujarat.	60-100	UR 0.7	30.0	42-45	1.40-1.50	Paddy	2,964-3,705	-	-
								Wheat	2,223-2,717		
								Groundnut	2,100-2,717		
[2]	1985-86	North Kheda, Gujarat.[1]	< 10	UR 0.7	14.0	15.0	1.07	Tobacco	1,606	15	1.89
								Sugarcane/ Banana	5,878		
								Cotton	2,347		
								Wheat	1,927		
								Groundnut	2,223		
[3]	1985-86	Charutar, Kheda, Gujarat.	15-25	UR 0.7	21.2	25-28	1.16-1.32	Cotton	1,630	-	-
								Sugarcane/ Banana	9,174		
								Wheat	1,482		
								Bajra	1,561		
[4]	Around 1884	Hoshiarpur, Punjab.	0-12	FR -	7.5	5.0	0.67	Wheat	358	8-10	1.3-1.7
[5]	Around 1883	Meerut, Uttar Pradesh.	< 12	FR 21.5	7.5	4.5	0.53-0.67	Wheat	326	8-10	1.3-1.7
[6]	1985-86	Allahabad, Uttar Pradesh.	2-10	FR 21.5	5.0	4.5	0.80-1.00	Wheat	371-395	10-12	1.7-2.0

Table 12.2. Water Market Environment and Water Rates, India. (Continued from previous page)

Source No.	Year	Location (District and State)	Water Table Depth (m)	Method and Level of Power Charges[2] (Rs)	Electric Pumps: Average Pumping Capacity (hp)	Electric Pumps: Water Rates (Rs/ hour)	Electric Pumps: Water Rates (Rs/hp/ hour)	Electric Pumps: Crops	Electric Pumps: Water Payment (Rs/ha)[3]	Diesel Pumps: Water Rates (Rs/ hour)	Diesel Pumps: Ratio of Price to Costs[4]
[7]	1985-86	West Godavari, Andhra Pradesh.	10-15	FR 4.0	8.6	3.0	0.35	Sugarcane	983	7.3-7.7	1.2-1.3
								Banana	1,428		
								Tobacco	595		
								Paddy	657		
								Groundnut	618		
[8]	Around 1985	Karimnagar, Andhra Pradesh.	10-20	FR 4.0	5.0	4.5-5.0	0.90-1.00	Paddy	1,112-1,259	16-21	2.7-3.5
								Ragi	494		
[9]	1984-85	Madurai, Tamil Nadu.	10-20	FR 6.0	5.0	4-6	0.90-1.00	Paddy	2,302	16-21	2.7-3.5
								Ragi	526		
								Sorghum	622		
								Groundnut	813		
[10]	1991-92	Vaigai Basin, Tamil Nadu.	10-25	Free	5.0	3-20	0.60-4.00	Paddy/ Sugarcane	225-3,411	15-20	2.5-3.3
[11]	1988-89	Nadia, West Bengal.	10-15	FR	5.0	4-5	0.80-1.00	Paddy	970	7.78	1.3

Sources: This is an updated version of tables in Shah (1993:75, 82-83 & 86-87). The original sources are: Shah (1985) for [1]; Shah and Raju 1988 for [2], [3], [7], and [8]; Jairath (1985) for [4]; Prasad (1984) for [5]; Shankar (1992) for [6]; Copestake (1986) for [9]; Janakarajan (1994) for[10]; and Kolavalli and Atheeq (1990) for [11].

[1] This is a water market in a canal region.

[2] While the unit rate (UR) is charged per kilo-watt-hour (kwh), the flat rate (FR), though charged per hp on an annual basis, is represented here on a monthly basis.

[3] These area-based water payments are based on the prevailing hourly water rates and the total number of hours of purchased water.

[4] These ratios are obtained by relating the water rates with the estimated hourly cost covering both diesel and repair costs. While the hourly costs for North Kheda is $7.95, it is $5.85 for other cases.

The hourly rates for diesel pumps are higher than those for electric pumps due to high diesel and repair costs. Regionally, the hourly rates in the Indo-Gangetic region are lower ($0.23 to $0.34) relative to those in the hard-rock region ($0.20 to $0.60). The rates in the Indo-Gangetic region are 1.3 to 2.0 times higher than operating costs, while in the hard-rock region, the rates are 2.5 to 3.5 times higher than operating costs. Finally, the regional pattern of area-based rates is very similar to that of hourly rates. They are generally higher for water-intensive and long-duration crops, such as sugarcane and banana, but lower for others. Similarly, commercial crops (e.g., cotton and tobacco) have higher water rates than food crops (e.g., millet and sorghum). Thus, the area-based water rates seem to reflect the effects of relative water consumption and water productivity.

Influence of Water Scarcity and Electricity Charges

Although it is clear from above that water scarcity and power charges influence water markets, the nature of their contribution needs further examination. Water scarcity influences not only the level but also the method of setting water rates. Unfortunately, it is difficult to isolate the exact amount of water payment that is due to scarcity. To understand this, let us decompose the water payment into the following four components: (a) pumping costs, primarily fuel, (b) labor costs in pump operation and monitoring, (c) a rental payment for the LIS, and (d) the value of water. As noted earlier, because most water sales occur without any reduction in sellers' irrigation, the opportunity cost principle is of little help in isolating the value of water. The same is true for the productivity approach. Water rates observed across states vary, more or less directly, with both crop productivity and water scarcity. But, cross-sectional comparisons yielding a relative result do not help estimate the productivity or scarcity value of water for a given area. For instance, if the area-based water payments in Table 12.2 are compared with the water productivity estimates used by the Vaidyanathan Committee (GOI, 1992b), in all cases, except for paddy and banana in Gujarat, the water payment is only a small fraction of the estimated water productivity levels.[6] However, if information is available on the first three components noted above, the value component can be obtained as the residual part of the water payment.

Although no empirical study provides such decomposed information on water payment, the residual approach, nevertheless, can be used in cases where the rates differ between water deliveries from stationary pumps (i.e., those permanently fitted on sellers' wells) and mobile diesel pumps that extract water from buyers' wells or from other common water sources. Such a differential pricing has been commonly observed with diesel pumps since the late 1960s. For instance, in Uttar Pradesh, the hourly rate for stationary pumps was $0.09 compared to $0.08 for mobile pumps (Patel and Patel, 1969). Similarly, in West Bengal, the hourly rate for water pumped from sellers' wells was $0.40 and that for water pumped from buyers' wells or from other common sources was $0.34 (Shah, 1993). Although the difference between the two rates is only a small fraction of the water payment, it represents an estimate of the value of water.[7] Although it is difficult to isolate the value in other contexts, it is clear that farmers in most cases should be attributing a certain fraction of the water payment to the water value.

Water rates are higher with variable electric power charges than with fixed power charges. A corollary is that since the flat rate system increases water sales, especially to small farmers, it has equity benefits. In terms of efficiency, however, the variable rate system is superior to the flat rate system. For instance, when the Gujarat government switched to the flat rate system for power charges in mid-1987, within four months, water charges in most parts of the state declined by 27 to 58 percent (Shah, 1993). The flat rates charged for power by different states varied from about $1.40 to $7.40/hp/year while the average water price varied only from $0.01 to $0.03/hp/hour (Shah, 1993). Although power charges--both level and method--certainly affect water rates (and even the pricing method), the changes in the level of power charges may have little effect on water sale and purchase decisions of farmers as long as water charges form only a fraction of the marginal value product of water (Saleth, 1996). This is because the combined effects of water and power constraints force many farmers to operate at a level where their marginal pumping costs are much lower than the corresponding marginal value product.

12.5 Institutional Dimensions and Behavioral Patterns

Water markets in India display wide variations in terms of organizational features and behavioral patterns. Those in northern Gujarat operate almost like agribusinesses with cash-based transactions complete with cash receipts and purchase records. In contrast, water markets in parts of Andhra Pradesh, Tamil Nadu, and southern Gujarat show shades of the feudal character involving 'water rent' and the provision for unpaid labor services. In between these two extremes fall those in the relatively water abundant Indo-Gangetic and deltaic regions that display a rather muted form of commercial character. Since the institutional aspects of water markets have important behavioral implications, some of these aspects are analyzed in detail below.

Size and Structure of Markets

Though the geographic locus of water markets is limited by the physical characteristics of water, their size is often enlarged by modern water conveyance technologies such as underground pipelines. Such market expansion is confined to a very few regions and can lead to an unbalanced market structure since more buyers are added to the market than sellers. Thus, most Indian water markets are both small and unbalanced, as indicated by the average number of buyers per seller. This number varies from 2 to 80 buyers per seller across regions (Shah, 1993). Typically, the number is higher for Gujarat, due to deep tubewells (up to 400 m in depth), large pumps (30 to 75 hp), and vast pipeline networks found in the state, but lower for other states. Another index of market size is the percentage of total water output traded. While this share is about 40 percent in Uttar Pradesh (Shankar, 1992) and up to 64 percent in West Bengal (Kolavalli and Atheeq, 1990), it is estimated to be over 80 percent in many parts of Gujarat (Shah, 1993).

In general, the sellers have more market power than buyers. The weaker bargaining position of buyers results from the unbalanced market with intense demand-side competition and because the benefits to the buyers are several times higher than those for the sellers. The opportunity cost of an exchange failure is much greater for the buyers. The sellers are usually large-scale farmers and the buyers are smaller scale farmers. There are exceptions to this pattern since small-scale farmers also supply water with a market share ranging from 30 to 45 percent (Shankar, 1992; Shah, 1993).

Since the behavioral patterns in markets where small-farm sellers dominate are likely to be distinctly different from those where large-farm sellers dominate, it is crucial to know the composition of water markets. Table 12.3 depicts a typical water sale-purchase matrix in Uttar Pradesh. The demand-side is dominated by smaller farms with less than 2 ha that account for 81 percent of the total irrigated area with purchased water. The supply-side is dominated by farms larger than 2 ha that account for 70 percent of the area irrigated with purchased water. Curiously, 75 percent of the area irrigated with water from small-farm sellers belongs to small-farm buyers, suggesting a kind of segmentation within the market where small-farm sellers deal mostly with small-farm buyers. This is because small sellers normally have lower capacity wells and pumps but higher irrigation needs per hectare (due to intensive land use and irrigation). Since they have less water to spare, their supplies better match the water demand of small-scale buyers. Furthermore, a market where the buyers and sellers have about the same sized farms and share a common socio-economic background is likely to be more equitable and less exploitative, as compared to a market where buyers and sellers have unequal farm sizes and belong to different social groups.

Non-competitive Behavior

Water markets in Gujarat and Tamil Nadu appear to have monopolistic or oligopolistic tendencies (e.g., Shah, 1993; Janakarajan, 1994), but those in Andhra Pradesh, Punjab, Uttar Pradesh, Orissa, and West Bengal appear to be relatively more competitive (e.g., Shah and Raju, 1988; Kolavalli and Atheeq, 1990; Shankar, 1992). Palmer-Jones (1994) argues that since most water markets are non-contestable, some forms of spatial monopoly become inevitable. The non-competitive character of water markets is based on two factors: water rates are higher than the pumping cost and the presence of price and non-price discrimination.

The fact that water rates are 2.5 to 3.5 times higher than pumping cost cannot always be an indication of monopoly behavior since this measure ignores fixed costs and the scarcity value of water and presumes marginal cost pricing behavior. While thin and unbalanced water markets often have monopoly potential, the sellers may not exercise their power because of social constraints as well as linkages between water markets and other rural markets. Moreover, the monopoly potential may be neutralized by state tubewells which not only bring down water rates, as in Allahabad district (Shankar, 1992) but also reduce water sales, as in Deoria district (Shah, 1993), both in Uttar Pradesh.

While markets with a single-rate structure are dominant, there are cases of various forms of both price and non-price discrimination. Yet price discrimination appears to be largely absent in mature water markets. For instance, Shah (1993) observes in Gujarat that even large water companies charge the same rate irrespective of whether the buyer is a member or not. In contrast, the markets in the Vaigai basin, Tamil Nadu, exhibit a multiplicity of prices (Janakarajan, 1994). Price discrimination is also observed in the case of in-kind payments. For instance, in Banaskantha district, Gujarat, sellers discriminate against certain buyers by requiring differential crop shares (25 to 50 percent), depending upon the conveyance distance and cropping pattern (Shah and Ballabh, 1993).

Unlike price discrimination, non-price discrimination, in the form of the quality and timeliness of irrigation service, is more widespread though difficult to observe. For instance, in the markets of Tamil Nadu, larger and more regular customers not only get a hidden price concession but also often receive priority service (Narayanamoorthy, 1991). Those with an unsatisfactory remittance history are either ignored altogether or given low priority (Janakarajan, 1994). In most contexts, both the price and non-price discrimination not only seem to have some economic basis but also appear to reflect the relative bargaining capacity of the buyers rather than the monopoly behavior of the sellers.

Table 12.3. Water Purchase and Sale Matrix by Farm Size, Allahabad District, Uttar Pradesh, 1987.

SELLERS	BUYERS					
(ha)	(ha)					
	< 1	1-2	2-3	3-4	4-10	Total[1]
	90.01	36.70	11.77	0.00	0.00	138.48 ha
< 1	65.00	26.50	8.50	0.00	0.00	100.00 %
	11.63	10.86	8.47	0.00	0.00	10.19 %
	134.84	69.52	35.36	19.07	13.62	272.41
1-2	49.50	25.52	12.98	7.00	5.00	100.00
	17.42	20.57	25.44	26.30	38.77	20.05
	108.08	62.71	35.93	16.26	9.29	232.27
2-3	46.53	27.00	15.47	7.00	4.00	100.00
	13.96	18.55	25.85	22.42	26.44	17.10
	75.87	39.74	18.79	5.78	4.34	144.52
3-4	52.50	27.50	13.00	4.00	3.00	100.00
	9.80	11.76	13.52	7.97	12.35	10.64
	365.28	129.34	37.12	31.40	7.88	571.02
4-10	63.97	22.65	6.50	5.50	1.38	100.00
	47.19	38.27	26.71	43.30	22.43	42.03
	774.08	338.01	138.97	72.51	35.13	1,358.70
Total	56.97	24.88	10.23	5.34	2.59	100.00
	100.00	100.00	100.00	100.00	100.00	100.00

Source: Shankar (1992:33).

[1] Figures in each cell are, respectively, the area irrigated in ha, and row and column-wise percentages.

Water Based Contracts and Conventions

There are a variety of water-based tenancy contracts (see Shah, 1993). For example, there are two distinct types of such contracts in Kheda district, Gujarat: (a) a two-party contract where the water seller provides irrigation water, shares 50 percent of the cash expenses (excluding labor costs), and claims 50 percent of the output and (b) a three-party contract where a water seller, a land owner, and a laborer share equally in the cash expenses as well as crop output. Similarly, in Karimnagar district, Andhra Pradesh, water sales occur within (a) labor contracts, (b) crop sharing contracts, and (c) crop and input sharing contracts. These contracts represent not only an institutional evolution of crop sharing within the context of water markets but also link them with other rural input/output markets.

Apart from these water-based tenancy contracts, the pricing methods are also accompanied by certain informal conventions and contractual obligations with considerable implications for water-use efficiency and risk-sharing (see Kolavalli and Atheeq, 1990; Palmer-Jones, 1994). The area-based pricing method, involving crop shares, provides less incentive for water conservation than the hourly method and allows risk-sharing between buyers and sellers. It also involves some informal contractual obligation for sellers to provide irrigation for the whole season. In the case of both the area-based and per-irrigation rates, there are mutually agreed upon conventions (e.g., the level or intensity of irrigation constituting 'full irrigation') to avoid conflicts and water overuse.

The functioning of water markets is affected not only by their economic linkages with other rural markets but also by non-economic linkages such as local customs, kinship, and social relationships. The water-based tenancy contracts discussed above show, in part, how water markets are linked with land, labor, and input markets. Moreover, water sellers are not only buyers of labor, credit, crop output, and land but also buyers of water in other parts of the village due to fragmented holdings and relative cost advantages (Kolavalli and Chicoine, 1989; Shah, 1993; Janakarajan, 1994). Besides these linkages, there is also a technical linkage between water markets and the rental market for LIS especially in the case of wells with permanently fitted pumps. In most contexts, since buyers and sellers have strong social bonds, it is unrealistic either to seclude water markets from their social setting or to consider them completely immune from local customs and conventions, and from personal and kinship relationships (Saleth, 1991).

12.6 Water Marketing: Efficiency, Equity, and Sustainability

Since water markets enhance the value of groundwater, they generate substantial efficiency and productivity gains. This is illustrated by comparing the area-based charges for groundwater with those for canal water. The canal rates vary from \$0.20 to \$28.60/ha, depending upon crop, season, and irrigation projects (Saleth, 1996) while the groundwater rates vary from \$6.40 to \$262.10/ha depending upon crop, season, and region (Table 12.2). The difference is still higher for specific crops, especially at the state level. In Gujarat, for instance, while canal water charges for paddy and sugarcane are respectively \$3.10 and \$23.70/ha, the groundwater charges for the same crops are \$84.70 to \$105.90 and \$170.80 to \$262.10/ha. Similarly, in Tamil Nadu, canal water charges for these two crops are in the range of \$1.40 to \$1.80/ha and the groundwater charges are in the range of \$6.40 to \$97.50/ha. The higher charges for groundwater not only induce private irrigation investment and fuller use of existing LIS capacity, but also could provide strong incentives for on-farm water-use efficiency.

Under the current institutional conditions, water buyers appear to use water more efficiently than sellers. This is shown in Table 12.4 where the relative performance of water sellers and buyers is compared in three hydro-geologically distinct regions of India. In all cases, water buyers have higher crop yields and a smaller number of irrigation applications. The buyers perform better because they, unlike the sellers, are mostly small farms known for higher cropping and input use intensities. In addition, buyers have better water-use efficiency because they face restrictions on their access to groundwater that the sellers do not face. Similar patterns of better or comparable performance of buyers are observed in other areas but this depends on factors such as farm size and cropping pattern. For example, in Uttar Pradesh, Shankar (1992) shows that the buyers obtain higher net incomes only when the comparison is confined to small farms. Similarly, Kolavalli et al. (1993) shows that the buyers have higher net incomes only for crops such as paddy, grams, and linseed.

If $1.38 billion/year is taken as a reasonable lower-bound estimate of the gains from water trade at the national level and the cost of water to buyers and sellers can be estimated, then the relative shares for buyers and sellers can be determined. If the average water payment is assumed to be $35/ha and the area irrigated with purchased water is assumed to be 6 million ha, the total payment to sellers will be about $210 million, leaving about $1.17 billion as the net gains to buyers. Similarly, if the actual cost of irrigation supply is assumed to be $14/ha, a reasonable figure under the flat-rate system, the net gain for sellers is about $126 million. This estimated total annual gain from water trade is sensitive to both cropping patterns and the extent to which areas are irrigated with purchased water. Hence, the actual magnitude of the gains can be either more or less than the estimated figure of $1.38 billion, depending upon factors such as rainfall, water and power availability, and crop prices. What is more important for the functioning of water markets is the way this total gain is shared among buyers and sellers. Although the relative share of gains can vary by crops, season, region, pricing method, and even by type of buyers, the buyers, as a group, gain many times more than sellers. This, plus the fact that water buyers are mostly smaller farmers, suggests the income distribution potential of water markets.

Table 12.4. Relative Input Use and Productivity Performance of Well Owners and Water Buyers, Andhra Pradesh, Gujarat, and Uttar Pradesh, 1976-77.

Items	Andhra Pradesh		Gujarat		Uttar Pradesh	
	Well Owners	Water Buyers	Well Owners	Water Buyers	Well Owners	Water Buyers
Crop Yield (kg/ha)[1]	2,429	2,649	1,494	2,815	1,564	1,627
Number of Waterings	5.40	4.42	4.35	2.11	2.50	2.40
Fertilizer Use (kg/ha)	107	119	54	69	44	27

Source: Satya Sai (1987:2-3).
[1] Since the crops studied differ by states (e.g., the main crops were paddy and sugarcane for Andhra Pradesh; paddy, wheat, and tobacco for Gujarat; and paddy and sugarcane for Uttar Pradesh), the crop yields were converted into rice equivalents to enable comparison.

The equity potential of water markets is further reinforced by the fact that, in addition to the direct output impact on small farms, there are also second-round employment and income benefits even for the landless. Such multiplier effects are particularly significant in areas facing water scarcity. However, these intra-generational equity benefits must be contrasted both with the intra- and inter-generational inequity effects of water markets. To understand the latter, let us distinguish two aspects of intra-generational equity, i.e., equity at the stage of *de facto* water control and equity at the stage of actual water use. As water markets benefit small farmers, they contribute to intra-generational equity at the stage of actual water use. Yet, if they reinforce and legitimize the *de facto* control of and the appropriation of rent over water by larger farmers, they contribute to intra-generational inequity among farmers at the stage of water control (Saleth, 1994). Unfortunately, the regulatory policies like well-spacing and licensing norms actually accentuate this intra-generational inequity as they protect the existing *de facto* control of rich farmers by restricting the entry of other farmers.

The inter-generational inequity effects of water markets are intimately linked with their effects on aquifer sustainability. Water markets can have both positive and negative effects on resource sustainability. For instance, in areas with serious depletion, efficient water markets with economic prices (which cover pumping costs plus the scarcity value of water) can improve resource sustainability in two ways. First, since they provide incentive for water conservation, water withdrawals will be lower when compared to the alternative scenario where every farmer has his own wells and ignores the scarcity of water. Second, an efficient water market could also reduce the over-crowding of wells causing effects similar to well-spacing regulations. This is because farmers have the option of buying water instead of digging a well right next to their neighbor's well.

Unfortunately, the economic features and geographical pattern of water markets suggest that their depletive effects tend to dominate over their efficiency and equity gains. Although it is unrealistic to attribute depletion problems primarily to water markets, water selling activities can exacerbate the problem in areas characterized by water scarcity and an economic rent on water that is higher than in areas with better water endowments. Due to this rent-seeking behavior and the existing institutional vacuum (i.e., the absence of legally enforced withdrawal limits), water sales occur by simply pumping additional water rather than by saving water through efficient use. As a result, the conservation imperatives remain limited for the seller and water withdrawals become excessive, causing increased aquifer depletion.

While the depletive potential of water markets is a forgotten aspect in most initial studies, a few studies do provide evidence that water trading can lead to groundwater depletion. In northern Gujarat which has more intense water marketing, the groundwater table declined by about 1.22 m during 1984-88. Notably, the Mehsana and Sabarkantha districts in this region witnessed a 3 m decline during the same period (Phadtare, 1988). Similarly, in Vaigai basin, Tamil Nadu, it is observed that, in the aftermath of intense water marketing, the proportion of wells with a depth of over 30 m has increased from 9 to 29 percent, suggesting a clear link between water trade and a secular decline in the water table (see Janakarajan, 1994). The depletion effects of water markets also foment secondary inequity effects as both farmers relying on traditional water lifting mechanisms and those unable to compete in the well-deepening race are often crowded out of the farming business.

12.7 Conclusion

Although water markets in India are localized, fragmented, and vary in terms of maturity, they are an evolving institution with substantial efficiency and welfare effects. Even with conservative estimates, they generate an annual gain of $1.38 billion, of which 70 to 80 percent accrues as net gains to buyers, who are primarily small-scale farmers. Since prices in water markets are several times higher than in public irrigation systems, they induce economic efficiency and disciplined water use. While the economic and equity contributions of water markets are tremendous, they also produce some negative effects. Groundwater depletion is a general problem of intensive agriculture and water markets do exacerbate the problem. Since individual water extraction is not restricted, markets, which enhance small farmers' access to water, simultaneously increase the sellers' *de facto* control over groundwater. Prevailing market prices are relatively more efficient and economically-oriented than those in public irrigation systems. However, since prices in many water markets do not reflect either region-specific scarcity or productivity, their resulting allocative role is limited. When the depletive and inequity effects of water markets outweigh their positive efficiency and equity benefits, the private benefits from water trade are likely to compensate for the social costs in most contexts. However, whether their efficiency and equity gains can compensate for their negative ecological and equity effects can be settled only within the context of each specific region given its resource endowments.

The root cause for the sub-optimality of Indian water markets lies not so much in their economic and organizational aspects as in the legal and institutional vacuum (i.e., the absence of mechanisms to quantitatively fix, enforce, and monitor individual and collective withdrawals) within which they operate at present. A legally instituted and locally managed water quota system defined within an ecologically consistent overall withdrawal limit could eliminate the negative effects of markets and magnify their positive efficiency and conservation benefits (National Academy of Agricultural Sciences, 1995; Saleth, 1994 and 1996). While the magnitude of benefits from observed water markets is tremendous, their contribution is only a fraction of the efficiency, equity, and sustainability gains possible from formal markets emerging within a well-managed water quota system. The prevailing institutional vacuum thus makes the currently observed water markets only a distant second-best option. Naturally, therefore, the sustainability of water markets as an economic institution in India (as in other Asian countries) hinges critically on the speed and effectiveness with which water-rights-centered institutional changes are implemented to set and enforce both individual and collective water quotas.

Endnotes

[1] Kautilya's *Arthasastra* dating back to 400 BC specifies the royalty to be paid for water, while south Indian inscriptions dating back to 1202 AD note the practice of selling surface water (Maloney and Raju, 1994).

[2] An exchange rate of 1 U.S. dollar to 35 Indian *rupees* was used to convert rupees to dollars in most of the text. Values in the tables are reported in rupees.

[3] There is some evidence for such pure pumpset rentals. In Nadia district, West Bengal, farmers rent diesel pumpsets to pump water from their own sources on payment of $34 to $43 per crop season to pumpset owners (see Kolavalli and Atheeq, 1990).

[4] The addition to output is based on the difference between the average irrigated and rainfed yields as reported by GOI (1992b).

[5] On the surface of it, the *de facto* system appears to resemble the appropriative rights system used in the western U.S. But, unlike the latter, the former is neither subjected to the 'beneficial use' doctrine nor does it guarantee any legal recognition.

[6] The water productivity is calculated as the difference between the average productivity levels of irrigated and unirrigated lands. The estimated water productivity

levels are: $104 for Gujarat, $96 for Punjab, $44 for Uttar Pradesh, $126 for Andhra Pradesh, $125 for Tamil Nadu, and $70 for West Bengal (GOI, 1992b).

[7] When water withdrawals are restricted, for example, by a water quota system, the estimated value for water will be a major component of water payment since the opportunity cost of water will be higher. The opportunity cost is higher because the sellers can sell only if they save a part of their quota either through efficient use or non-use.

References

Agro-Economic Research Centre, 1988. *An Evaluation of Shallow Tubewells in West Bengal*, Vishwa Bharati, Santiniketan, West Bengal.

Asopa, V. N. and B. H. Dholakia, 1983. *Performance Appraisal of Gujarat Water Resources Development*, Vol. I, Summary and Recommendations, Indian Institute of Management, Ahmedabad.

Copestake, J. C., 1986. *Finance for Wells in a Hardrock Area of Southern Tamil Nadu*, The National Bank for Agricultural and Rural Development, Bombay.

Government of India (GOI), 1976. *Report of the National Commission on Agriculture*, Part V: Resource Development, Ministry of Agriculture, New Delhi.

GOI, 1992a. *Model Bill to Regulate and Control the Development of Groundwater*, Ministry of Water Resources, New Delhi.

GOI, 1992b. *Report of the Committee on Pricing Irrigation Water*, Planning Commission, New Delhi.

Jairath, J., 1985. "Private Tubewell Utilization in Punjab: A Study of Costs and Efficiency," *Economic and Political Weekly*, 20(40).

Janakarajan, S., 1994. "Trading in Groundwater: A Source of Power and Accumulation," in Moench, M., ed., *Selling Water: Conceptual and Policy Debate Over Groundwater in India*, VIKSAT/Pacific Institute/Natural Heritage Institute, Ahmedabad.

Kolavalli, S., and L. K. Atheeq, 1990. *Groundwater Utilization in Two Villages in West Bengal*, Indian Institute of Management, Ahmedabad.

Kolavalli, S., and David L. Chicoine, 1989. "Groundwater Markets in Gujarat, India," *Water Resources Development*, 5(1).

Kolavalli, S., G. Naik, A. H. Kalro, 1993. *Groundwater Utilization in East Uttar Pradesh*, Indian Institute of Management, Ahmedabad.

Maloney, C., and K. V. Raju, 1994. *Managing Irrigation Together: Practice and Policy in India*, Sage Publications, New Delhi.

Narayanamoorthy, A., 1991. "Deep Borewell Water Business in Pudukkottai District, Tamil Nadu: Some Empirical Evidence," *Indian Journal of Agricultural Economics*, 46(3).

National Academy of Agricultural Sciences, 1995. *Agricultural Scientists' Perception on National Water Policy*, New Delhi.

National Sample Survey Organization (NSSO), 1984 and 1985. *Sarvekshana*, 8(2) and 9(1).

Palanisami, K., 1994. "Evolution of Agricultural and Urban Water Markets in Tamil Nadu," in M.W. Rosegrant and R.G. Schleyer (eds.), *Tradable Water Rights Experiences in Reforming Water Allocation Policy*, USAID, Arlington, VA.

Palmer-Jones, R. W., 1994. "Groundwater Markets in South Asia: A Truncated Discussion of Theory and Evidence," in Moench, M. (ed.), *Selling Water: Conceptual and Policy Debate Over Groundwater in India*, VIKSAT/Pacific Institute/Natural Heritage Institute, Ahmedabad.

Patel, S. M. and K. V. Patel, 1969. *Economics of Tubewell Irrigation*, Centre for Management in Agriculture, Indian Institute of Management, Ahmedabad.

Phadtare, P.N., 1988. *Geohydrology of Gujarat State*, Central Groundwater Board, Ahmedabad.

Prasad, D., 1984. *Resource, Poverty, and Technology-Enterprise Choice by Small Farmers: A Field Report based on Seven Village Studies*, Institute of Rural Management, Anand, Gujarat.

Saleth, R. M., 1991. "Factors Affecting Farmers' Decision to Buy Groundwater: Empirical Evidence From the Indo-Gangetic Region," *Indian Journal of Agricultural Economics*, 46(3).

Saleth, R. M., 1994. "Groundwater Markets in India: A Legal and Institutional Perspective," *Indian Economic Review*, 29(2).

Saleth, R. M., 1996. *Water Institutions in India: Economics, Law, and Policy*, Commonwealth Publishers, New Delhi.

Satya Sai, K. J. S., 1987. "Performance of Irrigated Farming Based on Water Purchased from Fellow Farmers," Paper presented in the Workshop: *Common Property Resources: Groundwater*, Roorkee University, Roorkee, February 23-25.

Shah, T., 1985. "Transforming Groundwater Markets into Powerful Instruments of Small Farmers Development: Lessons from the Punjab, Uttar Pradesh, and Gujarat," ODI Irrigation Network Paper 11d, London.

Shah, T., 1993. *Groundwater Market and Irrigation Development: Political Economy and Practical Policy*, Oxford University Press, Bombay.

Shah, T., and V. Ballabh, 1993. "The Social Science of Water Stress: An Exploratory Study of Water Management Institutions in Banaskantha District, Gujarat," Paper presented in the Seminar: *Water Management: India's Groundwater Challenge*, Ahmedabad, December 14-16.

Shah, T. and Raju, K. V., 1988. "Groundwater Markets and Small Farmer Development," *Economic and Political Weekly*, 22(13).

Shankar, Kripa, 1992. *Dynamics of Groundwater Irrigation*, Segment Books, New Delhi.

13 Groundwater Markets in Pakistan: Institutional Development and Productivity Impacts

Ruth S. Meinzen-Dick

13.1 Introduction

There is considerable interest throughout much of South Asia in water markets as a means of increasing access to and use of groundwater for irrigation (Kahnert and Levine, 1993; Moench, 1994). This interest arises not only from a recognition of the importance of groundwater resources for agricultural production, but also as an example of a spontaneously-developed institution for water management. In contrast to formalized water markets in countries such as Chile, groundwater markets in South Asia do not represent government-sanctioned or regulated sales of tradable water rights, but rather informal "spot markets" for the sale of water from private wells.

Although water sales from private wells are a long-standing practice, these informal arrangements have only recently been recognized and empirically examined (Chambers et al., 1989). Much of the literature on this subject is based on field studies conducted in India and Bangladesh (Pant, 1993; Wood, 1995; Kolavalli and Atheeq, 1990; Kolavalli et al., 1992; Shah and Raju, 1988; Mandal, 1993).

In most cases in South Asia, water sellers are farmers who sell surplus water after meeting the needs of their own fields. Tubewell water sales have become a profitable enterprise for small farmers in Uttar Pradesh, India (Shankar, 1992a), and even for the landless under the PROSHIKA program in Bangladesh (Ahmed, 1993; Wood and Palmer-Jones, 1990). In Gujarat State, India, where groundwater markets are highly developed, individuals and even private water companies are investing in wells primarily to sell water to others (Shah, 1985; Kolavalli and Chicoine, 1989).

The potential gains from groundwater market development lie in improving utilization of tubewell capacity, increasing access to irrigation water supplies (especially among farmers with small or fragmented holdings), and lowering water tables in areas of waterlogging (Chambers et al., 1989; Chaudhry and Young, 1990). By providing water to other farmers, tubewell owners can use a higher proportion of their well capacity than they would otherwise use on their own holdings, thereby reducing the need for other farmers to install their own wells. Because groundwater markets increase the use of installed pumping capacity, they can improve the economic efficiency of private tubewell irrigation.

Groundwater markets also make it possible for those without wells to use groundwater for irrigation.[1] This improves the equity of resource distribution because it is generally the smaller-scale farmers who do not own tubewells. In addition, the opportunity to sell groundwater can make it profitable for farmers to invest in wells even if their own holdings are too small to use the full pumping capacity (Shankar, 1992b). Even some farmers with wells purchase groundwater from others. This may be to irrigate fragments located away from their own well, or to take advantage of the more reliable water supply from wells located closer to the canals or in pockets of fresher groundwater. Those with diesel tubewells may prefer to purchase water from electric tubewells because the cost is less than pumping from their own (reserving their own well for times they cannot buy from others).

Some researchers on groundwater markets voice concern regarding who appropriates the gains from irrigation (Pant, 1991; Janakarajan, 1994). The prospect of exploitative "water lords" has been raised, especially where control over water through well ownership reinforces inequality based on land and other assets (Barah, 1992). Although legal definitions of property rights to groundwater are ambiguous in much of South Asia (Saleth, 1994), *de facto* ownership of the resource is accorded to the owners of the wells which lift the water. This, in turn, requires considerable investment in wells and pumps, as well as ownership of at least some land above an aquifer. Janakarajan (1994) argues, based on studies in Tamil Nadu, India, that as a result of the unequal access to resources, water buyers are dependent on tubewell owners, while the latter extract surplus from their control over groundwater.

Shah (1991) finds that well owners are most likely to extract monopoly rents from the sale of water where markets are not competitive. Since water transactions are restricted by topography and by the distance between source and field, market competition can be difficult to achieve. Shah suggests that the availability of groundwater resources and alternative irrigation supplies (especially canal water), a high density of wells, and the presence of lined conveyance structures can reduce the sellers' monopoly power and hence the price of water. Palanisami and Easter (1991) note that the price of water in wells from Tamil Nadu, India, was lower where groundwater recharge was higher. Others (notably Palmer-Jones, 1994) have argued that groundwater markets are contestable, especially where water tables are high, but market imperfections due to transaction costs, risk, and lack of access to credit affect the conditions of water sales.

Kolavalli (1989) and Wood (1995) note that transactions are not impersonal, but are part of multi-stranded linkages in which buyers may give preference to relatives or those with whom they have other relationships, either through lower water rates or priority for service. The markets for groundwater, land, labor, and credit are often interlinked (Janakarajan, 1993 and 1994). This can be a way of dealing with high transaction costs for water sellers, particularly where water is provided on credit (see Bardhan, 1984). Pant (1991) observed that the relative social and economic position of buyers and sellers affects water rates in Orissa State, India, where small farmers selling water to large landowners charged less than nearby large farmers selling to small landowners.

Empirical research and anecdotal evidence indicate a variety of contract forms and a wide range of prices in groundwater markets. Buyers may pay for water by providing labor, fuel, or a share of the crop, though the tendency is to move toward a cash charge per hour of water supplied as groundwater markets develop (Chaudhry, 1990; Shah, 1991). Several studies of groundwater markets have dealt with the price of water (Kolavalli and Atheeq, 1990; Shah, 1989).

Although the quality of irrigation service is acknowledged to be critical, analysis of the reliability of purchased private irrigation services and its impact on productivity has been missing from most studies of groundwater markets. Chambers et al. (1989) cite farmers' preference for purchased irrigation water above canal supplies as evidence of quality of service, but admit the difficulty of estimating adequacy, reliability, and other indicators. Freeman et al. (1978) and Renfro (1982) studied water trading and sales in Pakistan as a means of increasing farmers' control over irrigation supplies. The former used yields as a proxy indicator of quality of irrigation services; the latter used data on water and cash input use, cropping intensity, and gross income per unit area. Both studies found that, while water purchases increased productivity compared to canal irrigation alone, they did not have as great an effect on productivity as tubewell ownership because tubewell ownership provided a higher degree of control.

Increases in productivity that result from greater control of water can be defined as "irrigation surplus." While in its most basic form irrigation surplus refers to increases due to the change from rainfed to irrigated production, the concept can be further refined to capture differences in productivity between different sources of irrigation. Those types of irrigation which provide better quality irrigation service will not only provide for higher yields, but also allow farmers to shift to more lucrative crops – all of which increases the "gross irrigation surplus" (Shah, 1993). The extent of control over water supplies which groundwater markets provide to water purchasers and the extent to which purchasers are able to capture the irrigation surplus from controlled supplies remains one of the most important questions about the functioning of groundwater markets in South Asia.

This chapter addresses the issues of participation in groundwater markets and the extent of irrigation surplus, using empirical data from sample villages in two districts of Pakistan: Faisalabad District, in the heart of the canal-irrigated plains of Punjab, and Dir District, in the hilly areas of Northwest Frontier Province. The data are from a household survey on agricultural production and operation of water markets carried out in 1991-92 (for details, see Meinzen-Dick, 1996). The following section provides background information on the role of groundwater markets in Pakistan's irrigation sector. The third section examines the characteristics of tubewell owners and groundwater purchasers in the study areas, and the nature of groundwater contracts. The fourth section discusses the reliability of purchasers' access to groundwater. The implications for productivity of purchased groundwater (compared to public irrigation sources and own tubewells) are presented in the fifth section. Finally, the chapter concludes with suggested policy options for promoting equitable access to groundwater.

13.2 Groundwater Markets in Pakistan's Irrigation Sector

In the arid and semiarid climate of Pakistan, irrigation provides crucial water resources for agricultural production. Irrigated area accounts for over 80 percent of the gross cropped area and over 90 percent of the total value of agricultural production. The Indus Basin Irrigation System (the world's largest contiguous irrigation system) provides water to more than 15 million ha. Although some expansion of canal irrigation has been made possible by the development of dams and barrages, available surface supplies are limited and cannot meet the existing demand for intensive irrigation in Pakistan, given current low, state-determined water prices. Since the 1970s, groundwater irrigation has been the most rapidly-growing source of irrigation: it now serves approximately 25 percent of the irrigated area and provides over 36 percent of the irrigation water available at the farmgate (Pakistan Ministry of Food, Agriculture, and Livestock, 1994).

Under canal irrigation, water is allocated according to a rotational water delivery pattern, or *warabandi,* which provides turns to farmers in fixed proportion to their landholdings. The fixed turns of irrigation flow during a rotation cycle are to be used only on a farmer's own land. This system, which was established in the Canal and Drainage Act of 1873, aims at distributing scarce water over as large an area as possible. The low water availability and rigid allocation pattern are not adapted to meet the demands of the more intensive agriculture that came in the wake of the Green Revolution and increasing population pressure in Pakistan. This rigidity limits the productivity of public irrigation systems, a limitation which is especially apparent in comparison with privately-managed groundwater irrigation, where farmers have more control over water timing (Renfro and Sparling, 1986). Moreover, problems with operation and maintenance of the public systems mean that deliveries are unreliable, and often never reach the tail end of the system.

From the mid-1950s to the 1980s, government policy on groundwater development focused on public tubewells to provide vertical drainage in canal command areas with high water tables. Water from public tubewells was mixed with canal water and allocated in the same manner as canal water. Because the tubewell systems were tied to the same rigid *warabandi* as public canal systems, they did not increase the flexibility of irrigation. Rising operation and maintenance expenses for public tubewells, in conjunction with their poor performance in terms of timeliness and reliability of irrigation supplies, led the government to devolve responsibility for groundwater irrigation development from the public to the private sector.

There are now over 374,000 private tubewells in Pakistan, compared to 16,000 public tubewells (Pakistan, 1994). Despite this expansion in private tubewells, ownership of tubewells remains limited to 6 percent of total farms. Control over privately-managed groundwater irrigation requires investment in wells and pumping devices. To the extent that wealthy large-scale farmers are most likely to own tubewells and small-scale or poor farmers are unable to make the investment, the latter may be excluded from the benefits of highly productive groundwater resources. Further expansion of tubewell ownership is restricted not only by the cost of the investment required, but, more importantly, by limitations in total groundwater availability. Moreover, widespread private ownership leads to over investment in wells and pumpsets, particularly where holdings are small or fragmented. Institutional arrangements are needed to spread access to groundwater to other farmers, and to increase agricultural productivity and equity in the use of irrigation water resources.

Informal water markets provide an institutional mechanism for increasing access to irrigation with private groundwater. While groundwater markets are not officially recognized, the sale of water from private tubewells is a growing form of water allocation. Groundwater markets are reported in all provinces of Pakistan, but are most active in Punjab, where the greatest groundwater development has taken place. Based on data from a survey of well owners throughout Pakistan, NESPAK (1991) reported that 21 percent of well owners sold water, but the fraction of water sold was often low. Despite the potential importance of this institution for equitable and productive use of groundwater, groundwater markets in Pakistan have received relatively little attention from research or official government policy.[2]

13.3 Water Buyers and Sellers in Faisalabad and Dir Districts

The empirical results of the study of groundwater use among sample farmers in Faisalabad and Dir Districts indicate that the pattern of tubewell owners having larger landholdings--found in much of Pakistan (Johnson, 1989)--holds also in the study areas. Table 13.1 presents average landholding size and other characteristics of tubewell owners, water purchasers, and nonusers of groundwater. Farmers who own tubewells also own significantly more land than either groundwater purchasers or nonusers of groundwater, though nonusers' operational holding sizes are not significantly larger than those of water buyers. Nearly one-third of water buyers and nonusers of groundwater are landless tenants, whereas all tubewell owners own at least some land. Water buyers are from significantly younger households than those of tubewell owners, indicating that they may not be as well established.

Table 13.1. Average Characteristics of Sample Tubewell Owners and Water Buyers

Characteristics	Tubewell Owners	Tubewell Water Buyers	Nonusers of Groundwater	Total Sample
Land owned (acres)	8.21	3.46	4.64	4.83
	(5.85)[1]	(6.56)	(7.89)	(7.43)
Operational holding (acres)	7.72	6.04	5.93	6.20
	(5.47)	(5.16)	(7.66)	(6.83)
Percent pure tenants	0	33	31	27
	(0)	(48)	(46)	(45)
Percent owner-tenants	13	13	16	15
	(34)	(34)	(37)	(36)
Age of household head	59.50	50.13	56.51	55.29
	(13.31)	(14.25)	(13.44)	(13.93)
Sample size	24	45	108	177

Source: Meinzen-Dick 1996.
[1] Figures in parentheses are standard deviations.

Under conditions prevailing in most of Pakistan, tubewell irrigation water is not a commodity which can be transported far from the source to the area of application. Conveyance losses between the tubewell and the field restrict purchasers to buying from tubewells located in close proximity to their fields (and restricts sellers to buyers within a limited radius of the well). The distance over which it is feasible and economically viable to transport water depends on the soils, topography, and type of channel used to convey the water. In the present study, the average distance between the tubewell and purchasers' fields was 600 meters in Faisalabad District and 180 meters in Dir, which has more undulating topography. Use of lined watercourses or field channels, which have lower conveyance losses than unlined channels, increases the distance over which water can be transported.

Although some well owners sell water, this is not a primary enterprise in the study areas. Well owners remain farmers first, and only sell surplus water after meeting the needs of their own crops (see also Strosser and Kuper, 1994). This practice is linked to both the price and productivity of purchased groundwater.

Those who do sell water do not extract much rent from the sale: the cost of purchased groundwater was not much higher than the cost of pumping for well owners. The two most common ways of charging for groundwater are a flat charge per hour of pumping (ranging $0.57 to $3.27 per hour, depending on the pump type, capacity, and location) and a "buyer-brings-fuel" arrangement whereby the buyer supplies the diesel and motor oil for the pump, and pays an additional fee (of $0.16 to $0.24 per hour to the well owner to cover the wear and tear on the engine.)[3] Water sellers with diesel pumps are apparently only recovering their own costs under either type of contract.[4]

A larger sample of water sellers and purchasers under different ecological and socioeconomic conditions would be necessary to estimate the effect of physical, social, and economic factors on the cost of private tubewell water. However, the prices reported under either system do not represent a large profit to the tubewell owner.[5] The gains to the seller are more likely to come from interlinked transactions (such as expectations that the buyer will provide labor to the seller when it is needed) or through gains in prestige.

It is notable that under both the hourly charge and buyer-brings-fuel contracts, the price of water does not vary over the course of a season to reflect changes in its scarcity value. Wood (1995) explains that a fixed price reduces transaction costs, and it avoids perceptions that a seller is profiting from the misfortunes and water needs of the buyer (who is often a neighbor, or relative, or both). However, water may not always be available at this (or any other) price (Wood, 1995). At times of peak demand, tubewell owners may choose not to sell groundwater.

13.4 Reliability of Groundwater Markets

The major advantage of private tubewells lies in the ability of farmers to match irrigation deliveries to crop needs. Because tubewell water is not tied to a fixed *warabandi* rotation schedule, farmers can give more frequent irrigations during periods of peak demand. Tubewell water is also available throughout the year, except during periods of mechanical breakdown. In the IFPRI sample, tubewell owners reported that pump or engine failures made groundwater unavailable for an average of 2 weeks per season in Faisalabad and 1 week per season in Dir. This compares favorably with the reported unavailability of canal water for an average of 4 weeks per season in Faisalabad and 5 weeks per season in Dir. Although the reliability of irrigation service under private tubewells is generally higher than under public sources, such as canals and government tubewells, it is lower for water purchasers than for farmers with their own wells because tubewell owners sell surplus water after meeting the needs of their own crops. Thus, the deficits created by shortages of groundwater or energy supplies are not shared equally between owner and purchaser, but rather reduce groundwater availability to purchasers first. Such groundwater shortages compound groundwater unavailability due to mechanical failure of pumps, the latter affecting both owners and purchasers.

Tubewell water sellers and purchasers were asked whether water was always available when requested, as an indicator of reliability. Farmers responded that water is most likely to be unavailable for sale or purchase during periods of electricity shortage, and during periods of peak water demand. Not surprisingly, water buyers were more likely to identify problems with water availability than were the water sellers. Over a fourth of Faisalabad water buyers reported that they were unable to purchase water to meet crop needs during times of electricity shortage, but no sellers reported being unable to sell because of electricity shortage. Times of peak water demand are more problematic: nearly a fourth of all sellers and over half of all buyers reported that purchased tubewell water was not always available when needed during such periods.

What influences the reliability of purchased irrigation water? Table 13.2 presents the results of a logistic regression for reliability as a function of tubewell characteristics (source of power, diameter, and depth), buyers' characteristics (land ownership, age), relationship between buyer and seller (relative, landlord), and regional dummy variables (Jaranwala village for Faisalabad District, Dir District). Results indicate that the type of tubewell and status of the buyer (indicated by land ownership and age) had a significant effect, but the relationship between seller and buyer did not. Electric tubewells are more susceptible to power outages, and are therefore likely to be less reliable. Larger-capacity tubewells are more reliable than smaller tubewells, but deeper tubewells are not more reliable than shallow tubewells--perhaps because they are located in groundwater-scarce areas. Buyers with higher social status also have more reliable access to purchased tubewell water. If social ties influence reliability of water markets, farmers who buy water from close relatives or their landlords would be expected to receive more reliable irrigation service, but this was not found to be a significant factor. Thus, the small farmers from young households, who are less likely to be able to own tubewells, also face precarious access to tubewell water through private groundwater markets.

Sharecropping contracts, which are used under both diesel and electric tubewells in Dir, could also influence the reliability of irrigation service. It is noteworthy that sharecropping for tubewell water is practiced for crops such as tomatoes, onions, or rice, which are sensitive to moisture stress at critical periods. In sharecropping contracts, the water seller has a stake in the outcome of the crop, and a share in the risk if the water supply does not meet crop needs. Therefore, the seller has an incentive to supply adequate tubewell water in a timely manner. This is not as critical for crops less susceptible to the timing of irrigation or where alternative sources of irrigation are readily available. Further empirical research on the provision of tubewell water in different agroecological zones and for different crops is necessary to establish whether risk-sharing is a significant consideration in the choice of contract and timeliness of irrigation service, but similar findings on sharecropping for moisture-sensitive crops in Gujarat State, India, suggest that this is the case (Aggarwal, 1996).

Although not specifically included in the regression analysis, social norms influence both the pricing and access to groundwater in Pakistan. These factors play a greater role in this case than in classical markets, because transactions are not between anonymous buyers and sellers or conducted through a brokerage institution. Instead, sales must be negotiated between neighbors on a recurring basis. Islamic prohibitions against profiting from the sale of "surplus" water may also play a role. Raising the price to capture the full scarcity value of water, and making a large profit from the sale of water are, therefore, more difficult, especially where buyer and seller are related (e.g., when a father's land is divided among sons, with the tubewell remaining on one's holding). There is also an often-articulated norm that no one should be denied water needed to save a crop (Vander Velde personal communication). However, there is no requirement that tubewell owners sacrifice production on their own farms to provide water to neighbors.

Table 13.2. Logistic Regression Model for Reliability of Purchased Groundwater

Independent Variables	Coefficient	t-Statistic	Wald Statistic
Electric tubewell	-2.189 **	-2.934	8.617
Tubewell diameter	1.267 **	2.368	5.587
Tubewell depth	-0.026	-1.245	1.553
Land owned	0.072 **	1.986	3.949
Age of head of household	0.045*	1.844	3.398
Buy from relative	0.855	0.866	0.750
Buy from landlord	10.613	0.435	0.189
Jaranwala village	0.668	0.726	0.527
Dir district	4.209 **	2.397	5.745
Intercept	-5.929 *	-1.838	3.378
Model Chi-Square	58.200 **	with 9 degrees of freedom	
Number of observations	96		

Classification Table for Reliability of Purchased Tubewell Water

	Not Reliable	Predicted Reliable	Percent Correct
Observed			
Not Reliable	49	6	89.1
Reliable	9	32	78.5
Overall			84.4

Source: Meinzen-Dick 1996.
* Significant at the .10 probability level
** Significant at the .05 probability level

13.5 Impact on Production

With greater water control, the production risk declines and agricultural productivity and incomes increase because of three factors: 1) increased input use, 2) increased crop yields, and 3) switches to more profitable, but water-sensitive crops. Shah (1993) argues that the irrigation surplus is therefore directly related to the quality of irrigation service. Previous studies in Pakistan (Freeman et al., 1978; WAPDA, 1980) have shown clear productivity gains to farmers purchasing groundwater over those using only public canal or public tubewell supplies, but the gains were much less than those obtained by tubewell owners.

In part, water buyers may have lower cropping intensities and yields than tubewell owners because buyers choose to use less water due to the cost of purchased tubewell water. However, the price of water for tubewell water purchasers in the sample survey used here was not much greater than the cost to tubewell owners. Whether tubewell water purchasers use less groundwater based on an input allocation decision or due to supply constraints is unclear. There are numerous occasions in which tubewell water is not available to water buyers at any cost, despite their demand for it. The lower reliability of purchased tubewell water compared to owned tubewell water is likely to be a major contributor to any yield or income gap between tubewell owners and water purchasers. Renfro (1982) concludes that, in comparison with water purchasers, tubewell owners exert more control over water supplies, with favorable impacts on their productivity.

The examination of a production function for wheat allows us to control for the number of irrigation applications and to compare the effect of each source of water on yields. Wheat is a staple crop grown by nearly all farmers in both Faisalabad and Dir. In this model, yields per acre are a function of seed, fertilizer (elemental nitrogen plus phosphorus), and labor inputs per acre, soil characteristics (alkalinity, potassium, phosphorous, and salinity),[6] and the number of irrigations from each source. This specification distinguishes between water from canals, groundwater from own tubewells, and groundwater purchased from other farmers. Because canal water is generally of higher quality, canal applications should have a greater contribution than groundwater, while applications from owned and purchased groundwater sources should have approximately the same impact, unless there are differences in the degree of control and timeliness of irrigation applications from different sources.

Table 13.3 shows the results of this estimation, using a linear regression.[7] After controlling for soil characteristics and non-irrigation inputs, all three types of irrigation inputs had a significant positive effect on wheat yields. But the magnitude of the coefficients indicates that irrigation applications from own tubewells have the highest impact on yield, followed by purchased groundwater and canal applications.

The number of applications is an imperfect indicator of irrigation, because it does not control for the volume of water used per application (though the area of crop irrigated is controlled for) or for timing of applications. The volume of water per application is usually lower for tubewell than for canal applications, and therefore would not explain the higher productivity of groundwater irrigation. However, farmers have relatively little control over timing under *warabandi* rotations of canal systems. Tubewell water can be adjusted to the crop needs and growing cycle, and therefore has more impact on production.

The relationships identified in this production function may differ under other agroecological conditions, such as in areas with significant salinity problems. However, these results point to a productivity gap between the effect of own tubewell water, over which farmers have considerable control, and purchased tubewell water, over which farmers have less control.

While a yield gap between irrigation sources for such a major crop as wheat is important, it does not capture the full effect of differential water control on cropping patterns and incomes. The contribution of different sources of irrigation to gross margins provides an estimate of the irrigation surplus derived from greater control of water. The gross margin is computed by deducting all cash input costs (including the costs of irrigation) from gross crop revenues. This margin indicates the returns to land, family labor, and own capital. (Rental payments for land have not been deducted from the gross margins, to ensure comparability between landowners and tenants.) The present analysis uses a reduced-form equation to model differences in household gross margins as a function of season, household size, tenure, landholding, tractor ownership, soil salinity, and district.

Separate dummy variables are used for each source of irrigation or combination of sources, with unirrigated farms as the base. Those farmers using tubewell irrigation conjunctively with canal water have greater recharge and water availability than those who use groundwater outside the command of surface systems. Farmers' decision-making on cropping patterns and input use, and their returns from these decisions, are more likely to take into account the total water availability from all sources. Thus, the benefits derived from each source of irrigation are likely to be affected by whether it is the sole source of irrigation or used conjunctively with other sources.

Table 13.3. Effect of Irrigation Applications on Plot-Level Wheat Yields in Faisalabad and Dir Districts

Independent Variables	Coefficient	Standard Error	t-Statistic	Variable Mean
Seed (kg / acre)	3.60 **	1.82	1.98	40.93
Fertilizer (N + P kg / acre)	4.67 **	0.73	6.09	47.12
Labor (man-days / acre)	1.89 *	1.12	1.68	27.86
Degree of alkalinity	253.54 **	69.20	3.66	0.62
Soil K (ppm)	1.25 **	0.38	3.32	128.14
Soil P (ppm)	-4.50	4.82	0.93	10.07
Salinity dummy	-44.61	68.99	0.65	0.13
Canal irrigations	31.14 **	9.00	3.46	2.53
Purchased GW irrigations	44.58 **	16.66	2.68	0.62
Own TW irrigations	48.31 **	18.45	2.62	0.50
Intercept	-63.16	112.20	0.56	
Adjusted R Square	0.31 **			
Number of observations	263			

Source: Meinzen-Dick, 1996.
* Significant at the 0.10 probability level.
** Significant at the 0.05 probability level.

Results of this model are presented in Table 13.4. Gross margins are higher in *kharif* (mid-April to mid-October) than in *rabi* (mid-October to mid-April), in part because of the higher water availability and cropping intensity in that season. Neither household size, tenancy, nor the dummy variable for Dir District have a significant impact on gross margins, but soil salinity does reduce total income from cultivation. This may be due to depressed yields on saline fields and to farmers switching to less profitable, salinity-tolerant crops. There is a gross margin of approximately $26.50 per acre, after controlling for other factors. Tractor ownership has a substantial impact (nearly $408) on household gross margins. Like tubewell ownership, tractor ownership gives farmers greater control over the timing of agricultural operations which, in turn, increases returns to cultivation.

Both canal irrigation alone and purchased tubewell water alone have coefficients of approximately $94, although neither is significant. This is not to say that canal irrigation does not increase gross margins over rainfed cultivation. However, the variability among those sample farmers using only canal sources is too high for the coefficient to be significant. The few farmers dependent on purchased tubewell water alone also face highly variable returns from production. However, the effect of own tubewell water alone is twice as great as for canal or purchased tubewell water only – approximately $204. Access to purchased tubewell water in conjunction with canal irrigation has a stronger effect ($253) than own tubewell water only, but the effect of own tubewell water plus canal access was more than twice as great as any other type of irrigation ($565). This indicates particularly high returns to tubewell owners in areas with conjunctive use. Water purchasers, who have less control over tubewell applications, are not able to reap as large a benefit from the conjunctive use of groundwater.

Table 13.4. Regression Model for Household Gross Margins per Season

Independent Variables	Coefficient	Standard Error	t-Statistic	Variable Mean
Season	-1,813.61 *	1,109.41	-1.64	0.49
Household size	7.10	154.12	0.05	3.48
Tenure	-336.54	1,194.66	-0.28	0.45
Salinity dummy	-3,483.46 **	1,793.54	-1.94	0.14
Land holding size	653.69 **	90.36	7.23	6.82
Tractor ownership	9,946.28 **	2,515.10	3.96	0.07
Dir dummy	-288.82	1,580.61	-0.11	0.47
Canal irrigation only	2,315.23	1,603.18	1.44	0.41
Purchased GW only	2,277.10	3,761.10	0.61	0.03
Own TW only	4,959.79 **	2,476.14	2.00	0.10
Canal + purchased GW	6,190.24 **	2,206.98	2.81	0.23
Canal + own TW	13,853.20 **	3,704.35	3.74	0.04
(Constant)	-191.56	2,282.84	-0.08	
Adjusted R Square	0.37			
Number of Observations	329			

Source: Meinzen-Dick 1996
* Significant at the 0.10 probability level
** Significant at the 0.05 probability level

13.6 Conclusions

The informal groundwater markets in Pakistan do not function as classic markets, with prices reflecting the full scarcity value of water. Rather, they are bound up with many other social and economic relationships between neighboring farmers. Water taken from tubewells is controlled by the owner of the well, which in practice means that most groundwater is controlled by larger landowners who have the resources to invest in tubewells. Water markets do allow other farmers to purchase groundwater and have become a major means of spreading access to groundwater. As fresh groundwater resources become scarce, such institutional mechanisms for sharing the resource are likely to become increasingly important to ensure the productivity and equity of resource management.

Yet too few studies have attempted to assess the equity of access, the quality of irrigation service under groundwater markets and their impact on productivity. Evidence from study areas in Faisalabad and Dir Districts indicates that it is the larger, more established farmers who are likely to own tubewells and the smaller farmers from younger households who are likely to purchase water. Although the markets are not formally regulated, the price well owners charge does not extract a large "rent" for the water resource (though non-price considerations, such as links to land, labor, or credit transactions, or gains in prestige may be significant). However, water purchasers cannot always obtain access to groundwater, especially at times of peak demand, when tubewell owners use much of the well's capacity for their own fields. This unreliability of access limits agricultural productivity for those who depend on water from other people's wells.

Government regulation is not likely to improve this situation. One of the strengths of groundwater markets in South Asia lies in their spontaneous functioning. It is doubtful whether the government would have the institutional capacity to regulate sales among hundreds of thousands of private tubewells, and if it had such capacity, it is unclear what such direct intervention could achieve.

Among indirect policy tools for groundwater market development, energy pricing has received the greatest attention, particularly in India. In particular, flat rate electricity charges (under which tubewell owners pay a fixed fee per month, regardless of power consumption) have been advocated as creating incentives for farmers to sell as much water as possible (Shah, 1993). However, this does not work when electricity or groundwater (or both) are in short supply, as in much of Pakistan, especially during times of critical water demand. Farmers in Pakistan have the option of flat rate electricity charges, but these have not led to equitable sharing of groundwater. Strosser and Kuper (1994) found that those with electric tubewells sold *more hours* of water supply, on average, than those with diesel power (averaging 180 hours per electric tubewell, and 60 hours per diesel tubewell per year), but because electric tubewell owners also pump more water for their own fields they sold a *lower proportion* of water than did diesel tubewell owners (8 percent of total pumping hours for electric, versus 20 percent for diesel tubewells).

Another indirect policy tool could be the lining of conveyance structures. Pipes and lined field channels or watercourses contribute to the development of more competitive groundwater markets by allowing purchasers to obtain water within a wider radius of their fields, thereby increasing the number of potential suppliers (and permitting sales to a wider number of fields from each tubewell). Lined canals and pipes ensure that water purchasers receive more of the water they pay for from the tubewell and underground pipes can even, to some extent, overcome topographic limitations to water sales.

Despite these advantages of groundwater markets, there has been little private investment in lining or pipes in Pakistan. Lined (*pukka*) watercourses are more frequently used. Watercourse lining results from government and collective farmer investment, particularly under the On Farm Water Management Programs. Lining enables farmers to transfer water over longer distances, although the use of public watercourses for private groundwater is formally prohibited. Lifting such restrictions could facilitate the use of watercourse lining investments to reduce transmission losses for high-value groundwater.

The policy options to encourage groundwater markets to better serve the needs of small farmers are limited and unlikely to overcome the problems related to unreliable access. Expanding tubewell ownership to more small farmers may be preferable. In view of the high cost of tubewells and limited groundwater supplies, individual ownership by small and medium farmers may not be a viable option. Joint tubewells offer considerably more promise. There is a long-standing tradition of shared wells in Pakistan. One of the study villages in Faisalabad District, where canal irrigation was not available, had a high proportion of joint investment that enabled even small farmers to own at least a partial share of a well.

Moreover, jointly-owned wells are not incompatible with water markets. Water can be made available to non-owners, although usually at a different cost than to owners. Shared ownership can facilitate optimal location of wells in terms of tapping good quality groundwater and being able to serve a greater number of farmers (such as near the head of a watercourse). While joint ownership of irrigation infrastructure is likely to provide greater opportunities for groundwater use, they also entail higher transaction costs. The need for farmers to cooperate in initial investment, in allocating and distributing the water, and in maintaining and expanding the system establishes a significant barrier to the creation of common management practices, even among small groups at the local level. However, as fresh groundwater resources become scarce, such institutional mechanisms for sharing the resource are likely to become increasingly important to ensure that a wide range of farmers can take advantage of opportunities to improve their productivity and income.

Endnotes

[1] In examining inequality of irrigation distribution, Gill and Sampath (1992) note the effect of water trading and sales, especially in the *rabi* season (mid-October to mid-April) in promoting equity.

[2] For exceptions, see Renfro, 1982; Renfro and Sparling, 1986; Strosser and Kuper, 1994; Bajwa and Ahmad, 1991; WAPDA, 1990; World Bank, 1994.

[3] The average 1991-92 exchange rate was 24.50 Pakistan *rupees* to the U.S. dollar.

[4] Unfortunately, much of the information on price of purchased tubewell water comes from water buyers, rather than from the sellers. There are thus not enough data on tubewell operation costs and water delivery rates to determine the profit margin for water sales or the exact price per unit of water pumped.

[5] Strosser and Kuper (1994) report that the buyer-brings-fuel sales in their study areas do not include a charge for wear and tear on the engine. In such situations, sellers do not even recover their maintenance and depreciation costs.

[6] The degree of alkalinity is computed by subtracting 7 from the original pH value. Soil K and P levels are in parts per million (ppm). Salinity is represented by a dummy variable indicating whether or not measured electrical conductivity levels are greater than or equal to 4 millimhos per centimeter, the threshold level for average crop tolerance of salinity.

[7] Alternative functional forms, such as Cobb-Douglas and log-linear, were tested but did not fit the data as well as a linear regression. The large number of cases with values of 0 for one or more of the independent variables, notably the irrigation inputs, may account for the poor fit of the Cobb-Douglas equation. Quadratic and interaction terms were tested but found not to be significant and were therefore not included in the model.

References

Aggarwal, R. M., 1996. "Risk Sharing and Transaction Costs in Groundwater Markets: Evidence from Rural India." Dept. of Agriculture and Resource Economics, University of Maryland (mimeo).

Ahmed, Q. F., 1993. "Socialization of Minor Irrigation: a Strategy for Growth with Equity," in *Groundwater Irrigation and the Rural Poor: Options for Development in the Gangetic Basin*, (eds.) F. Kahnert, and G. Levine, 83-87. World Bank, Washington, D.C.

Bajwa, A. R., and C. B. Ahmad, 1991. *Trading of Canal and Tubewell Water for Irrigation Purposes*. Watercourse Management and Evaluation Directorate, P&I, WAPDA, Lahore, Pakistan.

Barah, B. C., 1992. "Emergence of Groundwater Market and Ecological Impact on the Drought Prone Areas." Paper presented at the International Conference on Prospects of Chinese Agricultural Development in the 1990s, August 20-25 Beijing, China.

Bardhan, P. K., 1984. *Land, Labor, and Rural Poverty: Essays in Development Economics*. Oxford University Press, New Delhi, India.

Chambers, R., N. C. Saxena, and T. Shah, 1989. *To the Hands of the Poor: Water and Trees*. Oxford and IBH Publishing Co, New Delhi, India.

Chaudhry, M. A., and R. A. Young, 1990. "Economic Impacts of Alternative Irrigation Water Allocation Institutions: Pakistan's Warabandi System," in *Social Economic, and Institutional Issues in Third World Irrigation Management*, (eds.) R. K. Sampath, and R. A. Young, 395-410. Westview Press, Boulder, Colorado.

Chaudhry, M. J., 1990. "The Adoption of Tubewell Technology in Pakistan," *Pakistan Development Review*, 29(3-4): 291-303.

Freeman, D. M., M. K. Lowdermilk, and A. C. Early, 1978. *Farm Irrigation Constraints and Farmers' Responses: Comprehensive Field Survey in Pakistan*. Colorado State University, Fort Collins, Colorado.

Gill, Z. A., and R. K. Sampath, 1992. "Inequality in Irrigation Distribution in Pakistan," *Pakistan Development Review,* 31 (1): 75-100.

Janakarajan, S., 1993. "Triadic Exchange Relations: an Illustration from South Asia. The Political Analysis of Markets," *IDS Bulletin,* 24(3): 75-82.

Janakarajan, S., 1994. "Trading in Groundwater: a Source of Power and Accumulation," in *Selling Water: Conceptual and Policy Debates over Groundwater Markets in India*, (ed.) M. Moench, 47- 58. VIKSAT, Pacific Institute, Natural Heritage Institute, Gujarat, India.

Johnson, Robert, 1989. *Private Tubewell Development in Pakistan's Punjab: Review of Past Public Programs/Policies and Relevant Research*. International Irrigation Management Institute, Colombo, Sri Lanka.

Kahnert, F., and G. Levine, (eds.) 1993. *Groundwater Irrigation and the Rural Poor: Options for Development in the Gangetic Basin*. World Bank, Washington, D.C.

Kolavalli, S. 1989. Comments at colloquium on "How to Reach the Poor Through Groundwater, Irrigation," organized by the World Bank, April 12-14, Washington, D.C.

Kolavalli, S., and L. K. Atheeq, 1990. *Groundwater Utilization in Two Villages in West Bengal*. Center for Management in Agriculture, Indian Institute of Management, Ahmedabad, India.

Kolavalli, S., and D. L. Chicoine, 1989. "Groundwater Markets in Gujarat, India," *Water Resources Development,* 5(1): 38-44.

Kolavalli, S., G. Naik, and A. H. Kalro, 1992. *Utilization of Groundwater in East Uttar Pradesh*. Center for Management in Agriculture, Indian Institute of Management, Ahmedabad, India.

Mandal, M. A. S., 1993. "Groundwater Irrigation in Bangladesh: Access, Competition, and Performance," in *Groundwater Irrigation and the Rural Poor: Options for Development in the Gangetic Basin*, (eds.) F. Kahnert, and G. Levine, 73-81. World Bank, Washington, D.C.

Meinzen-Dick, R., 1996. Groundwater Markets in Pakistan: Participation and Productivity. Research Report No. 105, International Food Policy Research Institute, Washington, D.C.

Moench, M. H., (ed.) 1994. *Selling Water: Conceptual and Policy Debates over Groundwater Markets in India.* VIKSAT, Pacific Institute, Natural Heritage Institute, Gujarat, India.

NESPAK (National Engineering Services Pakistan), 1991. *Contribution of Private Tubewells in the Development of Water Potential.* (Field Report). Government of Pakistan, Ministry of Planning and Development, Planning and Development Division, Islamabad.

Pakistan, Ministry of Food, Agriculture and Livestock, 1994. *Agricultural Statistics of Pakistan 1992-93.* Government of Pakistan, Islamabad.

Palanisami K. and K. William Easter, 1991. "Hydro-economic Interaction between Tank Storage and Groundwater Recharge," *Indian Journal of Agricultural Economics*, 46(2): 174-179.

Palmer-Jones, R. W., 1994. "Groundwater Markets in South Asia: a Discussion of Theory and Evidence," in *Selling Water: Conceptual and Policy Debates over Groundwater Markets in India*, (ed.) M. Moench, 11-46. VIKSAT, Pacific Institute, Natural Heritage Institute, Gujarat, India.

Pant, N., 1991. "Groundwater Issues in Eastern India," in *Future Directions for Indian Irrigation: Research and Policy Issues*, (eds.) R. Meinzen-Dick, and M. Svendsen, 254-258. International Food Policy Research Institute, Washington, D.C.

Pant, N., 1993. "Performance of the World Bank Tubewells in India," in *Groundwater Irrigation and the Rural Poor: Options for Development in the Gangetic Basin*, (eds.) F. Kahnert, and G. Levine, 119-130. World Bank, Washington, D.C.

Renfro, R. Z. H., 1982. Economics of Local Control of Irrigation Water in Pakistan: a Pilot Study. Colorado State University, Fort Collins, Colorado.

Renfro, R. Z. H., and E. W. Sparling, 1986. "Private Tubewell and Canal Water Trade on Pakistan Punjab Watercourses," in *Irrigation Investment Technology, and Management Strategies for Development*, (ed.) K. W. Easter, 193-210. Westview Press, Boulder, Colorado.

Saleth, R. M., 1994. "Groundwater Markets in India: a Legal and Institutional Perspective," in *Selling Water: Conceptual and Policy Debates over Groundwater Markets in India*, (ed.) M. Moench, 59-71. VIKSAT, Pacific Institute, Natural Heritage Institute, Gujarat, India.

Shah, T., 1985. *Transforming Ground Water Markets into Powerful Instruments of Small Farmer Development: Lessons from the Punjab, Uttar Pradesh and Gujarat.* Network Paper 11d. Overseas Development Institute, London, U.K.

Shah, T., 1989. *Efficiency and Equity Impacts of Groundwater Markets: a Review of Issues, Evidence and Policies.* Anand, India, Institute of Rural Management.

Shah, T., 1991. "Water Markets and Irrigation Development in India," *Indian Journal of Agricultural Economics*, 46(3): 335-348.

Shah, T., 1993. *Groundwater Markets and Irrigation Development: Political Economy and Practical Policy.* Oxford University Press, Bombay, India.

Shah, T., and K. V. Raju, 1988. "Groundwater Markets and Small Farmer Development," *Economic and Political Weekly,* 23(13): A23-A28.

Shankar, K., 1992a. *Dynamics of Groundwater Irrigation.* Segment Books, New Delhi India.

Shankar, K., 1992b. "Water Market in Eastern Uttar Pradesh," *Economic and Political Weekly*: 931-933.

Strosser, P., and M. Kuper, 1994. *Water Markets in the Fordwah/ Eastern Sadiqia Area: an Answer to Perceived Deficiencies in Canal Water Supplies?* IIMI Working Paper No. 30. International Irrigation Management Institute, Colombo, Sri Lanka.

Vander Velde, 1996. Personal e-mail communication. February 2.

WAPDA (Water and Power Development Authority), 1980. *Private Tubewells and Factors Affecting Current Rate of Investment.* WAPDA, Economic Research Section, Planning and Development Division, Lahore, Pakistan.

WAPDA, 1990. *Trading of Canal and Tubewell Water for Irrigation Purposes.* Watercourse Monitoring and Evaluation Dte. P & I Organization, Planning Division, Lahore, Pakistan.

Wood, G. D., 1995. *Private Provision after Public Neglect: Opting out with Pumpsets in North Bihar.* Occasional Paper 01/95, Center for Development Studies, University of Bath, Bath, U.K.

Wood, G. D., and R. Palmer-Jones, 1990. *The Water Sellers: a Cooperative Venture by the Rural Poor.* Kumarian Press, West Hartford, CT.

World Bank, 1994. *Pakistan Irrigation and Drainage: Issues and Options.* Report No. 11884-PAK. World Bank, Washington, D.C.

14 Economic Analysis of Water Markets in the Spanish Agricultural Sector: Can They Provide Substantial Benefits?

Alberto Garrido[1]

14.1 Introduction

In normal years, Spain's total rainfall that flows into rivers, lakes and aquifers amounts to 110,000 million m^3. Substantial structural efforts to increase and stabilize supplies that date back to the Roman Empire domination of the Iberian peninsula, currently allow the control of about 40 percent of that volume of water (45,000 million m^3). Until very recently, water policies were virtually limited to supplying water for residential consumption and the agricultural sector. Other objectives, such as streamflow benefits, were overridden by the need to meet those increasing demands. Presently, Spain devotes 80 percent of all available water to irrigate about 3.3 million hectares, and the remaining 20 percent to urban and industrial consumptive uses. Rainfall patterns across the country differ in such a magnitude that the northern basins get 18 times more water than the driest, southeastern basins. Moreover, the latter are the regions where irrigated agriculture produces the most valuable crops and where population demands for water during the tourist and driest season are two or three times larger than in winter.

The growth in water demand and maturity of the Spanish water economy has turned the water problem into a major political issue. Although the environmental deterioration of rivers and reservoirs has gained increasing concern, the agricultural sector still dominates most policy proposals due to the constraints it imposes on any water management strategy. The fact that a 10 percent reduction of the water consumed in irrigation is equivalent to all the nation's residential demand is frequently cited to illustrate agriculture's prominence within the Spanish water sector.

Overall water scarcity has led to a search for policies that induce greater efficiency in agricultural water use and the consideration of alternative strategies that would help alleviate future water shortages. Most of these options have already been used in other countries and have received considerable attention from researchers all over the world. Yet, one of the most frequently mentioned options, namely water reform, coupled with some degree of decentralization, has received little attention in Spain. This chapter attempts to determine whether water markets restricted to the agricultural sector would help improve the efficiency with which water is used to produce food and fibers. It tests empirically whether or not water markets could play a role in allocating water in four representative water districts. The analytical framework developed incorporates two features often neglected. One is realistic estimates of transaction costs, and the other is accounting for asymmetric water allotment reductions among water districts due to non-overlapping drought cycles.

14.2 The Water Crisis in Spanish Irrigated Agriculture

Spain's water policy is strongly influenced by the 1985 Water Law, which overruled the 1879 Water Law and established the public domain over all water resources including rivers, lakes, glaciers, and waterways. The 1985 water code imposes rigid conditions on farmers holding water-use rights in the sense that they are appurtenant to land-use rights and are not subject to trade even between farmers belonging to the same district. While it also assigns each Basin Agency the authority over water management plans and the distribution of water rights, these agencies are dependent on the central government's budget to cover their expenditures and investments, since user charges are insufficient to cover expenditures. The 1985 Law also ranked agriculture's rights as junior only to urban demands and, in doing so, constrained many water reform proposals and secured agriculture's rights against potential users with higher willingness to pay for water. In addition, the 1985 Water Law entailed the nationalization of water resources, and moved the Spanish water sector even further away from decentralized management.

One important action the 1985 Law failed to induce is the approval of a master plan to guide the nation's use of all its water resources. The severe four-year drought that ended in 1995 raised serious doubts concerning whether the proposed water plans at the national and basin levels would suffice to solve the anticipated scarcity problems. The last draft of the National Water Plan (MOPTMA, 1993) proposed in April, 1993, and rejected in Parliament in November, 1994, emphasized building interbasin aqueducts and adding more storage capacity. Hardly any attention was given to demand management policies and to the potential use of economic mechanisms to reallocate available water resources among competing users. More recently due to its failure in the legislative assembly, the government significantly reduced its ambitious water plan and switched the focus to rebuilding old and inefficient water districts. The new water-saving program, which aims to reduce water network losses, foster investment in more efficient technologies, and increase certainty of water allotments, received budgetary approval for 553 billion *pesetas* (4.4 billion U.S. dollars)[2] for the period 1996-2005 (MAPA, 1995). This new plan has been widely praised because it recognizes the need to improve infrastructure in existing districts rather than invest in new irrigation. The available funds will be allocated to water districts that voluntarily decide to set up an improvement project and share the cost. Hence, its success is conditioned on the voluntary response of farmers and districts. No information is yet available concerning the program's success.

The water crisis is illustrated by the significant number of irrigation districts that have not been allocated enough water to meet their legally established water-use rights in many of the past twenty years. The 1989 Farm Census (I.N.E., 1991), based on farm survey data collected during a normal rainfall year, shows that over 45 percent of the farms do not have adequate water supplies. In the southern and southeastern basins[3], which comprise 40 percent of the country's irrigated land, water scarcity is so pervasive and water allotments so insecure that there are serious doubts concerning whether structural solutions can solve the problem by focusing only on supply-side policies.

14.3 Political Approaches and Recent Experiences

The panoply of proposed solutions to alleviate the Spanish water crisis is a mixture of ideas and beliefs, rather than a body of evidence or a set of rigorous arguments. Two factors negatively influence the discussion and generate even more confusion. First, the economic interests at stake are so high that all groups that support them – green groups, farmers, power utilities with hydroelectric capacity, basin agencies or municipalities – have put together a number of reports and proposals that seem globally incompatible. Second, documented evidence obtained by Pérez-Diaz et al. (1996) concluded that water policy has become a negotiating instrument used by political parties to gain support for amendments and new bills on other political matters. Their political analysis also concluded that, once water policies were no longer perceived as a national issue but as a single element embedded in a set of interregional issues, the debate reached an impasse that precluded any progress towards a national agreement on water policy reform.

Thus, from 1993 to the end of 1996 three major political approaches have dominated the political debate on water policies. Up to mid-1993, the predominant line of thought was that Spain's water demand required important structural investments to increase water supply by about 8,300 million m^3 at a budgetary cost of $11.6 billion for the period 1993-2012. This was followed by the revised 1993 government proposed water plan which had two major shortcomings: first, it entailed a level of public spending inconsistent with other general macroeconomic objectives; and second, agricultural markets did not seem to offer an adequate outlet for the 400,000 ha of new irrigated land that the 1993 proposal would provide. In addition, this second approach suggested that all water users should pay more for their water services and thus help reduce the public deficit and induce farmers to use water more efficiently. The third and current proposal considers administered water prices as unlikely to be effective in reducing water consumption in a manner consistent with the relative scarcity of water across basins and drought cycles. The proposed solution emphasizes decentralizing water management and opportunity cost pricing to distribute scarcity signals among current or potential users.

As Maass and Anderson (1978) described in their analysis of the water economy of Valencia (East Spain), it is well documented that some Spanish water communities have been managing their water resources since the 10th century. Supported by favorable institutional arrangements, water markets involving fully privatized water rights have played an important role in Valencia, mostly for groundwater. These local markets exist due to the 75-year moratorium established in the 1985 Water Law for all wells registered before that date, so that the water pumped from them is considered private and tradable property. The mixture of institutional systems coexisting in Valencia, which includes intense intervention of some public agencies, private companies, and individual entrepreneurs, has resulted in a non-sustainable exploitation of the aquifer. For instance, Tobarra (1995) reports on a southeastern aquifer that has dropped about 14 meters per year in the last 15 years. The stress under which the Valencian water economy is evolving has generated significant responses in the private sector. Farmers with insecure rights over publicly controlled water resources have become actively involved in private ventures that finance well constructions to capture deep water and distribute it among shareholders. Shares can be acquired and sold through legal markets and access to private water can also be purchased through daily auctions. It is not uncommon to encounter farmers in Valencia who use public water at virtually no cost, and also own shares with prices that reflect capitalized water prices close to $0.50/m^3.

In the dry and volcanic Canary Islands of Spain, where the national 1985 Water Law does not apply, water resources are entirely in the hands of the private sector. Although market transactions are common, some of these are never reported or registered by any public agency. Aquifers, which together with desalinization plants provide all the water available on the islands, are exclusively exploited by the private sector. The lack of a public institution to correct for the classical market failure of unregulated aquifer mining has brought the islands' water economies to a condition characterized by the rule of capture (Domínguez Vila, 1996).

Water markets are almost nonexistent in the rest of the country. In spite of the explicit banning of water trading established in the 1985 Water Law, two specific agreements involving water trading illustrate different ways in which parties exchange water rights with the acquiescence of the basin agencies. While both are local arrangements with specific legal approval, they are frequently posed as good examples of strategies that solve scarcity problems without public intervention or further structural investments. The first is a water transaction between the city of Seville (Andalusia, Southern Spain) and one irrigation district that illustrates the type of contingent market arrangements proposed by Michelson and Young (1993) and Fisher et al. (1995). It was initiated just before the summer of 1994, when the city of Seville's water storage proved to be insufficient to cover its residential demand. A nearby irrigation district held, at the same time, water rights for a full reservoir which could not be appropriated by Seville's water utility. The district had, many years earlier, paid for the full costs of the dam and conveyance system. In financing its' storing facilities, the irrigation district had legally appropriated some of the subbasin's water resources. After thorough negotiations, both parties agreed on an emergency sale of 30 million m^3 at $0.056 per m^3, which met the city's deficit for the summer and reduced the farmers' water allotments by 30 percent. Farmers' proceeds from the sale represented about 25 to 30 percent of the average farm net returns, but it involved no risk.

The second experience also shows how increasing urban and industrial demands can be met through a mutually beneficial agreement in which water rights held by farmers become available to other users. The intense industrialization of the metropolitan area of Tarragona (Eastern Spain) during the late 1970s resulted in an excess demand of 126 million m^3 per annum. Instead of meeting it by capturing the equivalent volume from the River Ebro and jeopardizing the delta's ecosystem, a consortium agreed to finance the full cost of structural repairs for the irrigation district's conveyance system which saved the needed volume of water. The arrangement entailed no cost to the farmers and had an estimated cost for the consortium of $0.04 in 1981 U.S. dollars per m^3. However, since the agreement implied a change in the nature of the water rights, from irrigation to urban uses, it required a specific governmental approval through a 1981 decree.

Although limited to local arrangements, these experiences illustrate that the incentives to release water resources from irrigation to alternative uses could promote further investments in irrigation/conveyance technology. However, since the legal system requires new laws and regulations in order to alter any water rights, the transaction costs involved in the process operate against any pareto improving trade. It is precisely the lack of any scheme to appropriate potential trade benefits that makes farmers and irrigation districts reluctant to invest in better water management systems. As long as their water rights are preserved, they only face the risk of allotment reductions imposed during drought periods. Due to the legal constraints imposed on water trading by the current water code, few economic studies have been done to estimate potential gains from market reallocations. González and Rubio (1992) estimated the aggregate welfare improvements that could be attained with various pricing policies imposed on users who would benefit from interbasin water transfers. However, their results should be taken as merely preliminary because of two major shortcomings. One is that water demands are assumed to be inelastic; thus, water deficits in some basins are taken as given. The other is the assumption that surplus basins can store and convey the needed water at no cost other than those strictly related to the construction of dams and aqueducts. What these authors do demonstrate is that water transfers at a national scale are potentially welfare improving. The research reported in this chapter is certainly narrower in geographical scope but it offers a more in-depth analysis of the demand for irrigation water. It also shows what can be achieved through the decentralization of water allocation systems in the agricultural sector.

14.4 The Analytical Framework

Two models were formulated to analyze differing schemes of water trading. The first is a conventional individual farm model to determine how farmers allocate their land and water resources. Since water availability constrains a given farmer's options, a private marginal valuation can be identified with the dual value associated with the water-constraint maximization problem. The second set of models simulates water markets among farmers – Model 2 – and among irrigation districts – Model 3. The market models are built on various surplus functions which, in turn, depend on the water marginal valuations obtained from the first model.

The Farm Model

The farmer's problem is represented by the following constraint maximization problem (Model 1 hereafter):

$$\max \pi(l_j, x_j, w_j) = \Sigma_j[\, l_j\, (p_j\, f_j(x_j, w_j) - r\, x_j - p_w\, w_j) - c_j\, (l_j)\,] \tag{1}$$

subject to:

$$\Sigma_j\, l_j \leq L \tag{2}$$

$$\Sigma_j\, w_j\, l_j \leq W \tag{3}$$

where, $\pi(.)$ is the profit function; l_j is the land area allocated to crop j, (j=1,...,J); p_j is crop j's price; $f_j(.)$ crop j's yield response; x_j is a variable input allocated to crop j; w_j is the amount water allocated to crop j; r is the variable input's price; p_w is the variable cost of one unit of applied water including water institutional charges; $c_j(l_j)$ is the cost function of crop j excluding x_j and w_j; L is the farm's land area; and W represents the amount of water available per hectare allotted to each farm by the basin agency. Written in this way, Model 1 assumes that farmers are price-takers in both input and output markets, and have perfect knowledge about water availability and crop prices. Although unrealistic in some settings, the assumption of a deterministic supply of water is consistent with the evidence indicating that farmers are provided with reliable information about their water availability before the planting season starts.[4]

If a solution to the primal problem of Model 1 exists, then two dual variables also exist, denoted by μ and λ, which represent the shadow values of the farmers' water and land constraints. If normal curvature assumptions are made for the yield functions, it is straightforward to show that at the optimum, the shadow value of water does not increase as more water is available, i.e., $\delta\mu^*/\delta W \leq 0$. The relationship between the dual variable of water and the amount of water available is central in this research because it will be later used to obtain an inverse water demand function.

A Model of Intra-Community Water Markets

As stated above, the shadow price of water depends on the amount of water available to the farm, w_i, where index i hereafter denotes an individual farmer; hence, there exists a function $\mu_I = \mu_i^*(w_i;\ \theta_i)$ relevant to each individual farmer (θ_i, representing his specific parameters). Moreover, the integral of that function up to the value of w_i is a measure of the farmer's surplus. Formally, each farmer's surplus, S_i^F, is given by:

$$S_i^F\ (w;\theta_i) = \int_0^{w_i} \mu_i^*(w;\theta_i)\, dw_i \tag{4}$$

Thus defined, an irrigator community's (IC)[5] hypothetical planner's objective is to reallocate water resources among the member farmers through a decentralized system in which none is made worse off, and the IC's aggregate surplus is maximized. This is expressed by Model 2 below:

$$\max I^{CAS}(b_i, s_i, p_{mar}) = \Sigma_i\, S_i^F(W_i;\theta_i) - t\, \Sigma_i\, (b_i + s_i) \tag{5}$$

subject to:

$$W + b_i - s_i \geq w_i \qquad i=1,...,I \tag{6}$$

$$\Sigma_i\, (b_i) - \Sigma_i\, (s_i) = 0 \tag{7}$$

$$S_i^F(w_i;\theta_i) - t\ (b_i + s_i) + p_{mar}\, (b_i - s_i) \geq S_i^F\ (W;\theta_i) \qquad i=1,...,I \tag{8}$$

$$p_{mar} - \mu^*_i(w_i;\theta_i) \geq 0 \qquad i=1,...,I \tag{9}$$

$$p_{mar},\ b_i\ ,\ s_i \geq 0$$

where $I^{CAS}(b_i, s_i, p_{mar})$ is the IC's aggregate surplus; $S_i^F(w_i;\theta_i)$ is farmer i's surplus (with $S_i^F(W;\theta_i)$ being farmer i's surplus under the baseline scenario – in which trade is not allowed); W is the initial water allotment, which is equal for all IC's farmers; w_i, b_i ,and s_i, denote water used after trading and the volume of water purchased and sold by farmer i, respectively; t is the unitary transaction cost; and, p_{mar} is the equilibrium market price.

In Model 2, equation (6) represents each individual farmer's water constraint; equation (7) is the IC's water market equilibrium; equation (8) is each farmer's incentive compatibility constraint that prevents any farmer from being made worse off after a water transaction; and equation (9) forces the market price to reflect the opportunity cost of water externally assigned to the IC. The w_i's are not choice variables, since they result from each seller's (s_i) or buyer's position (b_i). Moreover, the market model expressed in equations (5) through (9) requires that any water transaction render sellers better off and buyers no worse off. The existence of transaction costs does not guarantee that there is an optimum solution to Model 2, but if one exists, then the equilibrium price, p^*_{mar}, will be equal to the highest willingness to pay among the set of buyers. This market price will make at least one buyer equally well off, since this particular buyer has a willingness to pay for water equal to the market price and the shadow value. Hence, while equation (9) will be binding for one or several buyers, for sellers, equation (9) will be slack, which explains why sellers always gain from trading. This assumption has been imposed based on the empirical evidence available from water markets showing that sellers are more reluctant to trade water rights than buyers, especially when those rights cannot be considered strictly private (Howitt, 1993). Jacques (1992) also provides substantial evidence of the differing valuations of willingness-to-pay and willingness-to-accept compensation for holding resource rights.

A Model of Inter-Community Water Markets

Although water markets among farmers belonging to the same IC might be efficiency-improving, they would fail to distribute true scarcity signals for the whole region. This is because ICs face non-overlapping drought periods and differ in their irrigation technologies, basic allotments, and farm production systems. Therefore, allowing water trading among ICs would provide a better estimate of the opportunity cost of water used in agriculture than would separate markets developed for each IC.

The hypothetical regional planner is confronted with a problem (Model 3, hereafter) slightly different from the one that the IC planner faces, although with a similar structure:

$$\max R^{AS}(b_k, s_k, P_{mar}) = \Sigma_k I^{CAS}{}_k(w_k;\theta_k) - t\, \Sigma_k (b_k + s_k) \quad (10)$$

subject to:

$$W_k + b_k - s_k \geq w_k \qquad k=1,\ldots,K \quad (11)$$

$$\Sigma_k (b_k) - \Sigma_k (s_k) = 0 \quad (12)$$

$$I^{CAS}{}_k(w_k;\theta_k) - t\,(b_k + s_k) + P_{mar}(s_k - b_k) \geq I^{CAS}{}_k(W_k;\theta_k) \quad (13)$$

$$k=1,\ldots,K$$

$$P_{mar} - \mu^*{}_k(w_k;\theta_k) \geq 0 \qquad k=1,\ldots,K \quad (14)$$

$$P_{mar},\ b_k,\ s_k \geq 0$$

where, index k represents an irrigation community; $R^{AS}(.)$ denotes the regional aggregate surplus; W_k is IC_k's specific water allotment; θ_k are the specific parameters of each IC; P_{mar} is the regional water market equilibrium price; and $I^{CAS}{}_k$ are each IC's surplus function, and result from:

$$I^{CAS}{}_k(w_k;\theta_k) = \int_0^{w_k} \mu_k^*(w_k;\theta_k)\,dw_k \tag{15}$$

Equations (10) through (14), and Model 2's equations (5) through (9) have the same economic meaning. However, there are two important differences between Models 2 and 3. First, while the $\mu^*{}_i(.)$ that entered S_i^F in equation (4) are the shadow value functions of water relevant to each individual farmer, the $\mu^*{}_k(.)$ represent the shadow value of water of irrigation district k. The shadow value functions of each IC are hypothesized to be dependent on whether water at a district level is centrally assigned or reallocated through a market mechanism. In other words, it is, in principle, reasonable to assume that the $\mu^*{}_k(.)$ resulting from the horizontal summation of the farmers' shadow value functions that belong to the same district will differ from the $\mu^*{}_k(.)$ that are obtained from the equilibrium market prices obtained from Model 2. If intra-IC markets are not permitted, then each IC's willingness to pay for water at each level of water allotment, w_k, corresponds with the horizontal summation of its member farmers' individual willingness to pay or shadow value:

$$\mu_k^*(w_k;\theta_k) = \Sigma_i\, \beta_{ik}\, \mu_{ik}^*(w_k;\theta_{ik}) \tag{16}$$

where β_{ik} is the weighting index of farmer i in community k; and $\mu_{ik}^*(.)$ are the dual value functions which relate water availability of farmer i in community k and the dual value of the water constraint in Model 1. Equation (16) establishes a correspondence between each IC's shadow value of water with its level of water allotment which, through equation (15), provides a measure of each IC's surplus as a function of w_k.

Alternatively, if water trading reallocates water among the farmers belonging to each IC, its water demand function can be elicited from the market price equilibrium that would occur for any level of water allotment. As shown in Model 2, there is one equilibrium price for each level of water allotment. In addition, we can obtain a correspondence between that equilibrium price, which varies for different w_k, and the amount of water available for each IC, formally written as follows:

$$\mu_k^*(w_k;\theta_k) = p^*{}_{k,mar}(w_k;\theta_k) \tag{17}$$

where $p^*{}_{k,mar}(.)$ is the water market equilibrium price in community k for a given w_k.

The second difference is that, whereas in Model 2 all potential participants have the same initial water allotment (measured in volume of water per hectare), in Model 3, regular allotments differ and are subject to non-overlapping reductions due to locally differential climatic conditions. This last point becomes critical in the design of strategies to overcome the stress imposed by below average rainfall. Intuitively, the probability that the regional planner will find an optimum solution (s^*, b^*, $P^*{}_{mar}$) to Model 3 is greater with uncorrelated w_k's than with those that have the same pattern of drought occurrence. In other words, as the w_k's differ widely across ICs for a given year, it is very likely that their $\mu_k^*(w_k;\theta_k)$ will also be quite different before trade. This is precisely the environment in which markets would permit intense water reallocations, rendering all participants better off or, at least, no worse off.

Empirical Estimates

The empirical version of Model 1 has two features that make it different from an otherwise conventional farmer's model. First, the yield response functions for four of the five crops – comprising 80 percent of all region's irrigated land – have been obtained from simulated data of an agronomic model, EPIC, calibrated for the soils, weather, and farming practices common in the region[6] (Williams et al., 1990). Simulated yields for different levels of nitrogen and irrigation were used to obtain input-response quadratic functions for wheat, corn, oilseeds, and cotton. Since sugar beets, the fifth crop, is not properly simulated by EPIC, its response function was obtained from *in situ* experimental research conducted by Morillo-Velarde (1991). Second, crop and farm specific cost functions were obtained following the method developed by Howitt (1995), using farm census data to perform the calibration procedure. The estimated parameters for the yield response functions as well as the rest of the details can be found in Garrido (1995).

Four ICs were selected for the study, all located in the Guadalquivir Basin and within a distance of less than 50 km from each other. The four were selected to be representative of the basin's ICs, but they differ from each other in several respects. Two – denoted by IC-1 and IC-4 – are more than thirty years old, use flooding irrigation systems, and are located on the best soils in the flood plain. The other two – denoted by IC-2 and IC-3 – were recently developed and are equipped with piped networks and sprinklers but have lower quality soils. Each IC has its own storage facility, catching water originating in separate sub-basins. Because all were publicly developed, their water allotments were initially evaluated based on their major crops' water requirements and the prevalent irrigation and conveyance technologies. This explains why the two old ICs were given water rights double the allotments assigned to the new ones (see Table 14.1).

Thirteen representative farms were selected based on the farm census data (I.N.E., 1991) and each IC's annual reports. For each one of these, a specific cost function was calibrated for each of the five crops. Thus, thirteen different versions of Model 1 were calibrated in order to obtain a water demand curve specific for each representative farm, based on the amount of water available. For each farm and water charge scenario, the corresponding farm surplus function is given by:

$$S_i^F(w_i) = \frac{-k_i}{b_i} \times (1 - e^{w_i b_i}) \qquad (18)$$

where i indexes farms; w_i is the amount of available water; and k_i and b_i are the demand parameters estimated by regressing the dual values of water against W, in Model 1, assuming that $\log \mu_i = k_i + b_i W$, with $k_i > 0$, and $b_i < 0$.

The empirical model outlined above was estimated for the four ICs comprising 13 representative farms. In addition to the scenarios imposed by considering two levels of institutional water charges, (the current ones and another with a surcharge of \$0.04/m^3 to approximate basin charges required to achieve cost-recovery), two scenarios involving different transaction costs (U.S. \$0.008 and \$0.024 per exchanged m^3) were included. The data used to calculate the transaction costs was supplied by Saura Martínez (1995) and is based on the investments that would be needed in different irrigator communities to install individual measuring devices and more efficient conveyance systems. Moreover, the estimated transaction costs fall in the same ranges of the transaction costs estimated by Hearne and Easter (1995) from actual water market transactions in Chile.

The empirical version of Model 3 analyzes potential water exchanges between ICs. While in general it is closely related to the empirical version of Model 2, it allows differing rainfall and water availability among the ICs. Based on evidence from the statistical analysis of past water reservoir releases, a set of eight water reduction scenarios were considered (denoted by A1 to A8 in Table 14.1).

Table 14.1. Water Reduction Scenarios for the Inter-IC Water Market

IRRIGATORS COMMUNITIES	Regular allotment[a]	Water reduction scenarios in % over regular allotments							
		A1	A2	A3	A4	A5	A6	A7	A8
IC-1(Bembézar)	9000	100	100	50	50	20	50	20	10
IC-2 (Fte-Palmera)	4300	100	50	100	50	50	20	20	10
IC-3 (Gen-Cabra)	4300	100	50	100	50	50	20	20	10
IC-4 (Guadam)	7000	100	100	50	50	20	50	20	10

[a] Measured in m^3/ha

In addition, since the surplus functions are dependent on whether water is centrally or market allocated within each irrigation community, Model 3's implementation will consider another set of eight scenarios, as shown in Table 14.2.

Table 14.2. Inter-IC Water Market Scenarios

	Scenarios							
	1	2	3	4	5	6	7	8
Water institutional charges[a]	Current	Current	+ 4	+ 4	Current	Current	+ 4	+ 4
Intra-IC market ?	N[d]	N	N	N	Y	Y	Y	Y
Transaction costs[b] (intra-IC mkt)	-	-	-	-	0.8	0.8	0.8	0.8
Transaction costs[c] (inter-IC mkt)	2.4	0.8	2.4	0.8	2.4	0.8	2.4	0.8

[a,b,c] All values are expressed in U.S. cents per cubic meter; [d] N= no, Y=yes

14.5 Market Simulation Results

Intra-Community Water Markets

As indicated above, Model 2 was run independently for each one of the scenarios which resulted from combining two levels of institutional charges for water with two levels of transaction costs. In addition to these four institutional settings, three runs were made for each setting to see how the market would work when farmers receive 100, 60, or 20 percent of their regular water allotments. Table 14.3 summarizes the results for each IC's closed water market. After trade, welfare gains for the most favored farm and for the whole IC appear in columns 4 through 9. The welfare gain for one individual farmer is measured by:

$$\text{Welfare gain}_i = 100 \times \{[S_i^F(w_i;\theta_i) - t\ (b^*_i + s^*_i) + p^*_{mar}\ (b^*_i - s^*_i) - S_i^F\ (W;\theta_i)]/\ S_i^F\ (W;\theta_i)\} \quad (19)$$

where $S_i^F(.)$, whose value depends on the amount of water used by farmer i, is the surplus function from equation (18). The welfare gain for the whole IC is the summation of its farmers' welfare gains.

Table 14.3. Intra-IC Market Results (Model 2)

Irrigator Communities	MARKET EQUILIBRIUM PRICE IN $/m³			% SURPLUS GAIN FROM TRADE OVER CURRENT SYSTEM: Most favored farm after trade			% SURPLUS GAIN FROM TRADE OVER CURRENT SYSTEM: The whole IC		
	(simulated/regular water allotments)[a]								
	20%	60%	100%	20%	60%	100%	20%	60%	100%
[Market scenario: current institutional prices and transaction costs = 0.8 cents/m³]									
IC-1	-[b]	0.24	0.19	-	10	21	-	2	4
IC-2	-	0.18	0.15	-	2	3	-	1	1
IC-3	0.27	-	0.16	4	-	1	3	-	< 1
IC-4	0.18	0.11	0.07	34	15	19	12	7	5
[Market scenario: current institutional prices and transaction costs = 2.4 cents/m³]									
IC-1	-	0.23	0.18	-	6	17	-	2	3
IC-2	-	-	-	-	-	-	-	-	-
IC-3	-	-	-	-	-	-	-	-	-
IC-4	0.21	0.14	0.10	7	16	17	6	3	2
[Market scenario: current inst. prices+ 0.04 $/m³ and transaction costs = 0.8 cents/m³]									
IC-1	-	0.21	0.15	-	14	28	-	3	6
IC-2	0.25	0.18	0.12	1	2	7	<1	1	2
IC-3	0.21	-	0.13	6	-	1	5	-	1
IC-4	0.17	0.10	0.06	87	71	63	26	25	26
[Market scenario: current inst. prices+0.04 $/m³ and transaction costs = 2.4 cents/m³]									
IC-1	-	0.19	0.14	-	9	23	-	3	5
IC-2	0.26	-	-	1	-	-	< 1	-	-
IC-3	-	-	-	-	-	-	-	-	-
IC-4	-	0.10	0.07	-	64	62	-	20	20

[a] Regular allotments (100%) are: IC-1: 9,000 m³/ha; IC-2: 4,300 m³/ha; IC-3: 4,300 m³/ha; IC-4: 7,000 m³/ha [3].
[b] indicates scenarios in which no trade would take place.

One striking result of the intra-IC market simulations is the number of scenarios in which no trade takes place. Putting aside, for the moment, the important impact of transaction costs, it is clear that the intra-IC markets are more likely to produce water reallocations when farmers are assigned their regular allotments than when allotments are reduced as a result of water scarcity. This can be seen by comparing the number of cases in which there exists an equilibrium price under column 3 (100%) with those under columns 1 (20%) and 2 (60%).

Table 14.3 also shows that the welfare gains for the whole community are quite modest. Only in IC-4 for one scenario would surplus gains become greater than 25 percent over the current system. Except for this one, welfare gains are in the range of 1 to 12 percent for the ICs. The assumptions imposed on Model 2 allow significant surplus improvements for the farmers who sell water to other members of the IC. Thus, as columns 4 to 6 in Table 14.3 indicate, the distribution of surplus gains from the market model makes some farmers significantly better off, though the aggregate improvements are much more modest.

The equilibrium prices resulting from the markets range from \$0.07 to \$0.27 per m^3. Equilibrium price differentials are partially explained by the differences in regular water allotments. As expected, ICs with larger allotments experience lower market equilibrium prices, since in these ICs water is relatively less scarce than in the ICs with smaller allotments such as IC-2 and IC-3. But other factors also explain the equilibrium price differentials. Perhaps one of the most important is the IC's conveyance efficiency. The relatively low efficiency of IC-1's conveyance system explains why its' equilibrium water prices are higher than in the other three ICs, despite the fact that IC-1 has the largest water allotment. Because lower efficiency implies a larger fraction of water lost through the system, farmers in IC-1 effectively receive at their farm gates similar volumes of water as IC-2 and IC-3. Moreover, the fact that IC-1 has better soils than IC-2 and IC-3 might also help explain the wider price spreads.

What seems to influence the gains from markets are transaction costs. If the exchanging system requires that farmers pay \$0.024 or more for each cubic meter to cover the marketing expenses, then no trade would take place in IC-2 and IC-3, independent of the level of the allotments. Even if trade occurs at high transaction costs, as it would in IC-1 and IC-4, the welfare gains are significantly lower. Both unambiguous results cast doubt on the degree to which market participants would exchange water rights in a closed environment involving a decentralization process completely financed within the IC.

Farmers' strategies to sell or buy water would bring about changes in the way irrigated land is allocated to alternative crops. Perhaps one of the reasons why some farmers are willing to sell water rights is the reform of the Common Agricultural Policy (CAP). Since the implementation of the CAP reform increased the subsidies per hectare for crops which are moderate water consumers, water selling farmers can reduce the amount of land devoted to cotton and sugar beets and increase the land devoted to CAP reform crops, such as wheat, corn, and oilseeds. Hence, water markets would likely reallocate the land devoted to industrial crops and to CAP reform crops in a manner closely related to the way water rights are exchanged among market participants.

Inter-Community Water Markets

Two dimensions are used to summarize the inter-IC market simulations in Table 14.4, in which equilibrium prices together with individual IC and regional surplus gains are reported. First, the aggregate water demand curve of each IC is dependent on whether a market exists that provides some degree of water reallocation within each community or if water is assigned following the current centralized approach. Second, non-overlapping drought cycles occur that could encourage ICs to engage in water trading in a regional market. To simplify the presentation, Table 14.4 has been divided into two sections. The market results pertaining to inter-IC trades in which each IC's surplus function is obtained from equation (16) are shown in the first part, and the market results which rely on ICs' surplus functions obtained from equation (17) appear in the second part of Table 14.4. Several conclusions emerge from the different scenarios.

First, if the ICs were guaranteed their regular allotments in most seasons, then the regional market would be rendered virtually useless. In other words, if the relevant and most probable allotment reduction scenario were A1, then only in three out of the eight studied cases would the market generate some degree of water trading. While the allocation of water rights in publicly developed water districts results from centrally planned decisions, it is estimated for each IC based on crop water demands and the actual level of irrigation technology. In a sense, the pattern of reallocation that a water market would generate has already been performed through the process of assigning initial rights to the ICs.

Contrary to general belief, the public sector in this case did not fail to distribute the initial rights in close accordance with the relative productivity of water for the ICs included in the study. The problem arises when those initial rights are subject to frequent downward revisions because of droughts and the growth of other sectors' water demands. Second, since the pattern of water allotment reductions observed from actual data on reservoir releases is clearly asymmetric, the relevant reduction scenarios would be A2, A3, A5 and A6 rather than A1, A4, A7 and A8 (see Table 14.1 for details). With the former scenarios, trade would reallocate water rights among the ICs much more intensively than with the latter scenarios. This result indicates that regional exchanges might help alleviate water shortfalls that only affect one or two ICs; the ICs short of water would buy some from the ones with larger allotments, making the IC sellers clearly better off. The converse is also true: if droughts impose equivalent reductions for all ICs, then the fact that the ICs do not differ very much from each other means that a market will generate only a few trades.

Third, equilibrium prices reach similar levels to those found for the individual IC markets. They also reflect the high or low institutional water charges: scenarios 1, 2, 5, and 6, generate higher prices than scenarios 3, 4, 7, and 8. Similar to the intra-IC market simulations, if the market equilibrium price reflects the true value of water, then market prices will be higher when the water users pay lower charges to the basin agency.

Fourth, similar to the intra-IC market simulations, both surplus gains and occurrence of trade are highly dependent on the level of transaction costs. However, a caveat should be made regarding the level of transaction costs in the regional market. Inter-IC markets have two specific features which would make the lower transaction

Table 14.4. Equilibrium Prices and Surplus Improvements from Inter-IC Water Trading (Model 3)

Scenarios without intra-IC Markets		Allotment reduction scenarios						
		A1	A2	A3	A4	A5	A6	A7
1. Current inst charges;	Price[a] ($)	-[b]	0.19	0.21	-	0.28	0.24	-
Trans costs: 2.4 cents/m^3	Region (%)	-	2	2	-	3	5	-
	IC-1(%)	-	<1	<1	-	<1	<1	-
	IC-2(%)	-	2	3	-	<1	15	-
	IC-3(%)	-	<1	2	-	1	1	-
	IC-4(%)	-	10	6	-	30	21	-
2. Current inst charges;	Price ($)	0.16	0.19	0.22	0.22	0.28	0.24	0.29
Trans costs: 0.8 cents/m^3	Region (%)	<1	2	3	2	4	6	5
	IC-1 (%)	<1	<1	<1	<1	<1	<1	<1
	IC-2 (%)	<1	3	5	<1	1	18	2
	IC-3 (%)	<1	<1	3	<1	3	3	<1
	IC-4 (%)	3	14	10	16	36	26	39
3.Curr inst char:+0.04$/$m^3$	Price ($)	-	0.12	0.17	-	0.24	0.20	-
Trans costs: 2.4 cents/m^3	Region (%)	-	2	2	-	5	4	-
	IC-1 (%)	-	<1	<1	-		<1	-
	IC-2 (%)	-	3	3	-	2	4	-
	IC-3 (%)	-	1	3	-	3	<1	-
	IC-4 (%)	-	11	9	-	40	26	-
4.Curr inst char:+0.04$/$m^3$	Price ($)	-	0.13	0.17	-	0.24	0.20	-
Trans costs: 0.8 Cents/m^3	Region (%)	-	3	3	-	6	5	-
	IC-1 (%)	-	<1	<1	-	<1	<1	-
	IC-2 (%)	-	9	5	-	4	8	-
	IC-3 (%)	-	9	4	-	5	<1	-
	IC-4 (%)	-	9	15	-	48	34	-

Scenarios with intra-IC Markets		Allotment reduction scenarios							
		A1	A2	A3	A4	A5	A6	A7	A8
5. Current inst. charges;	Price ($)	-	-	0.24	-	-	-	0.27	-
Trans costs: 2.4 cents/m^3	Region %)	-	-	4	-	-	-	4	-
	IC-1 (%)	-	-	1	-	-	-	<1	-
	IC-2 (%)	-	-	8	-	-	-	8	-
	IC-3 (%)	-	-	4	-	-	-	4	-
	IC-4 (%)	-	-	7	-	-	-	19	-
6. Current inst charges;	Price ($)	0.19	0.19	0.24	0.25	0.30	0.27	-	0.34
Trans costs: 0.8 cents/m^3	Region (%)	1	2	4	2	4	4	-	4
	IC-1 (%)	<1	<1	<1	<1	<1	<1	-	<1
	IC-2 (%)	1	3	9	<1	3	7	-	<1
	IC-3 (%)	<1	5	5	<1	3	<1	-	<1
	IC-4 (%)	3	5	9	14	30	22	-	33
7.Curr inst char:+0.04$/$m^3$	Price ($)	-	0.16	0.20	0.21	0.26	-	-	-
Trans costs: 2.4 cents/m^3	Region (%)	-	3	4	2	5	-	-	-
	IC-1 (%)	-	<1	<1	<1	<1	-	-	-
	IC-2 (%)	-	<1	<1	8	7	-	-	-
	IC-3 (%)	-	<1	3	<1	5	-	-	-
	IC-4 (%)	-	20	13	18	27	-	-	-
8.Curr inst char:+0.04$/$m^3$	Price ($)	0.13	0.15	0.20	0.21	0.27	-	0.28	0.31
Trans costs: 0.8 cents/m^3	Region (%)	2	4	5	3	8	-	6	8
	IC-1 (%)	1	<1	<1	<1	<1	-	<1	<1
	IC-2 (%)	<1	4	11	<1	11	-	2	4
	IC-3 (%)	<1	7	6	<1	8	-	2	8
	IC-4 (%)	13	21	18	25	34	-	39	44

[a] Equilibrium prices are quoted in U.S. $ per cubic meter.
[b] Indicates scenarios for which no trade would occur.

cost assumption more realistic. First, market transactions among ICs would result from negotiations among ICs' representatives, each of whom would act on behalf of several hundred farmers. Second, water exchanges would not necessarily require physical transfers of water but transfers of water rights for users downstream.

Finally, welfare gains for the region are fairly moderate. For some ICs, particularly for IC-4, the gains-from-trade are large. This results from the fact that IC-4, though it is more than thirty years old, has experienced an intense rate of technology adoption without having its regular allotment reduced to reflect its actual level of irrigation efficiency. Hence, it is an IC that can easily sell a part of its allotment without significantly reducing crop yields.

14.6 Discussion and Conclusion

The empirical analysis of water markets has shown that decentralization generates a range of short-run probable gains. If markets were limited to the farmer members belonging to the same irrigators community, surplus gains would not exceed 10 percent except for the old ICs. This is because in the older ICs differences among potential market participants have developed with the passage of time. Recently developed ICs encompass more homogeneous sets of farmers who select similar cropping strategies and use virtually the same irrigation techniques. Under these circumstances, markets only accessible to community farmers would have a limited role. Moreover, this research has shown that intra-IC markets' benefits are larger when farmers receive regular allotments than when droughts impose proportional reductions in farmers' water rights. Therefore, intra-IC markets may not be a great help in alleviating the impacts of droughts.

However, in interpreting these results it should be noted that the model results reproduce strict short-run effects. It is reasonable to assume that the shift from a purely centralized environment to a market system would generate indirect effects as well. Some of these long- or medium-term responses would imply changes in farmers' water management systems, farms' irrigation equipment, and farmers' attitude toward their IC's strategies for water conservation. Comparing the price equilibria obtained in different market settings and ICs with the investment required to introduce more advanced irrigation technologies, it becomes clear that a likely response to a market would be improved water-use efficiency. The extent to which this alternative is actually used by farmers remains to be studied in further research. But as Caswell and Zilberman (1985) demonstrated, the incentives to invest in technology decrease as soil quality and water availability increase, which is the case in the old ICs. Thus, the economic incentives operate against the desired goal of increasing the water-use efficiency in old districts with ample water allotments.

A larger scope for welfare-improving markets emerges at a regional level if ICs are allowed to exchange part of their allotments. This is particularly true if the ICs that exchange water have allotment reductions due to drought cycles which are not totally overlapping through time. This research provides evidence about the limited role of regional markets when ICs get their regular allotments or suffer proportional allotment reductions. Though not too far from each other, the four ICs included in this study get their water supply from reservoirs in different sub-basins. The fact that micro-climatic conditions may modify regional weather patterns increases the probability of asymmetric allotment reductions. This purely natural effect, together with a variant set of ICs, creates ideal conditions to establish a system of water exchanges among ICs. Although the estimated welfare gains obtained from the various market simulations were always less than 8 percent of the total gains created by the current system, if these gains can be achieved for a number of years, the scope for increasing these modest gains is promising. The fact that these results fall in the realm of short-term changes suggests that larger changes could occur in the medium- and long-term.

Finally, the research shows that both the intra- and inter-IC markets are very sensitive to the magnitude of transaction costs. Developing a water market *ex novo* requires not only new investment in facilities to permit water transactions but also the creation of a set of rules which are accepted by all potential participants. This implies that rights can be enforced and that conflicts can be solved at a relatively low cost. If the market's costs require users to pay a participation fee higher than about \$0.02 /m^3 exchanged, then water markets among farmers belonging to the same district as well as those in different districts would have few participants. An alternative solution would involve defraying some of the market's overhead by the public sector. But that would require that the public sector view water markets as a feasible option to the construction of expensive structural facilities.

Endnotes

[1] This work is based on results obtained in my doctoral dissertation. I am indebted to Prof. Carlos Romero for his comments and suggestions. Funding from the Spanish Ministry of Agriculture is also acknowledged.

[2] All figures throughout the text are denominated in U.S. dollars converted from pesetas at the average rate of 125 pesetas/\$.

[3] Southern and southeastern basins are (no. of hectares in parenthesis): Guadalquivir (443,000), Guadalete + Barbate (34,200), Jucar (380,000), Segura (235,000) and Sur (152,000). This research is based on irrigation districts located in the first basin.

[4] In virtually all water districts whose water allotments are subject to reservoir levels, district managers have sufficient information to anticipate minimum guaranteed water allotments before the planting season starts. If spring rainfalls are above average, they might raise the allotments correspondingly. This information was confirmed by the author through interviews with several water managers who have direct responsibility for water supplies pertaining to this study.

[5] Water districts are hereafter referred to as irrigator communities, since this is the literal translation of *Comunidades de regantes* in Spain.

[6] EPIC (Erosion Productivity Impact Calculator) has been calibrated to the region as a part of the E.U. funded project entitled: *Analysis of the socio-economic impacts of different policies in some European regions: competitiveness and environmental protection.* D.G. VI, Contract No. 8001-CT91-0306.

References

Caswell, M. and D. Zilberman, 1985. "The Choices of Irrigation Technologies in California." *American Journal of Agricultural Economics* 67 (2), 224-234.

Domínguez Vila, A., 1996. *"El Mercado de Aguas en Canarias." Paper presented at the VIth Meeting on Water Law*. Zaragoza, Spain, March.

Fisher, A., D. M. Fullerton, N. Hatch, and P. Reinelt, 1995. "Alternatives for Managing Drought: A Comparative Cost Analysis." *Journal of Environmental Economics and Management,* 29, 304-320.

Garrido, A., 1995. *La Economía del Agua: Análisis de la Asignación de Recursos Mediante el Establecimiento de Mercados de Derechos de Agua en el Valle del Guadalquivir.* Doctoral Dissertation, Universidad Politécnica de Madrid, Madrid.

González, A. and S.J. Rubio, 1992. "Optimal Interbasin Water Transfers in Spain." Report for the European Science Foundation's research program: *Sharing fresh resources in the Mediterranean region: An economic perspective.*

Hearne, R. R. and K.W. Easter, 1995. *Water Allocation and Water Markets. An Analysis of Gains-from-Trade in Chile.* World Bank Technical Paper Number 315, World Bank, Washington, D.C.

Howitt, R. E., 1993. "Empirical Analysis of Water Market Institutions: The 1991 California Water Market." Paper presented at the Conference: *Economic Aspects of International Water Resources Utilization in the Mediterranean Basin.* Fondazione Eni Enrico Mattei. Milan, Italy.

Howitt, R. E., 1995. "Positive Mathematical Programming." *Journal of Agricultural Economics*, 77(2):329-342.

I.N.E., 1991. *Censo Agrario 1989.* Madrid.

Jacques, S., 1992. "The Endowment Effect and the Coase Theorem." *American Journal of Agricultural Economics,* 74: 1316-1321.

Maass, A. and R. L. Anderson, 1978. *...and the Desert Shall Rejoice. Conflict, Growth and Justice in the Arid Environments.* M.I.T. Press, Cambridge.

MAPA, 1995. *Avance del Plan Nacional de Regadíos.* Secretaría General de Desarrollo Rural y Conservación de la Naturaleza. Ministerio de Agricultura, Pesca y Alimentación, October, Madrid.

Michelson, A. M. and R. A. Young, 1993. "Optioning Agricultural Water Rights for Urban Water Supplies During Drought." *American Journal of Agricultural Economics,* 75: 1010-1020.

MOPTMA, 1993. *Memoria del Proyecto del Plan Hidrológico Nacional.* Madrid.

Morillo-Velarde, R., 1991. *Respuesta de la Remolacha de Siembra Otoñal al Riego.* Doctoral dissertation. Universidad de Córdoba, Córdoba, Spain

Pérez-Diaz,V., J. Mezo, and B. Alvarez, 1996. *Política y Economía del Agua en España.* Círculo de Empresarios, Madrid

Saura Martínez, J., 1995. "La Modernización de Regadíos." *El campo* 132, 185-200.

Tobarra, P., 1995. *Estudio del Alto Guadalentin Desde la Perspectiva Economíca de la Gestíon del Agua Subterránea.* Caja de Ahorros del Mediterraneo, Murcia (Spain).

Williams, J. R., P. T. Dyke, W. W. Fuchs, V. W. Benson, O. W. Rice, and E. D. Taylor, 1990. *EPIC: Erosion Productivity Impact Calculator.* United States Department of Agriculture, Temple, Texas.

15 Welfare Gains from Potential Water Markets in Alberta, Canada

Theodore M. Horbulyk and Lynda J. Lo

15.1 Introduction

There are few jurisdictions where the topic of potential gains from water markets is as timely and central as in the province of Alberta, Canada. Following five years of public review and discussion, the Legislative Assembly of Alberta passed legislation in August, 1996, that will allow market transfers of the right to use surface and groundwater among competing users and uses (Alberta, 1996).[1] The new approach represents a major shift away from a legislative framework and a water management approach that had remained essentially unchanged for the previous hundred years. Historically, water rights have not been tradable among sites or uses, and there has not been an efficient rationing system to deal with water shortages. Whether or not this major change in water policy will be viewed as a success in the longer term is an open question. It is essentially a question about whether water markets will be able to generate significant welfare gains relative to the status quo, and relative to other policy options.

In practice, the 1996 legislative change approves only a new framework for water resource management. The technical provisions of the Act have not yet taken effect, and are not expected to come into force until 1998 at the earliest. Water management in the province is in a state of transition, waiting for the creation of necessary regulations, regional water management plans, and the approval, enforcement and appeal processes. Even so, water users are now in a position to start to plan their own futures under the new policy regime.

Although little is known about the many ways that the creation of water markets will change resource use and valuation in the province, there is a significant amount that can be learned from established knowledge of market behavior in general, and from observation of water markets elsewhere. Specifically, absent any current observations on the water markets that will be created, hypothetical or potential markets can be created and analyzed under a range of assumptions about how similar markets might operate in practice.

There are many important questions to be asked about the design, operation, and anticipated outcomes of water markets. When answers cannot be found by observing the experience of others, that is, when there are few so-called natural experiments in other similar jurisdictions, then computer simulation can be a valuable tool. Specifically, computer modeling can simulate the effects of a market where none currently exists. Modeling can describe expected changes in resource usage as well as the effects of these changes on the economic welfare of the water users. Moreover, where there are structural or procedural alternatives for the design of new markets, the effects of these alternatives can be portrayed in advance.

Envision, in what follows, the creation of a market in southern Alberta for transferable water entitlements where no such market currently exists. Some of the questions that may be asked include the following:

- How will the water be (re-)allocated under the new system?
- Who will win and who will lose, and by how much?
- What is the impact on resource users of varying the market's geographical and sectoral scope? (i.e., What should be the set of agents with whom each agent can trade?)
- What is the effect on resource users of varying specific property rights endowments when the new policy is introduced?

This chapter, then, considers the case of the southern region of the province of Alberta, Canada, where surface water drawn from river flows has traditionally been allocated using a command-and-control approach. The chapter examines short-run behavior in a simulated market, "as if" in a period of severe water deficit or drought.

The principal questions explored here concern the magnitude of changes in water use and associated welfare gains that might arise from short-term water reallocation, such as from rural to urban users and among rural users experiencing varying degrees of water deficit within a single growing season. The problem's market scope dimension refers to the design issue of whether or not water can be traded across sectors, across river sub-basins, or indeed, whether public entitlements to water, such as for instream flow needs or for interjurisdictional transfers, can also be traded in the markets (for example, by a public resource agency). The problem's property rights dimension refers to what (new or historical) rights or entitlements will be granted to (or preserved for) the existing agents when a new private market is established.

As an exercise in applied economic research, this chapter builds on a rich tradition of earlier papers by Samuelson (1952), Enke (1952), Takayama and Judge (1964), Flinn and Guise (1970), Vaux and Howitt (1984), Enright and Lund (1991) and Booker and Young (1994), all of which use static optimization and mathematical programming approaches to characterize optimal allocations of water under scarcity. In addition to its exploration of scope and property rights endowments, this chapter enriches the existing literature by incorporating in its model specific instream flow needs and exogenous demands for interjurisdictional transfer flows.

Three of the findings that will be presented are likely to generalize to other jurisdictions, and indeed, to other natural resource problems where economic instruments might be introduced. First, where the scope of potential markets is at issue, some of the largest gains might be realizable with even the smallest exposure to market forces. Thus policy makers might choose to "start small" and expand slowly. Second, by careful design of the "property rights" dimension of the new resource policy, almost any desired distribution can be achieved for the potential (monetary and non-monetary) welfare gains that the new policy will generate. Third, if the largest gains are to be realized where markets and market forces are at work, governments must be prepared to participate as active players in meeting their own (or society's) resource needs. At a minimum, governments should base their resource-use decisions on the information about resource values and scarcity that is generated by these new markets.

The five following sections of this chapter present an overview of the historical and emerging regulatory environment in Alberta, a description of the programming model, the scenarios analyzed, the welfare results compared, and the conclusions drawn.

15.2 The Alberta Water Economy

The province of Alberta is bounded on the west by the continental divide (the Rocky Mountains) – a distinguishing characteristic for its patterns of water supply and use (Figure 15.1). The province is bounded on the south by the state of Montana, with which Alberta shares relatively little area of common drainage or watershed and relatively little transboundary surface flow. Any surface drainage in the southern drainage basin of Alberta is to the east, toward the province of Saskatchewan with which a water-sharing (or so-called apportionment) agreement is currently in force.

There are four major connected river sub-basins that provide the main source of fresh water for economic activity in southern Alberta. There is considerable variability in the aggregated flow volumes in these four sub-basins over the years 1912 through 1986. The mean annual volume is 9.2 billion m^3 with a variance of 2.8 billion m^3 (Alberta Environment, 1988). Average annual precipitation levels range from 35 to 43 cm (14 to 17 inches) with values in the lower range as one moves south and east – that is, in the downstream direction of flow in the major river sub-basins. A historical interprovincial agreement requires in essence that, during each season, half of the total surface flows emanating from these four sub-basins in southern Alberta must be left in the South Saskatchewan River channel for the benefit of users in the downstream province (Prairie Provinces Water Board, 1969). Given the flow pattern of the rivers, Alberta, the upstream province, can choose which sub-basins are to supply this requirement (Alberta Environmental Protection, 1994). Although most of southern Alberta's water is fully allocated to existing users even in a year of average water levels, this chapter will examine the short-run reallocation of water in a year of extreme water deficit.

During the summer months about 90 percent of water use is for the irrigation of agricultural crops, including forages, grains, and vegetables. Most irrigators in southern Alberta gain access to irrigation water through membership in an irrigation district which holds the necessary water licenses. There has been a steady expansion of areas under irrigation within the boundaries of irrigation districts in southern Alberta over the years 1911 through 1993, reaching about 450,000 ha by 1993 (Alberta Agriculture, Food and Rural Development, 1995). Historically groundwater has not been an important source of water for irrigation use, accounting for only 3 percent of Alberta's annual consumptive water use (Alberta Environmental Protection, 1994).

Although water storage structures are commonly operated by irrigation districts or by the Province, water storage is less commonly practiced following diversion from the river by private (individual) licensees. Water storage is perhaps more important for hydroelectric power generation and for urban supplies, since the preferred periods for these uses are less closely matched (than are irrigation uses) to the periods of peak river flow.

Water has been used for irrigation in Alberta since the 1870s and has been subject to legislation since 1894. Until the 1996 Water Act comes into force, the legislation that governs water allocation (Water Resources Act, 1931) features non-transferable, site-specific and use-specific licenses without any significant pricing or fee-based allocation mechanism.[2] Water licenses are defined on the basis of (gross) annual volumes diverted from a watercourse or from groundwater, so that rates of return flow (i.e., non-consumptive use) are not credited or regulated.[3]

Figure 15.1. Southern Alberta Water Systems

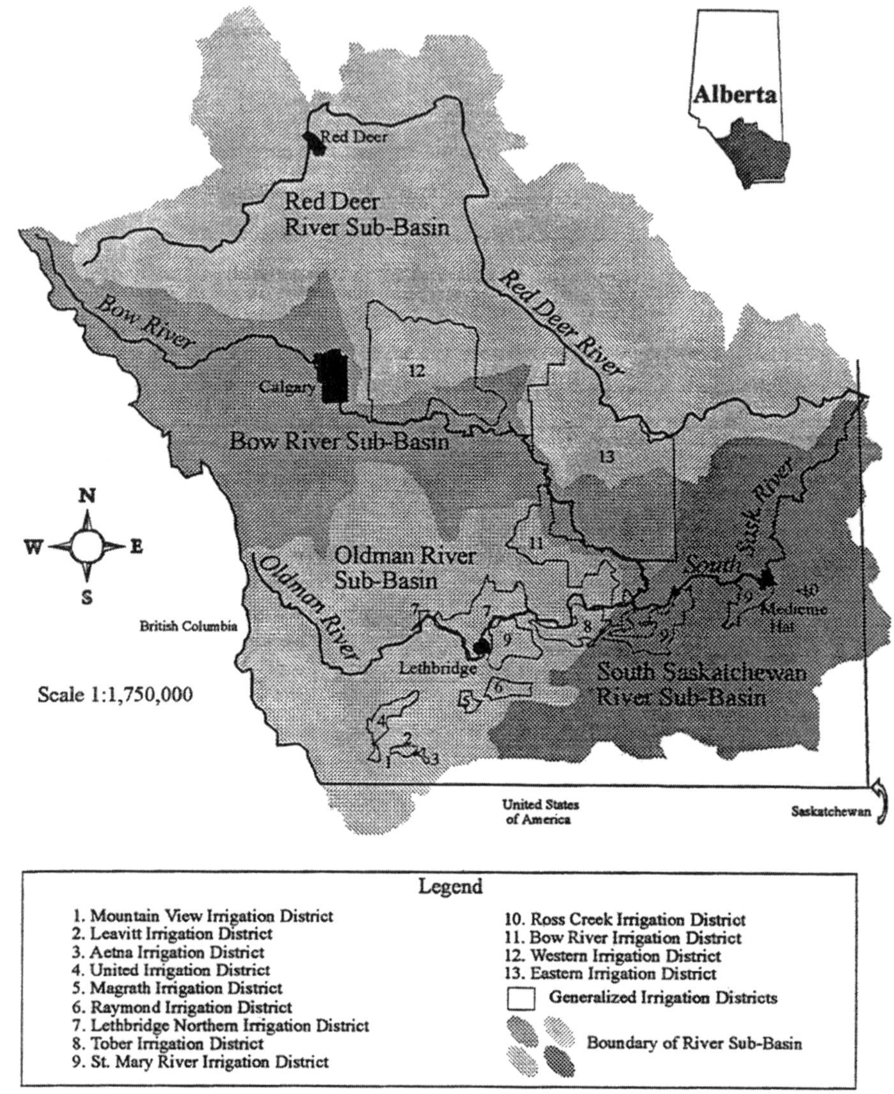

Cartographer: Steven Waters

Alberta's water "use permits" are legally tied to specific parcels of land, and historically have been transferred from land sellers to land buyers. There is evidence that these previous water transfers have been priced at (implicit) prices that reflect the (capitalized) future value of water use at a specific site as part of the total price paid for a parcel of land (Adamowicz and Horbulyk, 1996).

The existing licenses are held by cities, individuals, private corporations, and large irrigation districts, where the latter are allowed to redistribute water internally to individual members of the district. Each permit's seniority date (the rights were granted over time on a historical basis) defines its relative priority to draw all of its maximum allowable volume in the event of shortage. If quantity rationing should become necessary in a specific year, a junior right may be denied any water until all senior rights have been met (Veeman and Freeman, 1993).[4]

The existing appropriative water rights are valid indefinitely and are the rights that will become tradable under the new provincial water policy.[5] Historically, there have also been co-existing riparian rights and "license-exempt" uses of small quantities of water, such as for rural domestic use and for stock watering. Under the new regime, some "small-volume" uses will remain exempt and other "larger-volume" uses will be converted to equivalent appropriative rights with a seniority date defined by the year the current use actually started.

The 1996 Water Act is the product of a five-year public consultation process (see Alberta Environment, 1991; Alberta Water Management Review Committee, 1995). The legislation serves purposes other than setting up tradable rights. For example, it mandates the establishment of regional water management plans; it restricts transfers of water among major river basins and restricts international exports of water; it streamlines the licensing and approval process for water users; and it encourages new forms of regulatory enforcement and public consultation. Following early stages of public consultation, the government removed parts of the legislation that would have enabled the introduction of water resource royalties or water pricing, or that would have enabled the use of economic instruments other than trading.

As it relates to tradable water rights, the effect of the 1996 Water Act will be to transform historical licenses into marketable commodities. The market is expected to be free from other regulatory quantity or price limits on private trades. The government will oversee and approve temporary or permanent trades among users, provided these are not inconsistent with the regional water management plans that will be published. The government has allowed itself the privilege of holding back, for other public or private use, up to 10 percent of the water volumes proposed to be transferred in a private trade, provided that two relatively innocuous conditions are met. The "hold-back" must be in the public interest and must be made in pursuit of some water conservation objective. The government will also be able to buy and sell water rights on the private market, and may suspend or cancel other water rights (including those not offered for private sale), subject to various terms and conditions.[6]

Numerous changes in resource use are likely to occur as a result of the new legislation and its appeal to market forces. Some of these changes will be longer-term structural changes in the way that water is supplied and demanded, eventually including changes to the patterns, locations and sources of water used, the crops irrigated, the water application methods employed and so on. There might also be short-term changes that reflect responses to the newly-created opportunity to trade water on a season-by-season basis with others. Nowhere will the effect of this water market be more evident than in the situation where the creation of such a market coincides with an acute and largely unanticipated decrease in water supply during the irrigation season. That hypothetical event is examined in the analysis that follows.

15.3 The Programming Model

The programming model describes how, within a five-month irrigation season under a severe water deficit, a frictionless and perfectly functioning competitive spot market for surface water (i.e., a market with full information and without transaction costs) determines each user's price, quantity demanded, total payment, and associated consumers' surplus. These water allocations can be accomplished by maximizing the sum of producers' and consumers' surplus accruing to all participants in the market, subject to a series of constraints that define available water supplies, conveyancing technology, and instream flow needs. In a specific technical sense, the purpose of the exercise is to determine and compare static optimum allocations of surface water supplied by four rivers and demanded by spatially-separated water users within an irrigation season.

Southern Alberta's four river sub-basins are those of the Bow River, the Oldman River, the Red Deer River, and the South Saskatchewan River. The first two rivers drain into the fourth within the province of Alberta whereas the Red Deer River drains into the fourth near the provincial boundary of Alberta with the province of Saskatchewan (see Figure 15.1).

The within-season water supply behavior in these basins can be thought of as exogenously determined, and thus independent of seasonal demand or expected water prices. Aggregate water supplies come from natural flows, unaided by changes in storage, diversions or groundwater supplementation. (In practice, these activities could occur on-farm after surface flows have been pumped or diverted from river channels.) Moreover, for the purposes of guiding actual market behavior or of simulating it numerically in a deterministic model, much will be known about expected seasonal river flows in time to inform potential water traders prior to a water-trading spot market that precedes the irrigation season. This is because natural supplies are a direct and predictable outcome of winter precipitation levels, snowpack accumulations and other upstream hydrology. Following Flinn and Guise (1970), seasonal water supplies in each sub-basin are represented by an infinitely inelastic supply curve.

Numerous demands for consumptive and non-consumptive uses of water, represented here at the sub-basin level, are grouped into four broad categories: (1) agricultural demands, (2) urban and industrial demands, (3) instream flow needs, and (4) additional water required to meet the interprovincial apportionment agreement.

Following Enright and Lund (1991), agricultural, and urban and industrial demands are represented as linear inverse demand functions calibrated using data from Alberta and from other studies in the agricultural economics literature. Short-run factor demand curves for water use in agriculture have been derived from the work of Birch and Van Deurzen (1984). These authors use data from field studies conducted between 1949 and 1968 to estimate incremental returns to irrigation water on-farm, for a mix of the principal irrigated crops in each of the four river sub-basins. Their estimates may be thought of as representing short-run, region-specific, valuations of irrigation water per hectare for a range of foreseeable water application rates.

Short-run irrigation water demand at the sub-basin level was derived by summing (horizontally) these per hectare demands across all areas irrigated in each sub-basin, using 1993 as a representative or base year.[7] The set of linear demand curves thus derived was found to underestimate seasonal water use by agriculture in 1993 for a water price of zero. (Note that 1993 was not a year of extreme water deficit as envisioned in this reallocation problem.) To preserve the relative valuations of water for irrigation in each sub-basin, while at the same time representing aggregate water demand more accurately, the vertical intercept of each of these linear curves was increased, holding its slope constant. This "model calibration" proceeded until quantities supplied and demanded were equal for a price of zero in 1993. Drought conditions were simulated by choosing, for each sub-basin, a seasonal water supply value that was exceeded in 90 percent of the years between 1912 and 1986 (Alberta Environment, 1988).

Urban and industrial demand curves were derived for the principal city in each sub-basin by observing current quantities demanded and municipal water rates. A linear curve was extrapolated by applying published own-price elasticity of demand values from the western U.S. (Gibbons, 1986).

The demands for instream flow needs, and additional apportionment requirements are (by assumption in the former case) not price sensitive and, therefore, are represented by vertical demand curves. Instream flow demand is defined as a minimum quantity of water required to remain in the river channel in order to meet environmental and recreational needs. Apportionment requirements are defined as the extra volumes of water (beyond the minimum instream flow requirements) to meet the interjurisdictional transfer agreement at the Alberta-Saskatchewan provincial boundary. This approach gives rise to three demand groups in each sub-basin and four sources of supply (one in each sub-basin) plus one interprovincial apportionment demand.

The objective of the static optimization problem is to maximize the post-trade sum of producers' and consumers' surplus associated with a feasible equilibrium allocation. This choice of objective function parallels that of Enright and Lund (1991), although, in the present case, a richer set of constraints binds the set of feasible outcomes.

The first constraints describe the physical relationships among the four sub-basins, where any short-term trading or reallocation can only occur by using the free and costless gravitational downstream flow of each river.[8] Within limits, a downstream user can sell water to an upstream user on the same watercourse by reducing downstream consumption, and vice versa. Again, within limits, such trades might also occur between users on different interconnected watercourses if they have a common trading partner to act as intermediary. Therefore, it would not be necessary to construct additional aqueducts or conveyance infrastructure in order to implement the transactions described by the programming model. The second set of constraints requires the minimum instream flow volumes and apportionment flows to be met or exceeded. An algebraic representation of the programming problem is coded and solved using GAMS (1991) (Lo, 1995).

An important point in allowing inter-sectoral or inter-sub-basin trading is one's choice of units by which to measure water withdrawals and consumption. For the numerical analysis, all usages, supplies and demands have been measured on a net consumption basis, that is, by adjusting total diversions for the average return flows that result in each specific urban or agricultural usage.[9]

15.4 The Scenarios Analyzed

In characterizing the market-determined water transfers that might be allowed under the new water policy, it is instructive to consider a series of alternative scenarios that could be pursued, especially where the market scope of these scenarios varies. For example, there might be an increased social concern about potentially harmful externalities or third-party effects if trades are allowed to take place over large distances or between two different sub-basins. A scheme that allows only local trades might be seen as a less controversial way to introduce water markets. Similarly, there might be nationalistic reasons for excluding potential trades with another downstream jurisdiction (be that Saskatchewan or the U.S.).

In this study, a base case (no market) and three definitions of allowable market scope are used to portray the effect of spot markets that allocate seasonal surface flows within and among the four sub-basins. In Scenario 1, there is no scope for markets—no trade is allowed. This "base case" is modeled by restricting water allocations to historical values representing those that are typical of a drought year in the absence of pricing or markets for river water withdrawals.

The maximized value of total consumers' and producers' surplus (economic welfare, hereinafter) in Scenario 1 forms the benchmark for subsequent comparison with functional markets. In this scenario, all water prices are zero, and since supply curves are vertical, there is zero producers' surplus. Thus, to estimate welfare levels in Scenario 1, the areas are calculated under the rural and urban water demand curves up to the annual water volumes consumed in each region. This provides a measure of users' total willingness to pay for these water allocations. This measure is all consumers' surplus, since in Scenario 1, as in historical practice, water is unpriced in Alberta.

The scope of trade in Scenario 2 is between neighbors in a sub-basin (i.e., rural to urban) holding the historical total consumption constant in each basin and holding constant each basin's flow contribution to instream flows and interjurisdictional apportionment. Whereas in Scenario 1 there are no markets, in Scenario 2 there are four local ones.[10]

Scenario 3 broadens market scope to allow water trading across sub-basins, thus establishing one regional market. As in Scenario 2, Scenario 3 holds constant each basin's flow contribution to instream flows and interjurisdictional apportionment, allowing only agricultural and urban users to trade their previous (Scenario 1) "entitlements."

Scenario 4 relaxes a constraint present in the previous scenarios and allows an optimal reallocation of water supply for the purposes of instream and apportionment needs. This outcome is achieved "as if" the water regulator also entered the market, trading with other agents to reduce the public cost of meeting fixed instream and apportionment demands.[11]

From a policy perspective, Scenario 2 is motivated by the suggestion that, if there is to be trading, it should proceed on a sub-basin by sub-basin basis using each sub-basin's historical entitlement. Such an approach can be administered and monitored locally. Moreover, it could prevent the spatial transmission of any negative effects that might arise with unrestricted trading. For example, when water transfers are allowed across sub-basins, downstream users might sell water rights to users in an upstream sub-basin. This could result in considerably lower instream flows downstream from the new point of diversion. Scenarios 3 and 4 both allow such transfers, but Scenario 3 allows private transfers of only the historical water entitlements, holding constant the public authority's sourcing of water for instream flow needs and for interprovincial apportionment. For resource managers, Scenario 3 might be easier to implement, since, unlike Scenario 4, it does not require active participation by resource managers in these new markets, nor does it require the redesign of historical instream flow management practices.

Table 15.1 characterizes the pattern of reallocation of net surface flows within season in the four river sub-basins. There is an aggregate net supply of 4,147 million cubic meters per season, of which half must be allowed to flow to the downstream province in each scenario. Scenarios 2, 3 and 4 transfer increasingly more of the available supply from rural to urban users in times of drought. That is, 8.1, 8.2, and 10.4 percent, respectively, of the available supplies are transferred from rural to urban uses. Overall, this represents about 23 percent of rural users' Scenario 1 consumption, yet since urban consumption is so much smaller, this is equivalent to nearly doubling (190 percent increase) the water that would be made available to urban (residential and industrial) users.

Table 15.1. Reallocation of Available Supplies Under Three Alternative Scenarios (net surface flows in millions of cubic meters per season)

	Consumptive use of water by sub-basin and by type of use in:			
Water Users	Scenario 1 (Base Case)	Scenario 2	Scenario 3	Scenario 4
Bow River				
Rural	853	702	628	581
Urban	73	224	224	223
Instream	838	838	838	912
Sub-total	*1,764*	*1,764*	*1,690*	*1,716*
Oldman River				
Rural	686	540	535	504
Urban	84	230	230	228
Instream	835	835	835	851
Sub-total	*1,605*	*1,605*	*1,600*	*1,583*
South Saskatchewan River				
Rural	307	268	345	332
Urban	20	59	61	61
Instream	1,604	1,604	1,604	1,635
Sub-total	*1,931*	*1,931*	*2,010*	*2,028*
Red Deer River				
Rural	0	0	0	0
Urban	50	50	50	145
Instream	387	387	387	291
Sub-total	*437*	*437*	*437*	*436*
RURAL TOTAL	1,846	1,510	1,508	1,417
URBAN TOTAL	227	563	565	656
INSTREAM TOTAL[a]	2,074	2,074	2,074	2,074
SCENARIO TOTAL	4,147	4,147	4,147	4,147

[a] Since water left for instream uses in the Bow and Oldman Rivers is also available for instream or consumptive uses in the South Saskatchewan Sub-basin, the former amounts are subtracted from the total to avoid double counting.

Figure 15.2 portrays graphically the aggregate welfare gains that accrue from water allocation across the four scenarios.[12] The effect giving rise to the progressive changes across scenarios is the increasing scope of the proposed markets. More than 90 percent of the potential short-term welfare gains from water reallocation are due to the simplest introduction of (local) markets and market forces within individual sub-basins. This may be an important outcome if policy makers are concerned about other unspecified costs or problems associated with interbasin transfers.

Figure 15.2. Welfare Gains as a Function of Volume of Water Reallocated from Rural to Urban Uses (with Progressively Greater Market Scope)

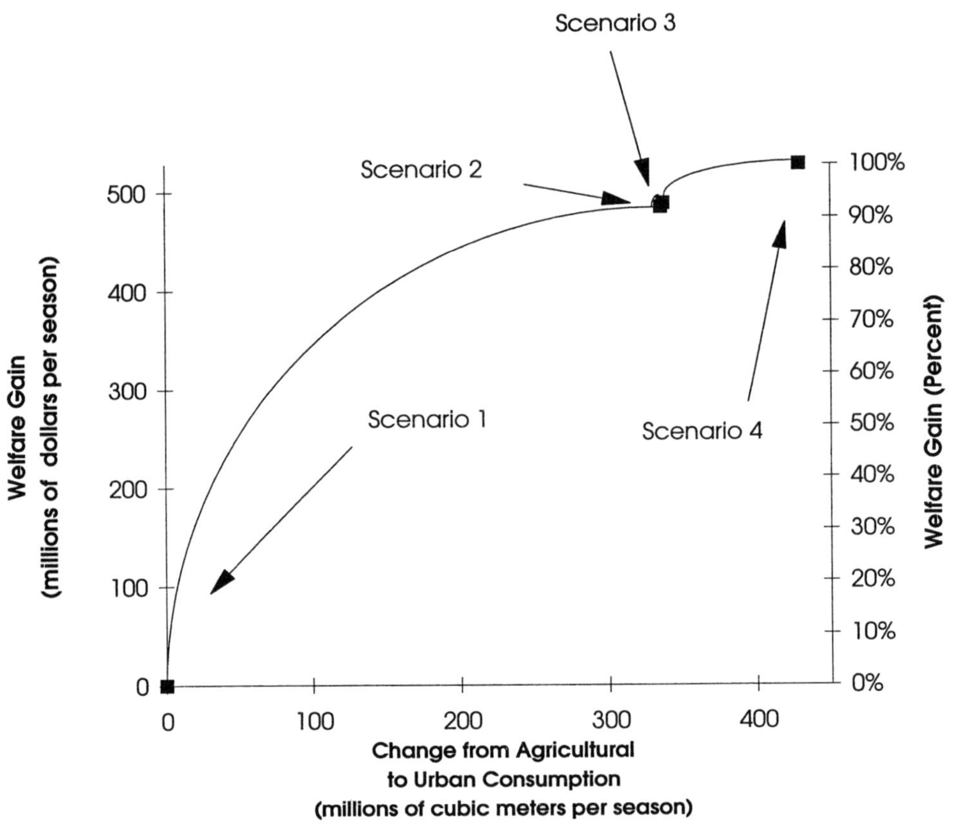

15.5 Resulting Welfare Comparisons

The distribution of welfare changes resulting from a move to markets as the principal water allocation mechanism will depend critically on the property rights or entitlements with which the existing agents will be endowed when they enter a new private market. In this analysis, the agents in question are the collectivity of rural (agricultural) or urban users in each sub-basin, plus the public water agency that has an opportunity to price or tax resource use in a way that transfers some benefits to other citizens and taxpayers, for example. Thus, when simulating a future move to markets, the analysis can also portray many of the alternative ways of distributing the economic welfare gains that derive directly from the resource reallocation.

Three alternative property rights regimes with equal (physical) water allocations are considered (those in the last column of Table 15.1). The pattern of redistributions of economic surplus across the three property rights regimes is shown in Table 15.2. This table shows the water allocations for Scenario 4, the single market with the broadest scope and with the largest total welfare gain from surface water usage. (Similar tabulations can be derived for the other scenarios as a guide to policy formation.)

Table 15. 2. Welfare Distributions Under Three Property Rights Regimes (in millions of Canadian dollars per season)[a]

Water Users	Scenario 1 (Base Case)	Scenario 4 Regime 1	Welfare Distributions Under: Regime 2	Regime 3
Bow River				
Rural	102.8	91.0	44.8	112.6
Urban	484.1	889.5	871.8	877.6
Sub-total	*586.9*	*980.5*	*916.6*	*990.2*
Oldman River				
Rural	98.2	90.3	50.3	104.8
Urban	87.9	152.5	134.4	141.2
Sub-total	*186.1*	*242.8*	*184.7*	*246.0*
South Saskatchewan River				
Rural	76.5	78.8	52.5	76.8
Urban	34.2	63.2	58.3	59.9
Sub-total	*110.7*	*142.0*	*110.8*	*136.7*
Red Deer River				
Rural	n/a	n/a	n/a	n/a
Urban	55.5	101.8	90.2	94.2
Sub-total	*55.5*	*101.8*	*90.2*	*94.2*
RURAL TOTAL	277.5	260.1	147.6	294.2
URBAN TOTAL	661.8	1,207.0	1,154.7	1,172.9
GOV'T. REVENUE			164.8	
SCENARIO TOTAL	939.3	1,467.1	1,467.1	1,467.1

[a] The exchange rate is about $.70 U.S. for $1.00 Canadian.

The first regime (Regime 1) supposes that water is reallocated "as if " by a market, but that, then, as now, no prices are actually charged.[13] Thus, it is an implementation of the "social planner's solution" yet no money changes hands. Predictably, when water is transferred – in aggregate – from rural to urban users without cost or compensation, the former are losers and the latter are winners. Importantly, as shown in Table 15.2, not all rural users lose since in this reallocation of water, the irrigators in the South Saskatchewan River Sub-basin gain the use of more water (332 million m^3 in Scenario 4, up from 307 million m^3 in Scenario 1) and profit accordingly.

Regime 2 is equivalent to an auction system requiring all users to pay the government for the privilege of drawing water at the equilibrium, spot-market-determined prices. The government collects all the revenue for the benefit of some unspecified third parties. For the agents whose consumption levels fall, the welfare effect is unambiguously negative, since they consume fewer units at a higher price per unit. For agents whose consumption levels rise, a gain or loss of welfare is possible, since there is a gain in consumers' surplus on the new consumption that can offset the agents' increased market expenditures on the existing consumption volumes.

As shown under Regime 2 the urban users all gain and the rural users all lose, but there is sufficient revenue raised by the water charge to more than compensate those on the losing end.[14] The tabulated results also portray a potentially important pattern of regional gains and losses, observable by comparing the sub-totals for each sub-basin to the corresponding base case. There are losses in some regions where rural water use is relatively more important than urban uses.

Regime 3 allocates or vests ownership in historical usage levels with all agents when markets are introduced. This regime most closely resembles the rights regime proposed under the new Alberta water law, provided that the ten percent holdback provision for approved transfers (without compensation) is not used. The new "resource owners" can use historical amounts without charge, pay for additional consumption, or gain from the sale of any water not used. Since all moves away from the status quo are voluntary market transactions, everyone's welfare rises from reallocation. As in Regime 2, there is an observable pattern to the welfare gains, but there is no direct source of resource royalties or revenues with which to effect any desired redistribution.

15.6 Conclusion

Not enough is yet known about the future operation of Alberta's new water markets to forecast or to predict which outcomes will be observed. All the same, some conclusions can be drawn. Water reallocation away from Alberta's historical command-and-control apportionments, through the use of well-functioning spot markets, can have a large short-term positive impact on the levels of economic welfare derived from use of this resource, especially in times of drought. Policy makers considering the move to such resource allocation mechanisms would be wise to heed the qualitative and quantitative lessons to be learned from economic simulations of these policy changes before the fact.

The welfare changes that would occur if market participants responded fully to a largely unanticipated drought are likely to be an upper bound illustration of the potential gains to be achieved from water markets in southern Alberta. In other situations, including those where the event is anticipated—and therefore crop choices or water management can also change—or where there are costs and delays in responding, the potential gain from markets would be lower. Even so, by showing the operation of a range of alternative market scenarios under extreme situations, the operational differences can be highlighted.

Where the scope of the markets to be created is at issue, either in geographic or sectoral terms, the varying effects can be illustrated. In the case of southern Alberta, large welfare gains arise from even the smallest introductions of market exchange. As with many policy changes, there is a possibility of infringing on existing property rights, be they legally established or socially perceived (see Horbulyk, 1995). Once again, models of the type employed here are capable of identifying potential winners and losers and the magnitudes of their welfare change.

To recap briefly, the ability to reallocate water optimally in times of an acute shortage, such as by using water markets, can increase water users' welfare significantly. For the model and data reported here, the aggregate consumers' surplus from water use increases by more than 56 percent ($528 million per season) in Scenario 4 compared to not reallocating (Scenario 1). Scenarios 2 and 3 present alternative, intermediate reallocations, such as due to markets with narrower market scope than in Scenario 4. Presumably these alternatives would be easier to implement, to administer and to monitor. These "constrained optimum" reallocations capture more than 90 percent of the $528 million per year in potential gains from water reallocation. Depending on whether or not historical property rights to unpriced use of water are recognized in the new Scenario 4 reallocation, Regimes 1, 2 and 3 describe some of the many ways that the $528 million of annual economic surplus would be distributed among sectors, regions, and the government itself. Importantly, various approaches cause all water users to be made better off.

Three of the findings presented here are likely to generalize to other jurisdictions. First, where the scope of potential markets is at issue, some of the largest gains might be realizable with even the smallest exposure to market forces, and thus policy makers might choose to "start small" and expand slowly. Second, by careful design of the "property rights" dimension of the new resource policy, almost any desired distribution can be achieved for the potential (monetary and non-monetary) welfare gains that the new policy will generate. Third, where markets and market forces are at work, if the largest gains are to be realized, then governments must also be prepared to participate as active players in meeting their own (or society's) resource needs.

Acknowledgments

Part of this chapter was written while Horbulyk was Visiting Associate Professor in the Department of Agricultural and Resource Economics, The University of New England, Armidale, Australia, and their research support is gratefully acknowledged. The authors also acknowledge financial support from Environment Canada through the University-Based Research Program on Economic Instruments. The views expressed here are not necessarily those of the Department of Environment, Canada. Any errors or omissions are the authors' alone.

Endnotes

[1] In Canada, natural resource policy falls mainly under provincial jurisdiction. Among the ten provinces, Alberta is the first to create water markets as part of its water law. Readers interested in other provincial and national policy approaches should see Canada (1988, 1990, 1991) and Karvinen and McAllister (1994).

[2] Some charges apply to the use of water stored or diverted by structures that are publicly operated but these are service charges designed to offset some related costs, not resource royalties per se. Irrigation districts that hold water licenses may divert surface

water without (wholesale) charge, but are then able to collect (retail) payment from individual users, especially where facility or personnel costs are incurred by the irrigation district (Horbulyk, 1997).

[3] There is little distinction in either the historical or the proposed water legislation between annual water volumes drawn from surface water versus groundwater sources Since most irrigators are equipped to exploit surface water flows, these are the water sources that most often require rationing in periods of water deficit.

[4] Since the current regime prevents private transfers, cooperation among users is encouraged by the government in times of acute shortage. When cooperative outcomes succeed, the actual rationing system will appear similar to the proportional rationing approach that is legislated in some U.S. states.

[5] The government reserves ownership of all water for itself, even though it grants (tradable) rights to use that water in various ways defined by its water policy.

[6] These features of the legislation might be needed if the government were to approve transfers of "sleeper" permits, that is, permits that were previously under-utilized but, following transfer, become fully utilized. This and other operational problems of tradable permit systems are described by Nichols (1992).

[7] Agricultural water demand for the Red Deer River Sub-basin is included with the Bow River Sub-basin to reflect the inter-basin transfers and aggregation which now occur in practice.

[8] A study of welfare gains in the long term would want to include other capital-intensive strategies including the use of storages, diversions, and available water-conserving technologies. There might also be increased exploitation of groundwater (Knapp and Olson, 1995). Since the welfare gains envisioned in the model all arise from reducing the differences in users' marginal valuations of water through market exchanges, other sources of such differences might also be captured in a long-term model. For example, there is likely to be considerable variation in the pattern of annual water supply and demand across regions over time that cannot be represented in a short-term model. Similarly, there might be variations in (net) demand within sectors in a sub-basin (so that intrasectoral trading is induced) yet this model allows only one representative water demander per sector, per sub-basin.

[9] Suppose that data are reported or derived for water demand on the basis of gross water consumption (i.e., total water diversions) per year, where some relatively constant portion ($1-\psi$) of water diverted returns to the watercourse as surface runoff and sub-surface drainage. Suppose that one wishes to report water prices or quantities on a net consumption basis. Return flows to the surface watercourse account for the difference between the gross and net consumption volumes and must be considered in modeling downstream water availability. Suppose that estimates are available (across uses and regions) of the ratio (ψ) of volumes of water actually consumed to the total amounts diverted.

One can re-scale (gross) water demand curves so that the vertical axis will represent dollars per cubic meter of consumptive use in the following fashion. If the (gross) inverse demand curve takes the form:

$$P = \alpha + \beta Q^D \tag{1}$$

where: P is price in dollars per m^3 diverted; Q^D is quantity demanded in millions of m^3 *diverted* per season, and ψ is the ratio of consumptive use to total diversions; then the (net) inverse demand curve takes the form:

$$P_n = \alpha/\psi + \beta/\psi^2 \, Q_n^D \tag{2}$$

where: P_n is price in dollars per m^3 consumed, and Q_n^D is quantity demanded in millions of m^3 consumed per season.

Since ψ will vary from region to region, care should be taken to convert values of P_n and Q_n^D back into their gross consumption equivalents if (gross) prices and quantities are to be used for administrative or comparative purposes, as in Alberta. In this model, agricultural water demands as reported in the units of equation (2) are:

Bow River Sub-basin	P_n = 0.23384-(2.6567 x 10^{-4}) Q_n^D	ψ = 0.74
Oldman River Sub-basin	P_n = 0.27896-(3.9600 x 10^{-4}) Q_n^D	ψ= 0.85
South Saskatchewan River Sub-basin	P_n = 0.39515-(9.5903 x 10^{-4}) Q_n^D	ψ=0.85

[10] The Red Deer River sub-basin does not have an effective market in Scenario 2 since there is no agricultural demand there, leaving one urban demander from a fixed supply, as in Scenario 1.

[11] In aggregate, instream flow needs and apportionment demands are treated here as being exogenously determined—they are represented by vertical demand curves in the allocation model. Even so, there is some flexibility to choose "least cost" sources to supply these needs. For example, within limits, consumption could be reduced in any of the Bow River, the Oldman River, or the South Saskatchewan River Sub-basins to meet the instream flow needs of the South Saskatchewan Sub-basin. Similarly, any sub-basin could reduce consumption to meet interprovincial apportionment demand. Scenario 4 endogenizes these choices and reports optimal patterns of supply that vary from those used historically by regulators.

[12] The welfare gains and losses of all agents (as derived from underlying changes in Marshallian consumers' and producers' surplus) are treated as commensurable, and are compared here in aggregate.

[13] This might be implemented by using a spot market or auction mechanism but then rebating any and all water charges paid on a lump-sum basis. As a description of property rights, this regime says users can only use those water volumes for which they hold a license (and that no price or fee needs to be paid). However, the state recognizes no property right or need for compensation when reallocating the historical license holdings.

[14] Provided that the water revenues raised are returned in some (lump sum) manner that does not influence water consumption decisions at the margin, the decision to charge a market-clearing price for water use could be implemented in a manner that is "revenue neutral" for the government. In practice, the introduction of any such pricing scheme has been excluded from the new Alberta water law.

References

Adamowicz, W.L. and T.M. Horbulyk, 1996. "The Role of Economic Instruments to Resolve Water Quantity Problems." *Canadian Journal of Agricultural Economics,* 44(4) (December).

Alberta Agriculture, Food and Rural Development. Irrigation Branch, 1995. Personal Communication. Lethbridge, Alberta Agriculture, Alberta.

Alberta Environment, 1988. *South Saskatchewan River Basin Historical Natural Flows: 1912-1978, 1979-1982 update, and 1983-1986 update.* Alberta Environment, Hydrology Branch, Technical Services Division, Water Resources Management Service, Edmonton, Alberta.

Alberta Environment, 1991. *Water Management in Alberta: Challenges for the Future.* Alberta Environment, Edmonton, Alberta.

Alberta Environmental Protection, 1994. *Alberta's State of the Environment: Comprehensive Report.* Alberta Environmental Protection, Edmonton, Alberta.

Alberta. Legislative Assembly of Alberta, 1996. *The Water Act.* Statutes of Alberta 1996, Chapter W-3.5. The Queen's Printer, Edmonton, Alberta.

Alberta Water Management Review Committee, 1995. *Report of the Alberta Water Management Review Committee.* Alberta Environmental Protection, Edmonton, Alberta.

Birch, A. and A.G.N. Van Deurzen, 1984. Irrigation Water Requirements in the South Saskatchewan River Basin: An Economic Perspective. Twelfth International Congress on Irrigation and Drainage. Fort Collins, USA; May 28-June 2. International Commission on Irrigation and Drainage. Volume 1(A): 325-341.

Booker, J.F. and R.A. Young, 1994. "Modeling Intrastate and Interstate Markets for Colorado River Water Resources." *Journal of Environmental Economics and Management,* 26(1): 66-87.

Canada. Environment Canada, 1988. *Federal Water Policy.* Environment Canada, Ottawa, Ontario.

Canada. Environment Canada. Canada's Green Plan, 1991. *The State of Canada's Environment.* Catalogue No. EN21-54/1991E. Minister of Supply and Services, Ottawa, Ontario.

Canada. Interdepartmental Committee on Water, 1990. *Federal Water Policy: A Progress Report.* Minister of Supply and Services Canada, Ottawa, Ontario.

Enke, S., 1952. "Equilibrium Among Spatially Separate Markets: Solution by Electric Analogue." *Econometrica,* 42(1): 40-47.

Enright, C. and J.R. Lund, 1991. "Alternative Water-District Organization: Screening-Level Analysis." *Journal of Water Resources Planning and Management,* 117(1): 86-107.

Flinn, J.C. and J.W.B. Guise, 1970. "An Application of Spatial Equilibrium Analysis to Water Resource Allocation." *Water Resources Research,* 6(2): 398-409.

GAMS (General Algebraic Modeling System), 1991. Student Version 2.25 with MINOS 5.3 Solver. South San Francisco: The Scientific Press.

Gibbons, D.C., 1986. *The Economic Value of Water.* Resources for the Future, Inc, Washington, D.C.

Horbulyk, T.M., 1995. "Integrated Water Management and the Treatment of Regulatory Takings: The Case of Canada's Water Allocation Reform," *Water Resources Update,* No. 100, (Summer): 48-54.

Horbulyk, T.M., 1997. "Canada," in Ariel Dinar and Ashok Subramanian (eds.) *Water Pricing Experiences: an International Perspective.* The World Bank Technical Paper No. 386, pp. 37- 45, Washington, D.C.

Karvinen, W.O. and M.L. McAllister, 1994. *Rising to the Surface: Emerging Groundwater Trends in Canada.* Queen's University, Centre for Resource Studies, Kingston, Ontario.

Knapp, K.C. and L.J. Olson, 1995. "The Economics of Conjunctive Groundwater Management with Stochastic Surface Supplies." *Journal of Environmental Economics and Management*, 28: 340-356.

Lo, L.J., 1995. "Water Scarcity and the Potential Gains from Water Trading in Southern Alberta." Unpublished M.A. Thesis. The University of Calgary, Calgary, Alberta.

Nichols, A.L., 1992. *Tradable Permits for Water Use: An Overview of Concepts and Experience.* Prepared for TransAlta Utilities. National Economic Research Associates, Inc, Cambridge, MA.

Prairie Provinces Water Board, 1969. *Master Agreement on Apportionment and Bylaws, Rules and Procedures: Consolidated Version May 1992.* Regina, Saskatchewan.

Samuelson, P.A., 1952. "Spatial Price Equilibrium and Linear Programming." *American Economic Review,* 42(2): 283-303.

Takayama, T. and G. G. Judge, 1964. "Equilibrium Among Spatially Separated Markets: A Reformulation." *Econometrica,* 32(4): 510-524.

Vaux, H.J., Jr. and R.E. Howitt, 1984. "Managing Water Scarcity: An Evaluation of Interregional Transfers." Water Resources Research, 20(7): 785-792.

Veeman, T.S. and R. Freeman, 1993. "Economic Aspects of Transferable Water Rights in Alberta," in David Manz (ed.) *Water and the Wilderness: Development, Stewardship, Management.* Proceedings of the 46th Annual Conference of the Canadian Water Resources Association, June 16-19; Banff, Alberta. Canadian Water Resources Association; 261-278, Cambridge, Ontario.

16 Development of Water Markets Using Experimental Economics[1]

Ariel Dinar, Richard E. Howitt, S. J. Rassenti, and Vernon L. Smith

16.1 Introduction

The large, complex water systems throughout the world, including the Western U.S., have traditionally required a centralized authority both to solve their fundamental coordination problems and to dictate solutions to their allocation issues. These problems are often exacerbated by factors such as the public good nature of water, externalities associated with transporting and consuming water, and the economies of scale in shared production facilities. In our world of rapid technological and environmental change, these centrally-managed systems tend to move too slowly to allow a society to adapt efficiently. One of the problems with proposals for substantial institutional change is that adaptation and irreversibility make the process slow, cautious, and costly to society. In this chapter, we will discuss both the recent development of 'smart', computer-assisted markets (McCabe et al., 1991) which provide the promise of developing decentralized solutions to these complex allocation problems, and the role that experimental economics can play in evaluating proposed institutional changes to help facilitate a more rapid and smooth adoption of these changes.

There is a substantial amount of evidence presented in previous chapters, which indicates that the development of water markets can improve water allocation by permitting the voluntary transfer of water to its highest-valued use. One such decentralized market mechanism is the so-called 'smart' market concept developed by McCabe et al. (1991), which allows for the pricing and allocation of resources in technologically interdependent environments. The essential idea is to combine the information and incentive advantages of decentralized ownership rights or responsibilities with the coordination advantages of central processing. The optimization data requirements for these 'smart' markets come in the form of willingness-to-pay demand, willingness-to-accept supply, and budget and capacity constraints, that are provided by decentralized decision-makers when ever price and allocation decisions are required. The required input data are available only from dispersed human decision-makers who best know their own circumstances and willingness to exchange, and benefit from revealing this information. The central program applies optimization algorithms to the bid-offer messages to determine the prices and allocations that maximize the gains from exchange.

Experimental economics yields a formal and replicable system for analyzing alternative market structures, such as the 'smart' market, before they are actually implemented. For example, a 'smart' water market can be developed and tested in the laboratory during years of sufficient water to simulate drought conditions. If the results are positive, such a market could be implemented during years of water scarcity.

The next section reviews experimental economics and its application to 'smart' resource-based markets. Then a prototype of the California water transfer model is presented, followed by a description of its incorporation into the experimental design scheme. Results are presented from preliminary experiments on the price efficiency achieved by the 'smart' market players when compared with a hypothetical set of perfectly competitive trades, and the rate of price discovery by the players.

16.2 Experimental Economics and Resource-Based Markets

Experimental methods have been applied to evaluate the performance of 'smart' computer-assisted markets in several industries. The first application was motivated by the question of how one might design an auction market for airport runway access rights to complement the deregulation of the airline transport industry. This led to the development of the experimental combinatorial auction mechanism[2] (Rassenti et al., 1982). This study did not lead to runway access markets, although there is now renewed interest in a more flexible market-based approach to defining priority for airport utilization space.

Subsequent applications of the 'smart' market technology have included natural gas pipeline networks and electric power networks. The structure of natural gas pipeline networks, consisting of well-head producers connected through pipelines to demand delivery nodes, closely parallels the physical features of water networks. In the gas network experiments, well-head producers enter location-specific offer schedules to supply gas at each production node. Wholesale bulk buyers submit location-specific bid schedules to buy gas delivered to their respective consumption nodes. Holders of pipeline transportation rights enter pipeline-specific offer schedules of transportation capacity. This decentralized information on willingness-to-pay and willingness-to-accept, expressed in the form of price-quantity step functions, defines a linear programming problem, that allows the computer dispatch center to compute prices and allocations that maximize the reported surplus. Since the reported individual demand (supply) functions need not, and generally do not, correspond to true willingness-to-pay (willingness-to-accept), the important research question is whether incentives to under-reveal demand (supply) have any significant impact on market efficiency. The natural gas network experiments show very high efficiency because, although intramarginal units are greatly under-revealed, marginal units are not, and this is all that is required to achieve efficient allocation in uniform price market mechanisms (McCabe et al., 1989, 1990, 1991).

Another feature studied in the context of gas pipelines is the use of competitive co-tenancy rules to introduce competition where only a single pipeline facility is needed. In such "natural monopoly" situations competitive policy requires the capacity of such facilities to be shared by multiple owners (which is commonly the case for large investments). The individual rights holders then submit separate competitive offer price schedules for their share of the capacity. This yields competitive outcomes in the short-run since the ownership structure is as if there existed parallel distinct pipelines with capacities equal to that of each shareholder. To obtain the long-run competitive outcome, one needs a property right rule stating that any co-tenant, co-tenants, or an independent outsider is free to invest in augmenting the capacity of the pipeline in return for the title to the increase in capacity rights. The point is that unique facilities need not imply natural monopoly as long as there are property right rules defining multiple rights holders who then must compete in using such rights to supply consumers with commodity or services. (See Rassenti et al. (1994) for a discussion of co-tenancy property right rules and an experimental evaluation of its efficacy for inducing competition on bottleneck pipelines.)

Based on what has been learned through natural gas research and other experiments dealing with institutional design, we consider the following mechanism properties desirable:

(1) *Self regulation.* Ultimately we seek to explore the feasibility of substituting a satisfactory market mechanism for rate of return and other regulatory constraints. The wide application of competitive auctions in financial markets, previous experimental results in electricity and natural gas networks (McCabe et al., 1991), and progress in the electricity industry in this field, internationally, suggests that a suitable extension and modification of this institution may have desirable properties in water network applications.

(2) *Non price-discrimination.* All price differences among wholesalers, and transportation rights holders in any given pricing period are justified only on the basis of marginal costs and opportunity costs. It is our interpretation that current law favors this property.[3]

(3) *Priority-responsive pricing for curtailments.* In the event of restrictions in water flow capacity (due, for example, to drought, link rupture, or pump station failure) there is an automatic restriction of deliveries giving preference to those wholesale buyers with the highest consumption priorities (and a corresponding restriction of supply schedules to give preference to those sellers with the highest priorities) as expressed in the bids (offers) each submits to the auction-dispatch center.

(4) *Value-responsive investment incentives.* A good mechanism should provide incentives for increases in transfer availability and transportation capacity where such increases have the highest value, given the level of sunk costs and lumpy investments already in place and capital market restrictions.

(5) *Simplicity and decentralization.* Although network systems may have inherent physical complications, it is desirable to have a price mechanism in which the participants can understand what is required of them and can readily fulfill these requirements. One of the advantages of a decentralized mechanism is that the participants can concentrate on judgments and actions based on the private information each is most directly informed about. Computer assistance is used where it is most appropriate and needed: to generate nondiscretionary, consistent, best, but strictly routine results based on the willingness-to-pay and willingness-to-accept judgments of dispersed individual participants.

(6) *Compatibility with existing institutions.* Present institutions emphasizing contract precommitments could continue to function simultaneously with a network auction. This could be accomplished by applying the auction to available capacities above the flows precommitted. Consequently the extent of the use of the auction would be determined by the decisions of the participants, and could grow in response to increased use without the necessity of a possible disruptive discontinuous switch-over. When the auction is used for spot trades, holders of long-term water rights could resell those quantities not needed for current consumption. In this case the auction would supplement and increase the efficiency of current institutions.

The deregulation of electric power has promoted interest in the use of experimental methods to examine market design, competitiveness, transmission pricing and other issues of performance and structure. A controversial question is whether and how bilateral contracting should coexist with a central computerized exchange. Due to externalities in the transmission system, bilateral contracts often impose external costs on all other traders. One means of avoiding this problem is to require all such contracts to be financial hedges on the spot price rather than physical contracts for delivery at a specified time. Thus a buyer's forward contract would be automatically "at market" in the spot exchange and buyers would submit bids to buy against their actual need for delivery in the spot market. Similarly, sellers would buy to offset their forward sales and go to the spot market with an offer to sell based on their current conditions. In this way each market participant locks in a future price, but the spot market is not constrained physically by the forward contract.

16.3 Modeling Interbasin Transfers in the San Joaquin Valley (SJV)

Market approaches to water allocation in California have been proposed for more than a decade. Among other studies, Vaux and Howitt (1984) evaluated interregional water transfers from agricultural to urban uses as a means to manage water scarcity. Their study compared projections of equilibrium prices for agricultural and urban water in 1995, with and without a market. They estimated annual benefits of about $240 million from improved water-use efficiency by 1995. The pollution reduction and water saving aspects of water markets were demonstrated by Dinar and Letey (1991). Their farm-level model estimated net social benefits in the range of $8 and $70 per acre as a result of the reduction in drainage pollution and water conservation when a water market was in place. Howitt (1994) estimated statewide net benefits from a drought water bank – a water transfer mechanism that took place in 1991 during the 1987-1992 drought in California – to be $106 million. Due to the political necessity for fixed administrative prices, the 1991 drought water bank did not clear the market at the established prices. However, it did demonstrate to the farmers and environmentalists that it is possible to reallocate water from less valuable uses to more valuable uses without destroying either the farm economy or the environment. While the drought water bank succeeded in the short-run, there is continuing interest in developing a market mechanism that provides faster price development and more timely notification of transactions to buyers and sellers, while still respecting distributional and environmental concerns.

In applying a 'smart' water market framework in the California context, a simplified approach was chosen. The approach is based on several components that include: (1) a water quantity/quality production function model of major agricultural crops (Dinar and Letey, 1996), (2) estimates of a demand function for urban centers, (3) environmental water quantity and quality requirements, and (4) conveyance cost between water supply and water consumption centers.

An agricultural region in the Central San Joaquin Valley was selected for the implementation of the 'smart' water market. It consists of several sub-regions in which potential buyers and sellers exist. The region is hydrologically linked with two urban areas, the San Francisco Bay area and the greater Los Angeles basin. These consuming regions are linked to three supply points in the Northern Sacramento Valley, groundwater in the San Joaquin Valley, and the Colorado River in the

South. The interbasin water transfers are effected by two major canal systems and several rivers. In most cases the transfer facilities have additional capacity that is under the control of independent agencies. In one case, the Sacramento Delta, control of flows through this critical junction depends on the environmental criteria. While the system is greatly simplified for this initial experiment, it contains the major elements that make the reallocation of water in California difficult, but potentially rewarding.

The Agricultural, Urban, and Environmental Sectors

Agricultural agents (buyers or sellers) are characterized by a set of physical conditions such as pan evaporation, severity of drainage problems, soil quality, surface water, and groundwater availability and quality, cropping patterns, and irrigation technologies. These conditions dictate a specific production function for each crop grown in each sub-region which, in association with input and output prices, allows the estimation of a water demand function.

Farmers are assumed to be profit maximizers. Regional water supplies from all sources have been used to grow a set of various crops in order to maximize net returns. The profit maximizing model includes various components. The basic features of the crop-water production model used are the production functions that are estimated for each crop in each of the agricultural regions in the study area.

Features involved in the calculation of the regional net return are an irrigation cost function that accounts for irrigation technologies and crops, and a set of coefficients needed for the estimation of the production functions. All relevant coefficients (such as pan evaporation, maximum attainable yields, etc.) are site/regional specific. The production functions are specified as nested CES functions that have a specific part (nest) for the relationships between water quantities, salinity levels and technologies.

The production functions are used as the pivotal instrument in a profit maximizing model, in addition to prices of inputs and outputs, and technical and administrative constraints. The profit maximizing model is calibrated to conditions (crop acreage) prevailing in the year 1995, using the approach suggested in Howitt (1995). By varying the quantities and qualities of water available for irrigation, a water demand function is estimated for each agricultural sector. This demand function is then incorporated into the market experiment.[4]

The two urban demand centers included in the analysis are: the Metropolitan Water District (MWD) that supplies water to consumers in the Los Angeles area in Southern California, and a hypothetical simplified water supply company for the San Francisco Bay communities. The two urban demands for water were estimated using updated urban water demands from Vaux and Howitt (1984)

Environmental demands at the critical Sacramento Delta junction are represented by the combined state and federal interagency group called (Cal-Fed). It purchases water for instream and outflow purposes to combine with the initial water endowments. To model a certain level of environmental safety, Cal-Fed was constrained to consume in certain nodes a minimum amount of water (for instream and outflow purposes).

The Regional System

The regional experiment refers to a simplified system that includes major agents only, at this stage, operating under simplified assumptions, for the sake of clarity. Initial allocations of water in this simplified setup are based on data in McClurg (1996) and represent the present situation. The present situation – the water endowments of the various agents that provide the initial values for the experiment – is based on water quantities available from various sources. Among these sources are the Sacramento River inflows, that feed into the Central Valley Project (CVP) and the State Water Project (SWP), the Colorado River, and San Joaquin Valley (SJV) and Kern County groundwater. The resulting initial allocation values are presented in Table 16.1.[5]

Link nodes are used here to allow water transfer between storage (surface water or groundwater reservoirs), through canals and rivers. It is assumed that link nodes, although associated with active agents, do not represent negotiating centers. That is, prices of water conveyance through these node links are fixed, and agents on both sides of the link cannot impact either the cost or the quantity of water to be transferred, except through their revealed demand. Cost of water conveyance are also modified from Vaux and Howitt (1984) adjusted for current values. Links in the experiment were constructed such that each agent had enough water rights at each link to secure deliveries of water demanded. Conveyance cost for links that are rivers were assumed to be zero, transport cost through canal links were assumed to be \$10-15/acre-ft, and in three links, (*lb*, *lh*, and *li*) lift costs were also added.

Figure 16.1 presents the system of potential buyers, potential sellers, and links through which water can be transferred. Each agent is assigned initial water allocations at some potential supply node (*w.*) and/or values to consume water at a potential buying node (*b.*). Table 16.1 lists the values, costs and capacities of the agents in our design. The column labeled Node gives the various locations in the network of Figure 16.1. Each of the agent has a set of rights to trade at the locations at which they are active. For example, agent SJV can sell up to 3.1 million acre-ft/period at a cost of \$0/acre-ft at location *w1*, while at location *b3*, SJV can buy and resell up to 2.5 million acre-ft/period at a price of \$54/acre-ft, and 4.5 more million acre-ft/period at a price of \$31/acre-ft. The chosen mix of agents reflects the current condition in the state in which there are both independent water right owners and "captive" sources owned by, or under contract with, a major consumer.

The representative environmental player in the model, Cal-Fed, is currently promulgating criteria to protect threatened fish species and other environmental indicators. These will probably take the form of a set of water flow and quality requirements that vary by the time of year. In this initial simplified model, environmental restrictions are characterized by a minimum flow rate through the Delta comprising two sets of flows. First, there is a larger flow from the Sacramento River that originates from the Shasta and Oroville dams. These flows travel through link ld (the Sacramento River) and can be diverted beyond the Delta to the demand sites. Through link lf (the Delta-Mendota canal) to the San Joaquin Valley or down the links lg and lh (the California Aqueduct) to demand nodes in Kern County (*b4*) and the Los Angeles Metropolitan Water District (*b2*). The second link, whose flows are regulated by the actions of Cal-Fed, is (*lc*) (the San Joaquin River) which conveys water from the San Joaquin Valley to the San Francisco Bay Area. The minimum flows are modeled as 3.3 million acre-ft through the Delta from the supply node (*w1*). This is achieved by valuing the first 3 million acre feet at zero to Cal-Fed, which essentially makes this amount of flow required, and allowing Cal-Fed a high price of $100/acre-ft for a further discretionary amount that is 10 percent of the base that can be purchased at (*b5*). The link (*la*) denotes the upper Sacramento River, and Cal-Fed has an interest in maintaining minimum flows of 0.7 million acre-ft in this stretch. The San Joaquin River link (*lc*) has a required amount by Cal-Fed of 0.2 million acre-ft and a further discretionary amount of 0.2 million acre-ft that is valued at $40/acre-ft.

Figure 16.1: Schematic Presentation of Potential Water Sellers, Buyers, and Links in the Study Area

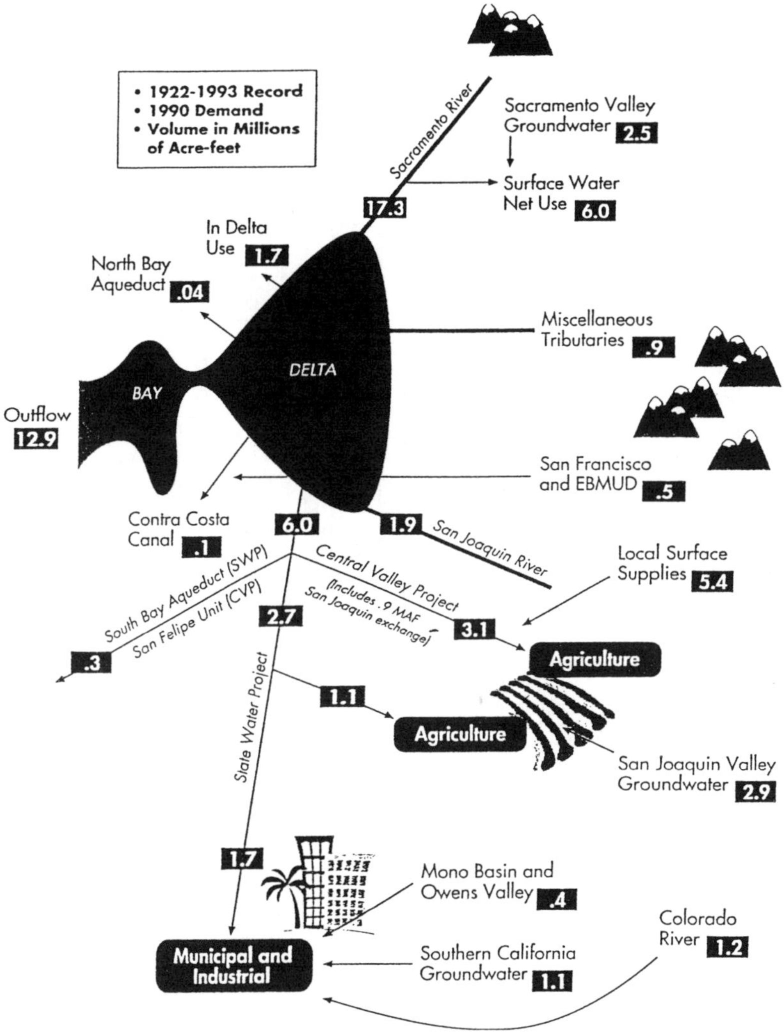

Source: McClurg (1996)

Table 16.1. Initial Allocations, Demand Specifications and Link Capacities, and Conveyance Costs in the Study Area

	SJV		Kern		MWD		SF		Cal-Fed		Robot[a]		Colo.	
Node	Q(MAF[b])	P($/AF)	Q(MAF)	P($/AF)	Q(MAF)	P($/AF)	Q(MAF)	P($/AF)	Q(MAF)	P($/AF)	Q(MAF)	P($/AF)	Q(MAF)	P($/AF)
w1	3.1	0	1.1	0	1.7	0	.9	0	3.0	0				
b5									.3	100	2.9	200		
b1							.3 .6	148 100						
w2	5.4	8												
b4			1.1	55										
b2					2.0 3.0	213 180								
w3					2.7	0								
b3	2.5 4.5	54 31											1.2	70
la									.7	0	3.4	0		
lb											.9	40		
lc									.2 .2	0 40				
ld	3.1	10	1.1	10	1.7	10								
le											5.4	15		
lf											3.1	10		
lg			1.1	10	1.7	10								
lh											1.7	80		
li											3.9	70		

[a]Robot is not an actual player but, as explained in the text, an accounting procedure to allow for adding/subtracting conveyance cost, or meeting constraint requirements.

[b]MAF= million acre-ft; 1 acre-ft = 1,235 cubic meters

Note: *w1* is initial allocation at the Delta; *w2* is initial allocation to the SJV from the CVP and SWP; *w3* is initial allocation to MWD from SWP and from the Colorado River; *b1* is San Francisco demand; *b2* is Los Angeles Metropolitan Water District demand; *b3* is SJV agricultural demand; *b4* is Kern agricultural demand; *b5* is Cal-Fed environmental demand; *la* is Upper Sacramento River link; *lb* is Robot; *lc* is San Joaquin River link; *ld* is Sacramento River link; *le* is Robot; *lf* is Delta Mendota canal; *lg* and *lh* are California Aquaduct links; *li* is Robot.

16.4 The Experimental Design for the SJV Water Market

A set of laboratory experiments was designed to study the efficiency and price performance characteristics of a uniform price double auction mechanism[6] for the simultaneous allocation of water and transportation capacity rights among buyers, transporters, and sellers. A computer coordinated auction market – *WaterMart* – maximizes total gains from exchanges, based on the submitted bids and offers and determines allocations and nondiscriminatory prices at all node.

Some of the salient features of *WaterMart* are as follows.

(1) Consumption centers (consisting of primary water buyers such as local distribution companies and industrial consumers) are connected to water sources (supply points of primary water right holders) by a capacity-constrained transportation network. At this initial stage the network we study experimentally is thin, and weakly competitive. In this sense it is intended as a worst-case example for any price mechanism.

(2) The auction market uses 'smart' computer support to process location-specific bid schedules from water buyers, location-specific offer schedules from water right holders, and segment-specific offer schedules of transfer capacity from transportation right holders. This means that each of the three types of participants is required to make judgments that reflect its own private circumstances, i.e., its willingness to buy water, sell water or sell transportation on each network link.

(3) The resulting prices in the market are nondiscriminatory. In other words, all sellers at the same location receive the same price, all buyers at the same location pay the same price, and all links connecting a given source location with a given receiving location obtain the same price. Other pricing mechanisms, such as double auction (without uniform prices) and posted pricing do not yield nondiscriminative prices except in full equilibrium.[7] *WaterMart* yields nondiscriminatory prices in each price period.

(4) Differences in the price of water between any two locations reflect differences in the marginal supply price of transportation and/or link capacity restrictions.

(5) The allocation of delivered water among buyers, of injected water among sellers, and of the resulting transportation requirements maximizes the total system aggregate gains from exchange (surplus) given the schedules of buyer bids, seller offers and transportation offers. In this sense, *WaterMart* guarantees a maximally efficient, competitive market given the submitted bids and offers of all participants: its first priority is to consume water from the lowest cost sources and deliver it to the highest value wholesale users using the lowest cost transportation routes.

The experiments we report evaluate the price and efficiency characteristics of *WaterMart* as a mechanism for the exchange of rights. The rights themselves are not labeled or interpreted in terms of the possibility of implementing a given trade in the water industry. Thus the rights actually traded in an implementation of *WaterMart* might be for hourly, weekly, or monthly spot water commitments, or for forward commitments next winter (firm or interruptible). Our primary research task is to evaluate *WaterMart* as a price discovery mechanism applied to any well-defined set of contractual rights.

Technically, *WaterMart* is an extension and generalization of the competitive (uniform price) sealed bid, and double sealed bid-offer, auctions. As a price mechanism, its distinguishing feature is that all accepted bids to buy are filled at a price that is less than or equal to the lowest accepted bid price of buyers – a price that just clears the market by making the total number of units offered equal to the number demanded. This contrasts with the discriminative sealed bid auction where Q units are offered for sale and the Q highest bids are accepted, but each accepted bid is filled at a price equal to the amount bid. Consequently, different buyers pay different prices for the same commodity. Also, the discriminative auction provides an incentive for strategic underbidding of full willingness-to-pay as each bidder attempts to avoid having to pay more than the minimum necessary amount (the lowest accepted bid). This incentive is reduced in the competitive auction, and in special cases, it is, theoretically, eliminated.

This chapter reports on a single "proof of concept" series of experiments using the representation of the *simplified California* water transmission network illustrated in Figure 16.1 of the previous section. There were four experiments each using six inexperienced subjects, followed by four experiments each using the same six, now experienced, subjects. The inexperienced subjects were all trained in a perfectly symmetric six node circular environment (three buy nodes interspersed with three sell nodes) in which they each owned identical rights to buy, sell and transport water. The training lasted for only 7-10 periods of trading because of the reading time required to digest the instructions the first time through. All the trained subjects returned to be randomly assigned a role in the highly asymmetric SJV system. The results of the training experiments were disregarded, as is typical in complex 'smart' market experiments.

Subjects were recruited from the undergraduate population at the University of Arizona. They were paid five dollars upon arrival for an experiment as an incentive to arrive on time. At the end of the experienced (second) experiment subjects were paid their total profit earnings in U.S. dollars. These payoffs varied from \$2 to \$70 dollars, U.S. currency, not including the show up fees. All experimental earnings are denominated in what are called "pesos," which are then converted into U.S. dollars at a pre-fixed individual rate. Using a conversion rate facilitates approximate uniformity and insures salience in individual rewards across experiments with different parameter designs and earning potentials.

All experiments were run on the NovaNet Computer system. This system was used: (1) to instruct subjects in the experimental institution; (2) to enforce the message rules which define the institution (and thus coordinate subject interaction); and (3) to record profits. Subjects then went through the instructions at their own pace. Experimenters were on hand both to assist subjects who had difficulty in understanding the rules and to enforce privacy. There were fifteen trading periods in each of the four experiments with experienced subjects. The subjects were not told when the experiment would end.

Each trading period lasted five minutes, with some agents responsible for submitting bids or offers to trade water for as many as six different rights. As the bid information was submitted to the electronic trade center, it was used to continuously resolve a linear network flow program.

Each arc (i) in the formulation below represents one bid or offer. If a buyer makes a two-part bid, then it is represented by two parallel arcs. Two-part offers by sellers are represented similarly, so each bid or offer is represented by the vector (s_i, e_i, d_i, u_i, c_i). In each case, s_i is its starting node, e_i its end node, d_i the least permissible flow on that arc, u_i the greatest permissible flow on that arc (determined by the bid or offer quantity entered), and c_i the bid value or offer price per unit of flow on that arc (bid values are negative costs).

For any set of bids and offers the revealed surplus in the network is maximized by solving the following:

(1) Maximize total surplus: $-\sum_i c_i \ f_i$;

(2) subject to $\sum_{i \varepsilon E_j} f_i - \sum_{k \varepsilon S_j} f_k = 0$ ($\forall$ nodes : j); and

(3) $d_i \leq f_i \leq u_i$ ($\forall$ arcs i),

where: f_i is the flow on arc i, E_j is the set of arcs which end at j, and S_j is the set of arcs which begin at j. Note that constraint set (2) maintains the balance of flow at each node j. Constraint set (3) ensures that the flow on each arc does not exceed the stated bounds.

Solving the linear programming problem (1)-(3) yields not only the optimal flows (and production and consumption pattern) but also the set of shadow prices π_j for all nodes in the network. Since the shadow prices are relative to one another, the difference in shadow prices at the start and end nodes of an arc gives us the value of the marginal unit of flow on that arc, hence the price associated with transportation. For example, if $\pi_7 - \pi_4 = 24$ then the owner of the line segment between nodes 7 and 4 receives \$24/acre-ft to transport water between locations 4 and 7. All buyers and/or sellers at a particular node would pay or receive the shadow price for that node. With the linear programming solution, the budget is always balanced since the physical flow on each segment represents an activity which is exactly cash compensated.

16.5 Laboratory Experiments for the California Water Market

For the four experiments analyzed and coded as 7x, 8x, 9x, and 10x (no inexperienced sessions are reported), Table 16.2 provides the average prices and quantities traded at each node. They can be compared with the competitive equilibrium (CE) which is the last column of Table 16.2. This CE solution is based on full revelation of demand by buyers and of supply by producers and transporters, and is computed by solving the linear programming problem with all bids at marginal redemption value and all offers at marginal cost. Of course, the interesting thing to note is that buyers and sellers do not need to reveal their infra-marginal values and costs to achieve the CE outcome. For example, SF need not reveal it is willing to pay at least \$100/acre-ft for all the water it consumes since the CE price at node *b1* is only \$40/acre-ft. Indeed the realized market prices range between \$62 and \$84/acre-ft, as SF has trouble negotiating free transportation from Cal-Fed.

There is a significant inter-experiment difference in the realized average mean prices and quantities. This is largely due to the relative negotiating capability of the various agents that varies from group to group. The more the overall quantities traded are shy of the fully revealed amount, the more flexible are the alternative price sets which can be observed to satisfy the revealed equilibrium conditions. It is apparent in these results that San Joaquin, as a consumer at *b3*, is the agent with the greatest consistent shortfall of quantity.

In any individual experiment, after about the first five periods prices tend to become quite stable, or to change very slowly. This result is a consequence of the bid-offer behavior of the subjects. Over time they tend to settle into a pattern in which bids to buy delivered water are at, or somewhat above, their respective node prices. Similarly producers offer at, or somewhat below, their node prices, while transporters offer at, or somewhat below, the difference between the upstream and downstream node prices. Consequently, a small reduction in a buyer's bid, or a slight increase in a seller's offer, is likely to cause agents to be cut out of the market or at least have their allocation reduced. This will have a relatively small effect on prices given this bid-offer pattern. The primary impact is on allocations and therefore efficiency and the distribution of surplus.

Table 16.3 contains the average per period earnings in dollars for each of the agents, expressed as a percentage of that expected under the competitive equilibrium scenario given in the first column. For example, SF (who is bargaining over the transportation fee) is earning between 47 and 88 percent of equilibrium earnings, depending on which group is observed. It is obvious that with the parameterization chosen, Cal-Fed was the only player with an easy bargaining situation. The ability to earn consistently much more than the competitive share is somewhat exaggerated by the fact that Cal-Fed retained only a few pivotal units under their bidding control. The rest of the water, for which they were not liable, was openly declared at margin by the robot. Thus a small absolute gain, which was easy in this bargaining position, became a large relative improvement over the CE solution.

Similarly, for Colorado, a small absolute gain is infinitely more than the CE profit, $0. But Colorado has limited success in an almost purely bilateral bargaining game with MWD. The agent who most closely and uniformly fits the CE "prediction" is the robot, yet this is the only non-strategic agent in the market. It simply declares its limit prices and trusts the revealed outcome to be profitable: it is!

These markets were relatively thin. Netting out what agents could deliver to themselves did not leave much potential transfer capacity. Yet they performed remarkably efficiently, 87 to 96 percent of overall potential gains realized by the perfectly competitive solution. This was despite the fact that, by current experimental standards, these agents were relatively inexperienced. They had not seen this network before and they only had fifteen periods to fathom its intricacies. Under such a complicated scenario we would expect improvement in gains from exchange through three visits (experimental sessions) and more than sixty trial periods. In fact, in the more detailed data not displayed here, overall efficiencies improved constantly for the duration of these experiments.

Based on these laboratory experiments with reward-motivated subjects, and other experience in network experiments, we conclude that water networks using a

Table 16.2. Average Prices and Quantities Traded at Each Node

	Experiment 7x		Experiment 8x		Experiment 9x		Experiment 10x		Competitive Equilibrium	
Node	Q (MAF)	P ($/AF)	Q (MAF)	P ($/AF)	Q (MAF)	P ($/AF)	Q (MAF)	P ($/AF)	Q (MAF)	P($/AF)
W1	6.4	20	5.9	18	5.8	19	7.1	11	8.5	0
B5	3.1	44	3.2	26	3.2	25	3.2	22	3.2	0
B1	.8	84	.9	66	.7	65	.9	62	.9	40
W2	3.1	26	2.6	21	5.4	15	4.4	16	5.4	5
J1	0.0	32	0	38	0	31	0	22	0	10
B4	.7	55	.7	67	.6	59	1.1	46	1.1	55
B2	4.5	145	4.5	151	4.0	157	4.5	137	5.0	140
W3	2.8	75	3.1	81	2.4	87	2.8	67	3.3	70
B3	3.3	41	2.3	36	5.1	30	4.6	31	7.0	20

Table 16.3. Average per Period Earnings and Percent of Competitive Equilibrium (CE)

	Experiment 7x		Experiment 8x		Experiment 9x		Experiment 10x	
Agent	Earn ($)	% CE	Earn ($)	% CE	Earn ($)	% CE	Earn ($)	% CE
SJV	151	94	1,186	73	180	112	1,415	88
Kern	369	96	303	79	362	94	377	98
MWD	4,178	81	4,011	78	3,066	60	4,663	91
SF	322	47	602	88	514	75	584	85
Cal-Fed	825	223	790	214	678	183	696	188
Robot	5,481	93	5,469	93	5,583	95	5,721	97
Colo.	32	1/0	118	1/0	207	1/0	7	1/0
Total [a]	11,358	90	12,481	89	10,590	87	13,464	96

[a] Totals may not agree with summation due to rounding.

computer coordinated 'smart' allocation mechanism will yield essentially competitive outcomes. Several institutional assumptions are inherent in the current experiments: first, alternative transportation paths have comparable costs; second, single paths have shared ownership or are common carriers at fixed rates (though we do not favor the latter because of its inadequacy in delivering appropriate price signals for new investment); and third, transport capacity is adequate. Such institutions appear to uniformly discipline the agents' attempts to manipulate market outcomes with strategic bidding because of the clear and endogenous signals built into the experimental structure.

16.6 Extension of the Approach to Other Countries and Water Situations

Water market initiatives can be observed, at various levels, in many developed as well as developing countries. Observations of the experience in various countries suggests that the process of water market development requires a long and difficult process of learning and implementation. The 'smart' market approach described in this chapter may reduce the learning cost involved in the development of a water market.

Use of the 'Smart' Market Approach in Developing Countries

Developing countries are more sensitive to water scarcity since their infrastructure is less flexible. Allocation of scarce water resources among competing demands in urban, industrial, agricultural, and environmental uses is becoming increasingly difficult and cumbersome, using existing institutions. The result is often misallocation, inefficiency, and negative third-party effects.

Markets for water or tradable water rights have the potential to increase efficiency in all these cases. However, in developing countries the transaction costs may be too high because (1) adequate information may not be available to all potential parties, and (2) the water infrastructure serves many small users, which makes it difficult to properly manage the transaction records.

In many developing countries, well defined property rights to surface and groundwater do not exist. However, in South Asia the sale of water from private tubewells is quite common. For example, Wood and Palmer Jones (1990) describe a market for groundwater trade in Bangladesh. Renfro and Sparling (1986) describe water trade in the Punjab region of Pakistan where farmers sell their water within the canal system. Meizen-Dick in Chapter 13 and Saleth in Chapter 12 have described groundwater trading in Pakistan and India, respectively. In Latin America, interest in water markets has lately arisen. For example, Chile has a water law that establishes a system for tradable water rights (see Hearne and Easter, Chapters 9 and 10). In Mexico, leases and sales of water among farmers for seasonal water use have existed even if it was illegal. In 1992, the new Mexican water law establishes water concessions or use rights on a volumetric basis (Thobani, Chapter 3). Many other countries such as Colombia, Peru, and Brazil are considering water trading.

The application of the model to the conditions prevailing in California allows us to calibrate the model to data that does not exist elsewhere. Having developed the approach, the application to conditions existing in other countries is straight forward. The model developed in this project, after being modified to the conditions in a particular country, could be used to estimate the total benefits from water trade. By applying the model to conditions in countries that do not yet have a water market, information can be provided that may be crucial for further development. For example, results from the 'smart' water market model may suggest construction of a canal to convey water from one region to another. The cost associated with such investment may have prevented the local authorities from doing it. The ability of the model to socially optimize allocations and predict potential outcomes will provide decision makers with an operational tool.

Use of the Approach to Test the Potential for a Particular Water Market

Several chapters in this volume evaluate the potential of water markets in particular regions in Spain and Canada (Garrido, Chapter 14; Horbulyk and Lo, Chapter 15). Other examples include regions in Australia (Musgrave, 1997), and the consideration of trade in water rights in Israel (Yaron, 1997). As in many other analyses, the evaluation procedure for the economic potential of the market mechanism may take several forms, including an optimization model that may be bound to *a priori* transfer patterns. In such cases, the use of a *WaterMart* approach may provide additional insights and allow the use of a market structure that reflects the actual relationships between existing agents.

16.7 Conclusions

In this chapter we discuss the application of a 'smart' computer-assisted market approach to development and evaluation of alternative institutional arrangements for water markets. Experimental economics has the ability to analyze alternative approaches to water markets using the 'smart' market software and quantitative models. The approach certainly has great promise for institutional testing and market familiarization, and probably can form the basis of an operational market. This approach may have significant advantages over conventional bilateral markets in that it may save society costs associated with actual trial and error in the stage of evaluation and establishment of the market. Initial water markets are likely to be thin, and to require low transaction costs to become established. The 'smart' market approach and the two-sided, sealed bid systems have advantages under these situations.

The analytical framework was applied in the California context using a simplified setup consisting of agents from the agriculture, urban, and environmental sectors. The preliminary experimental finding allowed us to determine that, because of the relatively thin nature of the market and other characteristics of the setup, certain agents have more bargaining power over the transportation and initial water allocation than do other agents. However, the performance of many of the agents was very close to the perfectly competitive equilibrium solution, with some (those with the bargaining power), realizing profits in excess of the perfectly competitive equilibrium. The convergence of the agents in the experiment to the optimal values was faster than expected, reducing the establishment cost of the market through fewer trail sessions.

The overall conclusion from the experiments with reward-motivated subjects, and much other experience in network experiments, is that water networks using a computer coordinated 'smart' allocation mechanism will yield essentially competitive outcomes and, therefore, should be further tested. The next stage of the work will be to apply the approach to a set of actual agents who represent the players in each of the sectors and to compare the market solution in that experiment with the market solution with the trained student experiments.

Acknowledgment

The authors acknowledge the assistance by James Murphy in establishing the market structure in the SJV. This chapter is based on the research project "Environmental Change and Adaptive Resource Markets: Computer Assisted Markets for Water Allocation," funded by an NSF grant # SBR-9513406.

Endnotes

[1] The views expressed in this chapter should not be attributed to the World Bank.
[2] In an experimental combinatorial auction mechanism, buyers and sellers trade budles of elemental goods. An example is a series of takeoffs and landing slots to route one plane for one day.
[3] There may be several exceptions, on a local basis, that include domestic uses – especially for low income users, and instream environmental uses.
[4] The regional production functions were calibrated using the data on crop yields, costs and returns from the Central Valley Production Model (CVPM) CH2MHILL (1997) adjusted for water quality and technological alternatives (Dinar & Letey 1996 Chapters 7 and 8).
[5] Simplification of the demand functions for agricultural producers in SJV and Kern regions are modified from Vaux and Howitt (1984) with proper adjustment for current values. They are introduced through a separate downward two-step demand function for SJV and a one-step function for Kern (see values in Table 16.1, buyer nodes *b3* and *b4*, respectively). In this initial experiment, the quality aspect of irrigation water, as could be obtained from the production function model, is not considered.
[6] A uniform price double auction is a market which allows buyers and sellers to make and change bids and offers while a clock counts down. As prices are submitted, the bids are sorted from highest to lowest, the offers from lowest to highest, and a uniform market clearing price for all trades (the intersection of supply and demand) is computed. However, trades are not final until the clock strikes zero.
[7] See Ketcham et al. (1984) for a comparison of the double auction and posted price exchange institutions.

References

Dinar, A. and J. Letey, 1991. "Agricultural Water Marketing, Allocative Efficiency and Drainage Reduction," *Journal of Environmental Economics and Management*, 20(2):210-223.

Dinar, A. and J. Letey, 1996. *Modeling Economic Management and Policy Issues of Water In Irrigated Agriculture*. Preager, Westport, CT.

CH2MHILL, 1997. *Central Valley Production Model: Supporting Document and Data*. Prepared for the Central Valley Project Improvement Act Programmatic AIS, U.S. Bureau of Reclamation and U.S. Fish & Wildlife Service, June.

Howitt, R. E., 1994. "Empirical Analysis of Water Market Institutions: The 1991 California Water Market," *Resources and Energy Economics*, 16:357-371.

Howitt, R. E., 1995. "Positive Mathematical Programming," *American Journal of Agricultural Economics,* 77, May, 329- 342.

Ketcham, J., V. L. Smith and A. Williams, 1984. "A Comparison of Posted Offer and Double Auction Pricing Institutions," *Review of Economic Studies*, 51, October, 594-614.

McCabe, K., S. J. Rassenti, and V. L. Smith, 1989. "Designing Smart Computer-Assisted Market: An Experimental Auction for Gas Networks," *European Journal of Political Economy*, 5:259-283.

McCabe, K., S. J. Rassenti, and V.L. Smith, 1990. "Auction Design for Composite Goods: the Natural Gas Industry," *Journal of Economic Behavior and Organization*, 14, September, 127-149.

McCabe, K., S. J. Rassenti, and V.L. Smith, 1991. "Smart Computer-Assisted Markets," *Science*, 254, October 25, 534-554.

McClurge, S., 1996. "Delta Development," *Western Water*, Jan/Feb., 4-13.

Musgrave, V., 1997. "Australia," in *Water Pricing Experiences, An International Perspective*, Dinar, A. and A. Subramanian (eds.), World Bank Technical Paper No. 386, Washington, D.C..

Rassenti, S. J., V.L. Smith, and R. L. Buflin, 1982. "A Combinatorial Auction Mechanism for Airport Time Slot Allocation," *Bell Journal of Economics*, 13, Autumn, 402-417.

Rassenti, S. J., S. S. Reynolds and V. L. Smith, 1994. "Co-tenancy and Competition in an Experimental Auction Market for Natural Gas Pipeline Networks," *Economic Theory*, 4, 41-65.

Renfro, R. and E. Sparling, 1986. "Private Tubewell and Canal Water Trade on Pakistan Punjab Water Courses," in *Irrigation, Investment, and Management Strategies for Development*, K. W. Easter (ed.), Westview Press, Boulder, CO.

The World Bank, 1993. *Water Resources Management*. A World Bank Policy Paper, Washington, D.C.

Vaux, H. J. Jr. and R. E. Howitt, 1984. "Managing Water Scarcity: An Evaluation of Interregional Transfers," *Water Resources Research*, 20:785-792.

Wood, G. D. and R. Palmer Jones, 1990. *The Water Sellers: A Cooperative Venture by the Rural Poor*. Kumarian Press, West Hanford, CT.

Yaron, D., 1997. "Israel," in *Water Pricing Experiences, An International Perspective*, Dinar, A. and A. Subramanian (eds.), World Bank Technical Paper, No. 386, Washington, D.C.

17 The Future of Water Markets: A Realistic Perspective

K. William Easter, Mark W. Rosegrant, and Ariel Dinar

17.1 Introduction

Although the experience with water markets is still limited, this situation is likely to change as more areas are faced with water scarcity problems and must look for better ways to reallocate this scarce resource. Under the appropriate conditions, it seems clear that water markets can be superior mechanisms for reallocating water to meet changing demand. A key reason to use water markets as the reallocation mechanism is because they give both potential buyers and sellers an incentive to conserve water and to participate in bringing about an equitable and efficient water reallocation. Markets also reduce the potential for rent-seeking by government officials. Trades can lead to permanent or temporary (during periods of drought) reallocations. If the reallocation of water rights is permanent, or between sectors, formal markets will need to be developed. In contrast, if the reallocation is only temporary and within the agricultural sector, then informal markets will usually work.

17.2 Conditions for Effective Formal Water Markets

As pointed out in Chapters 2 and 3, effective formal markets are dependent on some basic institutional arrangements such as tradable water rights or water-use rights (see Table 17.1). In many cases, institutional arrangements will also be needed to deal with third-party effects that result from changes in return flows or declining economic activity for the region selling water. Adequate management and infrastructure will be needed for trades that are not local in nature. Mechanisms may also be necessary to prevent monopoly control over water and to prevent the over-exploitation of groundwater caused by its open-access nature.[1] However, these latter two problems can be dealt with, by-in-large, through the appropriate design, quantification, allocation, monitoring, and enforcement of the water rights. How to do this effectively for groundwater is a particularly vexing problem in many developing countries as well as developed countries.

In many cases, governments will also need to reserve or buy some of the water rights in order to protect instream uses such as fish production, preservation of aquatic environments, and recreation that have strong public good characteristics. As Howe points out in Chapter 5, this has become a growing concern in the western U.S. as the demand for recreation and environmental services has grown. California has even reallocated water from irrigation to improve instream flows into the San Joaquin-Sacramento Delta (see Chapters 7 and 8 on California).

Another condition for efficient water markets is the provision of adequate information concerning water supplies and demands. In Chile, for example, water user associations (WUAs) were essential in providing such information (Chapter 9). The central government, generally, has a comparative advantage in obtaining water supply data while the users are better able to determine their own individual demands. WUAs with access to both sources of information can be important conduits for such information. In contrast to administrative pricing, asymmetric information is much less of a problem for water markets with strong WUAs. Yet even in markets, farmers may withhold information concerning their willingness to buy or sell water as they try to obtain a better price.

Table 17.1. Water Markets: Strategies for Mitigating Problems and Constraints (continued on next page)

A. Potential Problems in Unregulated Markets

Problem

**[a]1. Third-party impacts from a decline in output and employment in the water exporting area. (1, 2, 3, 6, 7)[b]
**2. Reduction or changes in return flows along with any changes in water quality. (1, 3, 4, 5, 6, 8, 9)
*3. Added incentive to overdraft open access groundwater stocks, damage the aquifer, and increase pumping costs. (3, 4, 10, 12, 13, 14)
*4. Increased costs of irrigation system for the remaining farmers. (3, 10)
*5. Drop in land values. (3, 11)
*6. Market power for large-scale buyers or sellers. (1, 3, 15, 16)

Mitigating Strategies

1. Require review and approval of transactions by public agency.
2. Establish a fund to compensate third-parties damaged in trading, financed by levies on water transactions.
3. Limit trades to a percentage of water rights in a given area or community.
4. Revise water rights downward.
5. Grant water rights to those using return flows.
6. Limit trading outside the river basin or sector to consumptive water use.
7. Open litigation to non-holders of water rights.
8. Tax or ban on trading from upstream to downstream users.
9. Set minimum instream flows to maintain aquatic ecosystems.
10. Require buyers to pay a fee for the costs imposed on the irrigation system from which water is transferred.
11. Require lending agency to clear permanent water sales.
12. Adjudicate groundwater rights.
13. Tax groundwater sales based on their scarcity value.
14. Limit trading in areas with rapidly declining groundwater stocks.
15. Provide for regulation of monopolies or expand supply options.
16. Aid small water rights holders with legal fees and registration.

Table 17.1. Water Markets: Strategies for Mitigating Problems and Constraints (continued from previous page)

B. Constraints that Raise the Transaction Costs of Water Trading

Constraints	*Mitigating Strategy*
***1. Incomplete or poorly defined water rights that are not separate from land. (1, 2, 8, 10)	1. Register and secure water rights.
**2. Inadequate infrastructure including conveyance and storage systems. (3)	2. Use proxies of water use (e.g., land area and use) to define water rights.
**3. Inadequate water management and /or the lack of WUAs. (4, 5)	3. Investment in infrastructure.
**4. Inappropriate initial allocation of water rights that causes conflicts among water users. (14)	4. Provide management training for irrigation agencies.
**5. Government agencies reallocate water among and within sectors without compensating original users. (1, 4, 16)	5. Provide water users with incentives to organize WUAs.
**6. Opposition from farmers and environmental groups (6, 10, 15). Also strategies 3, 6, 7, 8, 9, 10, 14, 15 from section A of the table.	6. Education programs explaining the benefits of water markets.
*7. Imperfect or asymmetric information on trading. (5, 9, 10)	7. Tax unused water rights
*8. The granting of more water rights than warranted by existing supplies. (11, 12, 13)	8. Keep up-to-date single basin-wide water rights registries.
*9. Sleeper or inactive water rights that might be sold and activated by water markets. (7)	9. Public agencies or WUAs act as clearing houses for trades.
	10. Aid small water rights holders with free legal protection and information.
	11. Encourage spot and option markets for water.
	12. Revise all water rights downward.
	13. Define two types of rights with one senior to the other.
	14. Base allocation of water rights on past use and employ an auction for any surplus.
	15. Use part of water traded to enhance stream flows.
	16. Base water rights on shares of the water supply rather than on absolute volumes.

[a] The frequency of a problem or constraint is indicated by the following:
* - infrequent; ** - frequent; *** - widespread.

[b] These numbers denote the mitigating strategy that is appropriate for the particular problem or constraint.

17.3 Informal Water Markets

If it is too expensive or too complicated to establish formal water markets, then informal markets may be an acceptable alternative. This is particularly true if there is a need for improved water allocation at the local level and the transaction costs of establishing formal markets are high. Besides allowing water to be sold to the most productive farmers, informal markets would give all farmers an incentive to use their water more efficiently.

Governments need to take steps to legalize water trading separate from land and to help better specify water rights. They should facilitate the establishment of WUAs and strengthen existing WUAs. Governments should also give WUAs more responsibility for water management and improve the certainty of water delivery schedules. The latter will involve improvements in the irrigation agency's water management capability and investments to improve the conveyance system. Most of these improvements need to be made even without the desire to establish water markets.

The scope of informal markets is likely to be limited. First, it will be difficult, if not impossible, to use informal markets to allocate water among sectors or among different river basins. For those types of trades, formal markets will be necessary as well as tradable water rights and mechanisms to address return flow and other third-party impacts. Second, it will also be difficult to make permanent sales of water rights separate from land through informal markets, except in a village or community where such trading is supported.

Finally, as pointed out by Thobani in Chapter 3, informal water markets are difficult to regulate. For groundwater markets, where the aquifer is being overdrafted, the inability to regulate informal water markets can increase the overdrafting and raise pumping and environmental costs, as discussed in Chapter 12. For surface water, informal markets can also cause environmental and third-party impacts that are difficult for governments to regulate. Therefore, in areas where water trading is likely to have substantial negative environmental impacts, it may be better to develop formal markets and to regulate these operations. Also, if intersectoral trading is important, it is better to establish formal markets like those in Chile, Mexico, or California (Chapters 7 through 11).

17.4 Potential Problems and Mitigating Strategies

Although water markets will play a bigger role in future water allocation, they certainly will not solve all water allocation problems (Garrido, 1997). There are good reasons why water markets have been limited in scope (see Table 17.1). Clearly, the expense of establishing water markets should not be incurred if water is not scarce or likely to be scarce in the foreseeable future. In Chile, where water markets have been encouraged, there is no trading in the southern part of the country because water is not scarce (Hearne and Easter, 1995). Water markets will only be active in regions with water scarcity.

Formal water markets are less likely to develop if they are constrained by high institutional development costs and by the absence of the management and infrastructure needed to implement trades. In areas where it is costly to establish, allocate, and enforce water rights, markets will be slow to develop. This has been an important factor in limiting market development in South Asia, where the existence of a large number of very small, fragmented farms makes it difficult to establish individual water rights. In such cases, water rights may have to be allocated to WUAs, as is being done in Mexico, or to villages, rather than to individual farmers. However, public agencies and farmers in South Asia will also have to take steps to improve irrigation infrastructure and management if water trading is to become widespread among farmers served by different canals.

In this region, there has also been a general aversion to markets, especially for allocating "basic" resources such as water and land. The belief is that markets will disadvantage low-income farmers. In practice, as Saleth and Meinzen-Dick show in Chapters 12 and 13, even "poor" farmers benefit from market exchanges. Clearly, if there are too few sellers, the buyers may be disadvantaged, but Saleth points out that social conditions in India, for example, tend to mitigate against such exploitation. Saleth finds little or no price discrimination in mature water markets or in markets where there are strong kinship and social relationships.

In the western U.S., Thompson (1997) identifies two traditional water paradigms that have limited the use and development of water markets. The first paradigm views water as a public resource that should be allocated broadly among society and carefully regulated. The second paradigm considers water as a local resource that should not be transferred among watersheds. The two paradigms have also been important in other countries. For example, both paradigms appear to have influenced Mexico's 1992 water law that established tradable 30-year water-use concessions, where the water is not owned by the owner of the concession and trades to users outside a water district are limited (See Chapter 11).

A directly related problem in establishing water markets is dealing with the issues associated with changes in return flows and water-related economic activities. This appears to be the case in California, as shown in Chapters 7 and 8. When water is transferred into other sectors or out of the river basin, governments will need to have mechanisms in place that require the traders to take into account these third-party effects. However, as Archibald and Renwick indicate in Chapter 7, these mechanisms must be carefully designed or they will foreclose many socially beneficial water trading opportunities.

Thus, for formal water markets to work, the government will have to take an active role in establishing the appropriate institutional and organizational arrangements. Clearly, a strong legal system is an asset in establishing such arrangements. A history of private irrigation development and strong WUAs are also key assets, as clearly illustrated in Chile and the western U.S.. Another important ingredient in establishing and effectively operating water markets is appropriate conveyance and storage systems. Hearne and Easter illustrate, in Chapter 10, how investments in storage capacity and a flexible infrastructure lower the transaction costs of water trading.

Yet, as shown in Chapter 11, too much government involvement in water reallocation decisions can prevent water markets from developing. Although Mexico has made a good start in establishing water markets, the National Water Commission (CNA) has broad authority to take water away from agriculture and transfer it to urban areas. In using this authority, CNA has foreclosed any possibilities of intersectoral market exchanges in several regions. A better strategy may have been for CNA to have been the broker in facilitating water trades between the two sectors. This would mean a major change in how CNA has functioned in the past and is not likely to occur without some strong political pressure from farmers and other interested parties.

Since the physical, institutional and technical conditions that affect the performance of potential water markets vary from area to area, it is very difficult to predict what will happen if water markets are introduced in a new area. Experiences from other countries or regions are likely to provide only partial insights into the potential performance of new water markets. The use of experimental markets, as suggested in Chapter 16, may be a low-cost first step for developing and evaluating alternative institutional arrangements for water market development. Although still not widely used for water resources, experimental markets can capture some of the complexities involved in water markets.

17.5 Conclusions

Contrary to the "nay sayers," water markets have worked and can be a superior mechanism for reallocating water. There are both formal and informal markets at work today. In addition, there are spot market sales and sales of permanent water rights. There are also different leasing arrangements that have been developed informally for water exchanges. These leasing arrangements, not surprisingly, are similar to leasing arrangements used for land, including crop sharing and cash rents (Chapter 12).

Where water is scarce and large amounts of the available water supplies were committed to particular uses many years ago, the economic benefits from water markets are likely to be large. In contrast, if the allocation was made fairly recently, based on the higher valued uses of water, and new opportunities are not available, then the gains will be much more modest, as is shown in the Spanish model in Chapter 14.

For markets to be effective, transaction costs must be kept low. To keep these costs low, the appropriate institutional and organizational arrangements need to be in place, as well as a flexible infrastructure and management. As pointed out above, the critical first step is to establish tradable water rights or water-use rights and have mechanisms to deal with third-party effects in place.

Yet if it is difficult to establish legally-enforceable water rights, a "thick" spot market may provide almost the same security as the ownership of permanent water rights (Chapter 8). In other words, the ability to buy the water needed at a reasonable price may provide enough security so that firms are willing to invest in enterprises that are dependent on this purchased water. A contingent or options market for water can provide additional security so that firms can be assured of a given volume of water at a set price. With only a spot market and no options market, firms may be subject to wide fluctuations in prices.

Such markets are probably a cheaper alternative to buying a substantial quantity of senior water rights in order to maintain adequate supplies during droughts. In Pakistan, the markets for groundwater have greatly improved the security of water supply, particularly in government irrigation projects. This security has allowed increased investments and increased production, as shown in Chapter 13. It will be important to see if spot and contingent markets have similar effects on the productivity of water.

17.6 Future Research Needs

There are a number of research issues that need to be addressed in the future. Well-conducted research can help guide governments in designing better institutional foundations for water markets. The first set of questions involves the type of water right required for formal and informal markets. Is a use right for 30 years adequate for most water transactions or is a 50-year or permanent use right significantly better? In addition, what type of water right is required for spot markets?

Water user groups have, in many cases, developed rules for allocating water among their members. Are these rules an adequate basis for establishing tradable water rights? How might these rights be protected from changes in upstream water uses? Does the formalization of water rights reduce transaction costs? In cases where water user associations have determined their own members' water rights, are transaction costs low enough for water trades?

Related to questions concerning water rights is the problem of enforcement of these rights. Does the type of right affect the ease of enforcement? Under what conditions are legally-based enforcement mechanisms more efficient than enforcement based on personal relationships (personal trust or fear)?

Can water markets be expected to operate effectively during periods of severe drought, or is it better for the government to do the reallocation under such conditions? If the answer is "maybe," then how severe does the drought have to be to make the water market ineffective?

In designing contracts for water, how complete do the contracts have to be? Does the answer to this question depend on the type of water rights or on the method of enforcement or both? Does the thickness of spot markets (short-term contracts) determine the level of assurance (security) provided to long-term investors concerning the availability of water? What can be learned from the development and operation of land markets that would help in establishing and improving water markets? Future research needs to address these issues in a comprehensive way so that governments can better design and implement water markets when necessary and desirable.

Endnotes

[1] Over-exploitation of groundwater is usually due to the lack of well defined water rights and water markets only accentuate the depletion problem.

References

Garrido, A., 1997. "Economics of Water Allocation and the Feasibility of Water Markets in Agriculture," paper presented at a workshop on Sustainable Management in Agriculture, Athens, Greece, Nov. 3-5.

Hearne, Robert R. and K. William Easter, 1995. *Water Allocation and Water Markets: An Analysis of Gains-from-Trade in Chile,* World Bank Technical Paper No. 315, World Bank, Washington, D.C.

Thompson, Barton H., Jr., 1997. "Water Markets and the Problem of Shifting Paradigms," in *Water Marketing – the Next Generation*, T.L. Anderson and P.J. Hill (eds.), Rowman and Littlefield Publishers, New York.

Authors Index

Subject Index